NON-NEWTONIAN FLUID MECHANICS

NORTH-HOLLAND SERIES IN

APPLIED MATHEMATICS AND MECHANICS

EDITORS:

J. D. ACHENBACH
Northwestern University

B. BUDIANSKY
Harvard University

W. T. KOITER
University of Technology, Delft

H. A. LAUWERIER
University of Amsterdam

L. VAN WIJNGAARDEN
Twente University of Technology

VOLUME 31

NORTH-HOLLAND – AMSTERDAM · NEW YORK · OXFORD · TOKYO

NON-NEWTONIAN FLUID MECHANICS

G. BÖHME

Hochschule der Bundeswehr Hamburg
Hamburg, F.R.G.

1987

NORTH-HOLLAND – AMSTERDAM · NEW YORK · OXFORD · TOKYO

ISBN: 0 444 70186 9

Translation of:
Strömungsmechanik nicht-newtonscher Fluide
©Teubner, Stuttgart, 1981
Translated by J.C. Harvey

Publishers:
ELSEVIER SCIENCE PUBLISHERS B.V.
P.O. Box 1991
1000 BZ Amsterdam
The Netherlands

Sole distributors for the U.S.A. and Canada:
ELSEVIER SCIENCE PUBLISHING COMPANY, INC.
52 Vanderbilt Avenue
New York, N.Y. 10017
U.S.A.

Library of Congress Cataloging-in-Publication Data

Böhme, G. (Gert), 1942-
Non-Newtonian fluid mechanics.

(North-Holland series in applied mathematics and mechanics ; v. 31)
Translation of: Strömungsmechanik nicht-newtonscher Fluide.
Bibliography: p.
Includes index.
1. Non-Newtonian fluids. 2. Fluid mechanics.
I. Title. II. Series.
QA929.5.B6413 1987 532 86-32937
ISBN 0-444-70186-9 (U.S.)

PRINTED IN THE NETHERLANDS

PREFACE

This book is a translation of the German textbook 'Strömungsmechanik nicht-Newtonscher Fluide' published by B.G. Teubner, Stuttgart, 1981. Because of its friendly reception by readers I was encouraged to bring out an English edition in order to reach a wider range of readers. Dr. J.C. Harvey of Yelverton, Devon, England has contributed to this project not only by translating the text, but also by producing the camera-ready copy, for which I am very grateful.

The book has its origin in lectures which I have several times given to engineering students after their intermediate diploma examination, and to co-workers in Hamburg who are interested in this subject. The book is intended for use in technical universities, and as a help to practising engineers who are involved with flow problems of non-Newtonian fluids.

The treatment of the subject is based throughout on continuum mechanics model concepts and methods. Because in non-Newtonian fluids the material properties operating depend critically on the kinematics of the flow, special attention is paid to the deriving and explanation of the adequate constitutive equations used. Thus I avoid as much as possible formal arguments, but instead use obvious arguments and I give detailed comment on the rheological law concerned, before a flow is analysed thoroughly, and the study of the chosen rheological process is brought up to definite results.

In order to ensure that the book can be read without reference to other sources it is necessary initially to consider some general principles of continuum mechanics. After this I begin with the study of simple motions, namely steady and unsteady shear flows, and I then proceed by degrees to kinematically more complex motions. Thus at the start I deal with the topic in greater detail and later I treat the topics rather more briefly. Problems of various degrees of difficulty at the end of each chapter invite active participation by the reader.

Numerous stimulating topics from the literature are considered in the book. I have however not allowed myself to present the problems dealt with in a more or less disconnected sequence, but have always been concerned to work out what is essential from a didactic viewpoint, to omit unnecessary ancillary details, and to bring the individual parts together into a unified whole. Hence many times merely the formulation of a problem and the result of a contribution were useful for my purpose. The solution method has always been harmonised with the methods described in the book. Therefore I have refrained from quoting the many journal articles which refer to the material being described and which I have found useful. It is impossible to give a complete reference list.

'Everything' cannot be included in a textbook, i.e. the author has to make a careful choice of his material. Thus I have considered only laminar flows throughout, and I have in the actual applications mostly assumed the fluid to be incompressible. This means no significant limitation when dealing with the highly viscous substances such as occur in plastics technology and processing techniques. It is rather different in the case of the strict limitation to one-phase flows. Multiphase flows with participation of a non-Newtonian fluid could not be considered for lack of space. Thermal effects had to be treated relatively briefly. The reader who has read the book thoroughly should however immediately be able to apply himself to a study of these and other topics in non-Newtonian fluid mechanics.

Hamburg, October 1986 G. Böhme

CONTENTS

1 Principles of continuum mechanics 1
1.1 Basic concepts 1
1.2 Material derivative 5
1.3 Deformation rates 7
1.4 Rivlin-Ericksen tensors 15
1.5 Strain tensor 17
1.6 Kinematics of steady shear flows 24
1.6.1 Plane shear flow 25
1.6.2 Poiseuille flow 26
1.6.3 Couette flow 27
1.6.4 Helical flow 28
1.6.5 Torsional flow 30
1.6.6 Cone-and-plate flow 30
1.7 Continuity equation 31
1.8 Stress and volume force 34
1.9 Equations of motion 36
1.10 Energy equation for fluid flow 38

2 Material properties occurring in steady shear flows 45
2.1 The flow function 50
2.2 The normal stress functions 60

3 Processes that are controlled by the flow function 69
3.1 Rotational viscometer 69
3.1.1 Effect of dissipation 72
3.2 Pressure-drag flow in a straight channel 81
3.2.1 Flow characteristics, pumping efficiency 86
3.2.2 Extrusion flow 94
3.2.3 Fluid dynamics theory of the roller 104
3.2.4 Flow in journal bearings 111
3.3 Radial flow between two parallel planes 119
3.4 Pipe flow 126
3.5 Helical flow 135

4 Effect of normal stress differences 141
4.1 Cone-and-plate flow 141

4.2 Weissenberg effect 146
4.3 Die-swell 150
4.4 Axial shear flow 156

5 Simple unsteady flows 161
5.1 Linear viscoelasticity 162
5.1.1 Sudden change in shear rate 167
5.1.2 Creep test and creep recovery 171
5.1.3 Oscillatory stress and deformation 173
5.1.4 Tuning a shock absorber 177
5.1.5 Flow in the vicinity of a vibrating wall 182
5.1.6 Rayleigh problem for a Maxwell fluid 184
5.1.7 Unsteady Couette flow 192
5.2 Non-linear effects in unsteady pipe flow 196
5.2.1 Constitutive equation for slow and slowly varying processes 204

6 Nearly viscometric flows 213
6.1 Shear flows with a weak unsteady component 213
6.2 Plane steady boundary layer flows 219
6.2.1 Stagnation point boundary layer 225
6.2.2 Modified lubricating film theory 230
6.3 Stability of plane shear flows 237

7 Extensional flows 245
7.1 Theoretical principles 245
7.2 Applications 250

8 Special rheological laws 259
8.1 Fluids without memory 260
8.1.1 Minimum principle for generalised Newtonian fluids . . 262
8.2 Integral models 265
8.2.1 Flow between eccentric rotating discs 271
8.2.2 Boundary layer at a plane wall with homogeneous suction 273
8.3 Differential models 276
8.4 Approximation for slow and slowly varying processes 281

9 Secondary flows 287
9.1 General theory 287
9 2 Rotational symmetric flows 292
9.2.1 Conical nozzle flow 294
9.2.2 Flow round a rotating body 303
9.2.3 Flow through curved pipes 309
9.3 Plane flows 312
9.3.1 Convergent channel flow 315
9.4 Steady flow through cylindrical pipes 320
9.4.1 Isothermal conditions 323
9.4.2 Effect of dissipation 324
9.4.3 Effect of a transverse temperature gradient 327
9.5 Periodic pipe flow 330

Appendix Set of formulas for special curvilinear coordinates . . . 335

Acknowledgements 340

References 341

Index 345

LIST OF THE MOST IMPORTANT SYMBOLS

Symbol	Dimension[1]	Meaning
$\mathbf{a}$	LT^{-2}	Acceleration vector
A	L^2	Area
$\mathbf{A}_n$	T^{-n}	n^{th} Rivlin-Ericksen tensor (n = integer)
b	L	Width, semi-axis of ellipse
Br	-	Brinkman number
c	$L^2T^{-2}\Theta^{-1}$	Specific heat
c	LT^{-1}	Wave speed
c	L	Semi-axis of ellipse
$\mathbf{C}_t$	-	Relative right-Cauchy-Green tensor
d	L	Diameter
$\mathbf{D}$	T^{-1}	Strain rate tensor
e	L^2T^{-2}	Specific internal energy
$\mathbf{e}$	-	Unit vector
$\mathbf{E}$	-	Unit tensor
f	-	Normalised flow function
$\mathbf{f}$	$ML^{-2}T^{-2}$	Force per unit volume
F	MLT^{-2}	Force
$\mathbf{F}_t$	-	Relative deformation gradient
g	LT^{-2}	Acceleration caused by gravity
G	$ML^{-1}T^{-2}$	Linear viscoelastic influence function
$G^*=G'+iG''$	$ML^{-1}T^{-2}$	Complex shear modulus
h	L	Height
I_1, I_2, I_3	-	Invariants of a symmetrical tensor
k	$ML^{-2}T^{-2}$	Pressure drop per unit length
K	-	Dimensionless pressure parameter
ℓ	L	Length
$\mathbf{L}$	T^{-1}	Velocity gradient tensor
M	ML^2T^{-2}	Torque
$\overline{M}$	-	Mean molecular weight
N_1, N_2	$ML^{-1}T^{-2}$	Normal stress functions
p	$ML^{-1}T^{-2}$	Pressure

---

1 Mass [M], length [L], time [T] and temperature [Θ] are basic quantities in the International System of units.

p	T^{-1}	Laplace transform variable
P	$ML^{2}T^{-3}$	Power
$\mathbf{q}$	MT^{-3}	Heat flux density vector
Q	-	Dimensionless discharge
r	L	Cylindrical coordinate, radius
R	L	Spherical coordinate
Re	-	Reynolds number
s	T	Time delay
S	-	Dimensionless velocity parameter
$\mathbf{S}$	$ML^{-1}T^{-2}$	Stress tensor
So	-	Sommerfeld number
St	-	Stokes number
t	T	Time
$\mathbf{T}$	$ML^{-1}T^{-2}$	Extra-stress tensor
u,v,w	LT^{-1}	Cartesian components of velocity
$\bar{u}$	LT^{-1}	Mean velocity
U	LT^{-1}	Constant reference velocity
$\mathbf{v}$	LT^{-1}	Velocity vector
V	L^{3}	Volume
$\dot{V}$	$L^{3}T^{-1}$	Volume flux
$\mathbf{W}$	T^{-1}	Vorticity tensor
We	-	Weissenberg number
x,y,z	L	Cartesian coordinates
α_1,α_2	ML^{-1}	Second order material coefficients
β	-	Angle
β	ML^{-1}	Second flow function
β_1,β_2,β_3	$ML^{-1}T$	Third order material coefficients
γ	-	Angle of shear
$\dot{\gamma}$	T^{-1}	Shear rate
δ^*	L	Boundary layer displacement thickness
Δ	-	Dimensionless displacement thickness
Δ	L^{-2}	Laplace operator
$\dot{\varepsilon}$	T^{-1}	Elongation rate
η	$ML^{-1}T^{-1}$	Shear viscosity
η_0	$ML^{-1}T^{-1}$	Lower Newtonian limiting viscosity, zero-shear viscosity

η_∞	$ML^{-1}T^{-1}$	Upper Newtonian limiting viscosity
η_*	$ML^{-1}T^{-1}$	Reference value for viscosity, in general = η_0
$\eta^*=\eta'-i\eta''$	$ML^{-1}T^{-1}$	Complex viscosity
$\bar{\eta}$	-	Pumping efficiency
$\eta_E, \eta_{EB}, \eta_{EP}$	$ML^{-1}T^{-1}$	Elongational viscosities
ϑ	-	Dimensionless temperature difference
ϑ	-	Spherical coordinate, angle
Θ	Θ	Absolute temperature
λ	$MLT^{-3}\Theta^{-1}$	Thermal conductivity
λ	T	Relaxation time, material specific time scale
ν_1, ν_2	ML^{-1}	Normal stress coefficients
Π	-	Dimensionless power parameter
ρ	ML^{-3}	Density
σ	-	Normalised shear stress = τ/τ_*
σ_{ij}	$ML^{-1}T^{-2}$	Total stresses, elements of **S**
τ	$ML^{-1}T^{-2}$	Shear stress
τ_*	$ML^{-1}T^{-2}$	Stress reference value
τ_{ij}	$ML^{-1}T^{-2}$	Extra-stresses, elements of **T**
φ	-	Cylindrical coordinate, angle
Φ	$ML^{-1}T^{-3}$	Dissipation function
Φ	-	Reduced discharge in a pipe flow
Ψ	L^2T^{-1}	Stream function for plane flows
Ψ	L^3T^{-1}	Stream function for rotational symmetric flows
ω	T^{-1}	Angular velocity
Ω	$ML^{-1}T^{-3}$	Potential of the extra-stresses of generalised Newtonian fluids

1 PRINCIPLES OF CONTINUUM MECHANICS

1.1 Basic concepts

Fluid mechanics is a field theory in the broad sense. It describes the observed phenomena by considering the material as a continuum. The true atomic structure of the material is not considered. This modelling assumes that the molecular dimensions are negligible when compared with the global dimensions of the flow field. Applications for which this assumption is not valid, for example in the case of the dynamics of highly rarefied gases, are not considered here.

The points from which the continuum is assembled are called *material points*. It is expedient in the representation of a material point to start out from a small particle of finite extent, an element of fluid which one may think of as being coloured. The smaller the spot of colour, i.e. the smaller the particle of fluid is chosen, the better is its approximation to a material point. The path of the coloured spot for any motion of the fluid which can be detected with the eye therefore approximates to the path travelled by a material point.

We understand by the term field a function of space and time. It may thus be assumed that a characterising condition for this point can be assigned to each point in a space filled by a liquid or a gas. There occur in continuum mechanics scalar functions of state like pressure and temperature, vector functions of state like velocity and acceleration, and second order tensor fields, particularly the stress tensor.

Because the processes which concern us here occur in the three dimensional space of the observation, we require three space coordinates for their description. In general we take as a basis a fixed reference system composed of Cartesian coordinates. It may however be more appropriate for certain applications to use other forms of spatial coordinates, especially cylindrical or spherical coordinates. We shall decide on this as the case may be in the treatment of actual examples. It is of course advantageous for the derivation of general statements to start from the Cartesian coordinates x, y, z.

We shall generally designate the Cartesian coordinates of a vector **n** by subscripts x, y, z and enclose them in a column matrix which represents the vector **n**, e.g.

$$\mathbf{n} = \begin{bmatrix} n_x \\ n_y \\ n_z \end{bmatrix}$$

We make an exception for the position vector **r**, whose components we shall designate x, y, z, and for the frequently occurring *velocity vector* **v**. In order to simplify the writing we shall not designate the Cartesian velocity components by indices, but by various letters, u,v,w:

$$\mathbf{v} = \begin{bmatrix} u \\ v \\ w \end{bmatrix}$$

A direction is assigned to each material point by the vector field of the flow velocity, which may possibly vary with time. The integral curves which make up this direction field at a certain time are called *stream lines*. The following applies for advance along a stream line dx:dy:dz = u:v:w (t constant). The stream lines consequently satisfy the following system of the ordinary differential equations (σ is the curve parameter for the stream lines; the time t remains constant):

$$\frac{dx}{d\sigma} = u\,(x, y, z, t); \qquad \frac{dy}{d\sigma} = v\,(x, y, z, t); \qquad \frac{dz}{d\sigma} = w\,(x, y, z, t) \tag{1.1}$$

Simultaneous integration of this system of equations yields a stream line for time t. The three constants of integration which arise can be considered as the space coordinates of a point through which the stream line will pass. Different stream lines are distinguished by various values of the constant of integration. A surface made up of stream lines only is called a *stream surface*.

One understands by the word *path line* the path traversed by a material point. One can determine the path lines for a given velocity field by integrating the set of equations

$$\frac{dx}{dt} = u\,(x, y, z, t); \qquad \frac{dy}{dt} = v\,(x, y, z, t); \qquad \frac{dz}{dt} = w\,(x, y, z, t) \tag{1.2}$$

Note that time plays the part of the parameter of the function. The constants of integration can be identified with the coordinates of that point which defines the position of the fluid particle singled out at a certain time (say $t = t_0$).

A flow is called *steady* when it has a time independent velocity field ($\partial \mathbf{v}/\partial t = 0$); otherwise it is described as *unsteady*. In the case of steady flows it is evident that according to equation (1.1) there is at all times the same stream line picture, and at the same time each stream line is simultaneously a path line. The following example shows that this can also occur with unsteady flows. However in general the stream line structure of an unsteady flow varies with time, and the stream lines and path lines no longer coincide.

For an example we now consider an unsteady flow with the velocity field

$$u = a(t)x; \quad v = -a(t)y; \quad w = 0 \tag{1.3}$$

Integrating equation (1.1) we obtain the parameter representation for the stream line passing through the point $x = x_0$, $y = y_0$, $z = z_0$

$$x = x_0 e^{a(t)\sigma}; \quad y = y_0 e^{-a(t)\sigma}; \quad z = z_0 \tag{1.4}$$

This is a planar curve (in the plane $z = z_0$). The curve parameter σ can obviously be easily eliminated. Hence the time dependent factor $a(t)$ also does not apply. The result

$$xy = x_0 y_0 \tag{1.5}$$

in all cases no longer contains time as a parameter, so that the stream lines according to equation (1.5) are rectangular hyperbolas, unchanging with time. The same holds for the path lines. One then obtains directly from equation (1.2) the representation for the path line which passes through the point $x = x_0$, $y = y_0$, $z = z_0$ at time $t = t_0$:

$$x = x_0 e^{\int_{t_0}^{t} a(\tau)d\tau}; \quad y = y_0 e^{-\int_{t_0}^{t} a(\tau)d\tau}; \quad z = z_0 \tag{1.6}$$

Here also the function parameter (that is to say t) can be easily eliminated. Hence the relationship (1.5) again follows. Stream lines and path lines therefore coincide and form a system of rectangular hyperbolas. The flow field under discussion has in the origin of the coordinates a stagnation point, where all the velocity components vanish. This therefore describes the events in the vicinity of a free stagnation point and hence is described as a plane *stagnation point flow*.

The expression (1.5) could also have been derived thus. Because $w = 0$ the stream and path lines are two-dimensional curves. The following applies for the two groups of curves

$$\frac{dy}{dx} = \frac{v}{u} \tag{1.7}$$

In the above case the right-hand side is independent of time (cf. (1.3)) and is equal to -y/x. The resulting differential equation dy/dx = -y/x is solved by equation (1.5).

Note that for any unsteady flows with the condition w = 0, not only the stream lines but also the path lines occur in the planes z = constant and under these conditions the stream lines can always be calculated from equation (1.7), but the path lines only when the right-hand side of (1.7) is independent of time.

We return again to the flow discussed above, having the velocity field (1.3), and examine the relative motion of three material points which were at the time $t = t_0$ located at $x = x_0$, $y = y_0$; at $x = x'_0$, $y = y_0$; and at $x = x_0$, $y = y'_0$ (in each case in the same plane $z = z_0$) (cf. Fig. 1.1). One can easily understand that those points which initially had the same x-coordinates and hence were located above each other in Fig. 1.1 have coincident coordinates at all times. Because the x-component of the velocity vector depends only on x, but not on y, both these points always move at the same velocity in the x-direction and therefore always have the same x-coordinate. The straight fluid line formed from the two material points as end points therefore always remains oriented parallel to the y-axis and is in the course of time simply compressed, but not rotated. Correspondingly a fluid element originally parallel to the x-axis is stretched but not rotated. The surface area of the cuboid volume whose sides are initially oriented parallel to the axes is therefore deformed as shown in Fig 1.1, whilst the height of the square remains constant. Because for such a volume the extensions (positive and negative) only occur in two directions perpendicular to each other, one also refers to this as a *two-dimensional elongational flow*. Moreover the volume of each fluid particle remains constant with time. The path lines of the material points initially occupying x_0, y_0 and x'_0, y'_0 are hence described by equation (1.6) and the corresponding relationships

$$x' = x'_0 e^{\int_{t_0}^{t} a(\tau)d\tau} \quad ; \quad y' = y'_0 e^{-\int_{t_0}^{t} a(\tau)d\tau} \quad ; \quad z' = z_0 \tag{1.8}$$

from which applying

$$(x' - x)(y' - y) = (x'_0 - x_0)(y'_0 - y_0) \tag{1.9}$$

the area equality of the rectangles shown and therefore the volume equality of the relevant square follow.

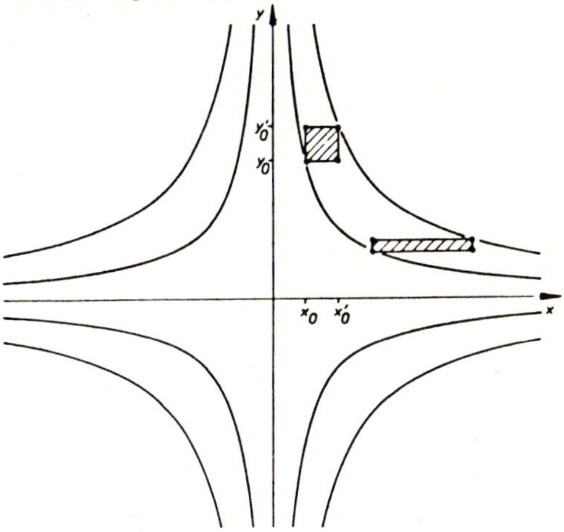

Fig. 1.1 Plane elongational flow

Flows in which the volume of each fluid particle remains constant are called *isochoric*. An elongational flow with the velocity field (1.3) has this property.

1.2 Material derivative

We consider a scalar field $\Phi(x,y,z,t)$, for example the temperature field of a flowing fluid, and investigate the changes with time of Φ which a stationary observer observes, and those observed by an observer who is moving with an individual fluid particle. The stationary observer obviously records the *local time derivative* $\partial\Phi(x,y,z,t)/\partial t$. On the other hand the change with time of a fluid particle is a *material substantial time derivative* and is designated by $D\Phi/Dt$. The relationship between the two time derivatives follows by consideration of the following. The material point which at time t is located at position x,y,z moves corresponding to its actual velocity in the short time interval Δt round the distances $u\Delta t$, $v\Delta t$, $w\Delta t$, and hence at time $t + \Delta t$ is located at $x + u\Delta t$, $y + v\Delta t$, $z + w\Delta t$. The material time derivative follows from the value of $\Phi(x + u\Delta t,\ y + v\Delta t,\ z + w\Delta t,\ t + \Delta t)$ found for this position at time $t + \Delta t$, the corresponding value of $\Phi(x, y, z, t)$ and the elapsed time Δt from

$$\frac{D\Phi}{Dt} := \lim_{\Delta t \to 0} \frac{\Phi(x + u\Delta t, y + v\Delta t, z + w\Delta t, t + \Delta t) - \Phi(x, y, z, t)}{\Delta t} \tag{1.10}$$

Hence it follows that

$$\frac{D\Phi}{Dt} = u\frac{\partial\Phi}{\partial x} + v\frac{\partial\Phi}{\partial y} + w\frac{\partial\Phi}{\partial z} + \frac{\partial\Phi}{\partial t} \tag{1.11}$$

The first three terms, which are designated as convective components of the material time derivative, can be considered as a scalar product of the vectors $\mathbf{v}$ and grad Φ, so that one can write equation (1.11) in a shorter form as

$$\frac{D\Phi}{Dt} = \mathbf{v} \cdot \text{grad}\ \Phi + \frac{\partial\Phi}{\partial t} \tag{1.12}$$

This form is also valid for the use of curvilinear coordinates, whilst equation (1.11) is for use with Cartesian coordinates only.

If one substitutes the set of velocity components u, v, and w for Φ in equation (1.11), then one obtains the components of the *acceleration vector* and its derivatives

$$\begin{aligned} a_x &:= \frac{Du}{Dt} = u\frac{\partial u}{\partial x} + v\frac{\partial u}{\partial y} + w\frac{\partial u}{\partial z} + \frac{\partial u}{\partial t} \\ a_y &:= \frac{Dv}{Dt} = u\frac{\partial v}{\partial x} + v\frac{\partial v}{\partial y} + w\frac{\partial v}{\partial z} + \frac{\partial v}{\partial t} \\ a_z &:= \frac{Dw}{Dt} = u\frac{\partial w}{\partial x} + v\frac{\partial w}{\partial y} + w\frac{\partial w}{\partial z} + \frac{\partial w}{\partial t} \end{aligned} \tag{1.13}$$

At this point we introduce the *velocity gradient tensor*

$$\mathbf{L} := \text{grad}\ \mathbf{v} \tag{1.14}$$

with the matrix

$$\mathbf{L} = \begin{bmatrix} \frac{\partial u}{\partial x} & \frac{\partial u}{\partial y} & \frac{\partial u}{\partial z} \\ \frac{\partial v}{\partial x} & \frac{\partial v}{\partial y} & \frac{\partial v}{\partial z} \\ \frac{\partial w}{\partial x} & \frac{\partial w}{\partial y} & \frac{\partial w}{\partial z} \end{bmatrix} \tag{1.15}$$

whose components are the nine spatial derivatives of the velocity field.

By using **L** the formulas (1.13) for the acceleration can be combined:

$$\mathbf{a} = \mathbf{L}\,\mathbf{v} + \frac{\partial \mathbf{v}}{\partial t} \tag{1.16}$$

Thus the vectors **v** and **a** are considered as column matrices and the product **Lv** can be calculated by the rules of matrix multiplication.

For some applications it is useful to know that one can also write this relationship in the following invariant form:

$$\mathbf{a} = \operatorname{grad} \frac{|\mathbf{v}|^2}{2} - \mathbf{v} \times \operatorname{curl}\, \mathbf{v} + \frac{\partial \mathbf{v}}{\partial t} \tag{1.17}$$

In this $|\mathbf{v}|$ gives the magnitude of the velocity vector, and the symbol $\times$ indicates the vector product of the two terms.

1.3 Deformation rates

We now look at the way in which for a given velocity field infinitesimally separated material points move relative to each other during the passage of time. For this purpose we consider two points '1' and '2' which at time t are located at **r** and **r** + **dr** respectively. In the time interval Δt they are displaced to a first approximation, i.e. disregarding components that vanish more quickly than Δt as $\Delta t \to 0$ along the distances $\mathbf{v}_1\Delta t = \mathbf{v}(x,y,z,t)\Delta t$ and $\mathbf{v}_2\Delta t = \mathbf{v}(x+dx,\ y+dy,\ z+dz,\ t)\Delta t$. Its relative position at time $t + \Delta t$ is therefore described by the vector

$$\mathbf{dr}^* = \mathbf{dr} + (\mathbf{v}_2 - \mathbf{v}_1)\Delta t + O(\Delta t^2) \tag{1.18}$$

(cf. Fig. 1.2). Now for the difference between the velocity components of two infinitesimally separated points

$$\begin{aligned} u_2 - u_1 &= \frac{\partial u}{\partial x}dx + \frac{\partial u}{\partial y}dy + \frac{\partial u}{\partial z}dz \\ v_2 - v_1 &= \frac{\partial v}{\partial x}dx + \frac{\partial v}{\partial y}dy + \frac{\partial v}{\partial z}dz \\ w_2 - w_1 &= \frac{\partial w}{\partial x}dx + \frac{\partial w}{\partial y}dy + \frac{\partial w}{\partial z}dz \end{aligned} \tag{1.19}$$

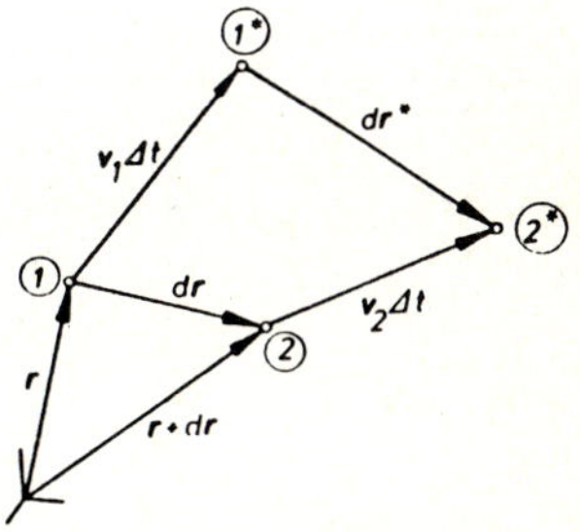

Fig. 1.2 Displacement of two infinitesimally separated points

By use of the velocity gradient tensor **L** these relationships can be written in abbreviated form

$$\mathbf{v}_2 - \mathbf{v}_1 = \mathbf{L}\,\mathbf{dr} \tag{1.20}$$

Hence from equation (1.18) we obtain

$$\Delta \mathbf{dr} := \mathbf{dr}^* - \mathbf{dr} = \mathbf{L}\mathbf{dr}\,\Delta t + O(\Delta t^2) \tag{1.21}$$

The term $\Delta\mathbf{dr}$ indicates how for an observer travelling with the material point '1' the distance vector varies with respect to the adjacent material point '2' during the element of time Δt. For the material time differential of the distance vector

$$\frac{D(\mathbf{dr})}{Dt} := \lim_{\Delta t \to 0} \frac{\Delta \mathbf{dr}}{\Delta t} \tag{1.22}$$

which we shall designate by a point for the sake of simplicity

$$(\mathbf{dr})^{\cdot} = \mathbf{L}\,\mathbf{dr} \tag{1.23}$$

therefore applies.

We now decompose the velocity gradient into its symmetric part

$$\mathbf{D} := \frac{1}{2}(\mathbf{L} + \mathbf{L}^T) = \begin{bmatrix} \dfrac{\partial u}{\partial x} & \dfrac{1}{2}\left(\dfrac{\partial u}{\partial y} + \dfrac{\partial v}{\partial x}\right) & \dfrac{1}{2}\left(\dfrac{\partial u}{\partial z} + \dfrac{\partial w}{\partial x}\right) \\ \dfrac{1}{2}\left(\dfrac{\partial u}{\partial y} + \dfrac{\partial v}{\partial x}\right) & \dfrac{\partial v}{\partial y} & \dfrac{1}{2}\left(\dfrac{\partial v}{\partial z} + \dfrac{\partial w}{\partial y}\right) \\ \dfrac{1}{2}\left(\dfrac{\partial u}{\partial z} + \dfrac{\partial w}{\partial x}\right) & \dfrac{1}{2}\left(\dfrac{\partial v}{\partial z} + \dfrac{\partial w}{\partial y}\right) & \dfrac{\partial w}{\partial z} \end{bmatrix} \tag{1.24}$$

and into its antisymmetric part

$$\mathbf{W} := \frac{1}{2}(\mathbf{L} - \mathbf{L}^T) = \begin{bmatrix} 0 & -\dfrac{1}{2}\left(\dfrac{\partial v}{\partial x} - \dfrac{\partial u}{\partial y}\right) & \dfrac{1}{2}\left(\dfrac{\partial u}{\partial z} - \dfrac{\partial w}{\partial x}\right) \\ \dfrac{1}{2}\left(\dfrac{\partial v}{\partial x} - \dfrac{\partial u}{\partial y}\right) & 0 & -\dfrac{1}{2}\left(\dfrac{\partial w}{\partial y} - \dfrac{\partial v}{\partial z}\right) \\ -\dfrac{1}{2}\left(\dfrac{\partial u}{\partial z} - \dfrac{\partial w}{\partial x}\right) & \dfrac{1}{2}\left(\dfrac{\partial w}{\partial y} - \dfrac{\partial v}{\partial z}\right) & 0 \end{bmatrix} \tag{1.25}$$

$\mathbf{L}^T$ is the transposed velocity gradient tensor, the matrix representation of which is obtained by interchange of rows and columns in $\mathbf{L}$. The expansion $\mathbf{L} = \mathbf{D} + \mathbf{W}$ transforms equation (1.23) into

$$(\mathbf{dr})^{\cdot} = \mathbf{D}\,\mathbf{dr} + \boldsymbol{\omega} \times \mathbf{dr} \tag{1.26}$$

Note that for each antisymmetric tensor $\mathbf{W}$ a vector $\boldsymbol{\omega}$ can be associated in such a way that the following holds for arbitrary vectors $\mathbf{n}$: $\mathbf{Wn} = \boldsymbol{\omega} \times \mathbf{n}$. In the case under discussion $\boldsymbol{\omega}$ is half the vorticity vector, as can be learned from equation (1.25)

$$\boldsymbol{\omega} = \frac{1}{2}\,\mathrm{curl}\,\mathbf{v} \tag{1.27}$$

Note that the kinematic vectors $\mathbf{D}$ and $\mathbf{W}$ are in general functions of position and possibly also of time. If $\mathbf{D} = 0$ holds for a certain position of the flow field, then the relative movement of the immediate surroundings of this point in accordance with equation (1.26) consists only of a rotation of angular velocity $\boldsymbol{\omega} = \frac{1}{2}\,\mathrm{curl}\ \mathbf{v}$. Therefore $\mathbf{W}$ is referred to as the *rate of rotation* or *vorticity tensor*.

For the interpretation of the *rate of strain tensor* $\mathbf{D}$ we consider the time change of the scalar product of two material line elements $\mathbf{dr}$ and $\delta\mathbf{r}$ (cf. Fig. 1.3).

$$(\mathbf{dr}\cdot\delta\mathbf{r})^{\cdot} = (\mathbf{dr})^{\cdot}\cdot\delta\mathbf{r} + \mathbf{dr}\cdot(\delta\mathbf{r})^{\cdot} \tag{1.28}$$

If one denotes the magnitude of the vectors $\mathbf{dr}$ and $\delta\mathbf{r}$ by ds and δs respectively, and the angle between them by $90^\circ - \gamma$, then the terms in brackets on the left-hand side can be written as $ds\delta s \sin\gamma$. The derivatives of the distance vector on the right-hand side can be eliminated with equation (1.23):

$$(ds\,\delta s\sin\gamma)^{\cdot} = \mathbf{L}\,\mathbf{dr}\cdot\delta\mathbf{r} + \mathbf{dr}\cdot\mathbf{L}\,\delta\mathbf{r} \tag{1.29}$$

By applying the mathematical law $\mathbf{Tn}\cdot\mathbf{t} = \mathbf{n}\cdot\mathbf{T}^T\mathbf{t}$ for a scalar product that is formed from two vectors $\mathbf{t}$, $\mathbf{n}$ and a tensor $\mathbf{T}$, the first term on the right-hand side can also be given in the form $\mathbf{dr}\cdot\mathbf{L}^T\delta\mathbf{r}$, hence all of the right-hand side becomes $2\mathbf{dr}\cdot\mathbf{D}\delta\mathbf{r}$. If one differentiates the left-hand side according to the product rule, and then divides through by $ds\delta s$, one obtains

$$\sin\gamma\left\{\frac{(ds)^{\cdot}}{ds} + \frac{(\delta s)^{\cdot}}{\delta s}\right\} + \dot{\gamma}\cos\gamma = 2\,\frac{\mathbf{dr}}{ds}\cdot\mathbf{D}\,\frac{\delta\mathbf{r}}{\delta s} \tag{1.30}$$

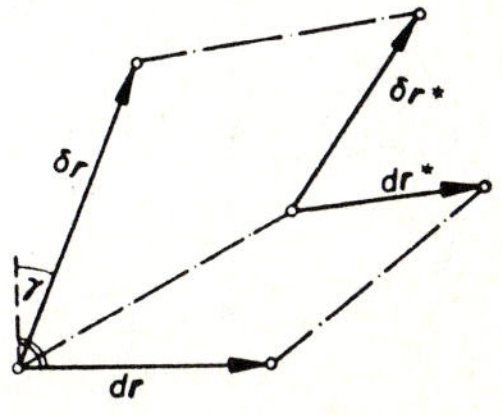

Fig. 1.3 Time change of two material line elements

The individual components of the tensor $\mathbf{D}$ can be easily discussed by means of this relationship. In order to eliminate from the right-hand side an element on the major diagonal, e.g. D_{xx}, we let the two line elements coincide ($\mathbf{dr} = \delta\mathbf{r}$, $ds = \delta s$), and choose their positions so that they are oriented in one coordinate direction, e.g. in the x-direction. Because under these conditions the angle between them vanishes, therefore $\gamma = 90^\circ$ applies, and equation (1.30) reduces to

$$\frac{(ds)^{\cdot}}{ds} = D_{xx} \tag{1.31}$$

This relationship indicates that the diagonal elements of the rate of strain tensor **D** define the *elongation rates* $(ds)^{\cdot}/ds$ of such line elements which are oriented in the coordinate direction.

In order to produce a secondary diagonal line element, say D_{xy}, on the right-hand side of equation (1.30) we choose two line elements perpendicular to each other, which lie parallel to the axes, e.g. in the x-direction and the y-direction respectively. Thus one obtains because $\gamma = 0$

$$\dot{\gamma} = 2\,D_{xy} \tag{1.32}$$

The non-diagonal elements of **D** are therefore half as large as the velocities at which the instantaneous right angle between the two line elements change, which are oriented exactly in the different coordinate directions. The tensor **D** therefore describes the elongation rates of the edges and the rates of change of the angles formed between the edges of an initially cuboid material element oriented parallel to the coordinate axes.

As an example we now consider the velocity field (1.3) of a two-dimensional elongational flow. Relative to the coordinate system taken as a basis the velocity gradient tensor obviously has the matrix form

$$\mathbf{L} = \begin{bmatrix} a & 0 & 0 \\ 0 & -a & 0 \\ 0 & 0 & 0 \end{bmatrix} \tag{1.33}$$

Because **L** is symmetrical the vorticity tensor **W** vanishes. The fluid particles accordingly do not rotate, which is also clearly indicated in Fig. 1.1. Therefore **D** = **L** holds, i.e. the deformation velocity tensor **D** is present in equation (1.33). The vanishing of all secondary diagonal elements signifies that the angles between line elements parallel to the coordinate directions do not vary with time. Hence the volume sketched in Fig. 1.1 remains rectangular at all times. The two non-vanishing diagonal elements show that the element lengthens in the x-direction with the elongation rate a, and contracts in the y-direction with the contraction rate -a.

As an illustration of a secondary diagonal element of the rate of strain tensor we consider a steady *simple shear flow* which has the velocity field

$$u = cy, \qquad v = 0, \qquad w = 0 \tag{1.34}$$

Hence this is obviously a parallel flow in the x-direction, so that the velocity u increases linearly with the coordinate y. Because all the fluid particles which lie in the plane y = constant move with the same velocity, there are no relative displacements between them with increase in time, so that each layer y = constant remains undeformed by the motion. This kinematic characterisation moreover is also correct when the flow velocity u depends if wished on y, and if the occasion arises, on time. The flow field is always characterised by the fact that an assemblage of two-dimensional layers, just like a stack of sheets of paper, slides one over the other. One therefore speaks of *two-dimensional shear flows*, and denotes conveniently the layers remaining undeformed as *shearing surfaces*. The simple shear flow with the velocity field (1.34) is a special case of two-dimensional shear flow.

We choose a rectangular volume of liquid from a simple shear flow, the edges of which at the actual moment in time run parallel to the axes (Fig. 1.4). Because the upper and lower enclosing surfaces travel at different velocities the rectangular volume becomes a parallelepiped, so that the deformation increases without limit with time. The shape of the volume of liquid after the lapse of time Δt is obviously completely defined by the angle Δγ between the tilted surface and the y-axis. As shown in Fig. 1.4 this angle is given by

$$\Delta\gamma = \frac{cy_2 \Delta t - cy_1 \Delta t}{y_2 - y_1} = c\,\Delta t \tag{1.35}$$

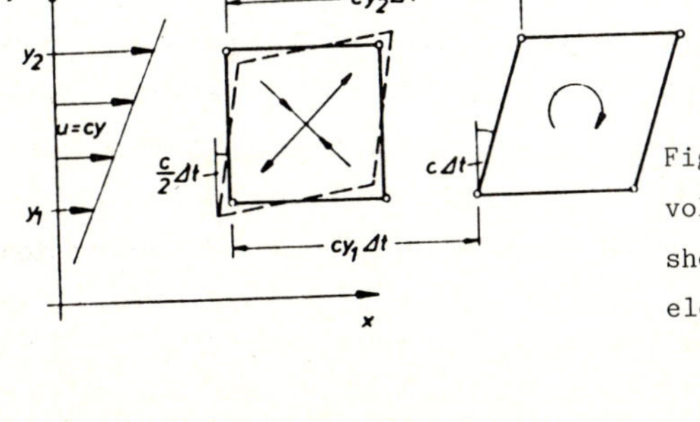

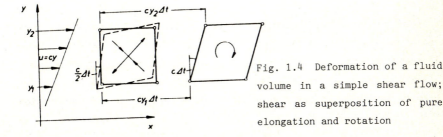

Fig. 1.4 Deformation of a fluid volume in a simple shear flow; shear as superposition of pure elongation and rotation

The limiting process accordingly yields the relationship $\dot{\gamma} = c$, or

$$\dot{\gamma} = \frac{du}{dy} \tag{1.36}$$

For a simple shear flow the rate of shear $\dot{\gamma}$ accordingly corresponds to the velocity derivative. Hence we have obviously once again derived equation (1.32) for this flow. A velocity gradient tensor belongs to the given velocity field, which one can easily write down for oneself, and the symmetrical part of it possesses the matrix representation

$$\mathbf{D} = \begin{bmatrix} 0 & c/2 & 0 \\ c/2 & 0 & 0 \\ 0 & 0 & 0 \end{bmatrix} \tag{1.37}$$

Of six independent elements only one chiefly occurs, namely $D_{xy}=c/2$. Because the double value c = du/dy given by equation (1.32) defines the rate of change of the right angle between two line elements in the x- and y-directions, equation (1.32) reduces to equation (1.36) in the case of simple shear flow. Equation (1.36) moreover holds not only for simple shear flow, but for all two-dimensional shear flows. The right-hand side is then of course no longer 3-dimensionally constant, so that the rate of shear depends on position and time. In the case of an unsteady two-dimensional shear flow it goes without saying that the right-hand side of equation (1.36) is a partial differential with respect to position for time = constant.

Now remember that one can transform every symmetrical tensor to the diagonal form, i.e. one can introduce a special system of coordinates, so that only the main diagonal elements of the matrix representing the tensor are non-zero. Hence the symmetrical tensor **D** has relative to the x-y-z coordinate system the form

$$\mathbf{D} = \begin{bmatrix} D_{xx} & D_{xy} & D_{xz} \\ D_{xy} & D_{yy} & D_{yz} \\ D_{xz} & D_{yz} & D_{zz} \end{bmatrix} \tag{1.38}$$

so there is always a rotated *principal axis system*, in which **D** takes on the diagonal form

$$\mathbf{D}^* = \begin{bmatrix} \lambda_1 & 0 & 0 \\ 0 & \lambda_2 & 0 \\ 0 & 0 & \lambda_3 \end{bmatrix} \tag{1.39}$$

The three elements λ_1, λ_2, λ_3 are called *eigenvalues* of the tensor **D**. They are found by manipulation of the eigenvalue expression

$$\mathbf{D}\,\mathbf{n} = \lambda\,\mathbf{n} \tag{1.40}$$

This homogeneous system of equations for the three components n_x, n_y, n_z of the desired eigenvectors **n** only has non-vanishing solutions if its determinant vanishes:

$$\begin{vmatrix} D_{xx}-\lambda & D_{xy} & D_{xz} \\ D_{xy} & D_{yy}-\lambda & D_{yz} \\ D_{xz} & D_{yz} & D_{zz}-\lambda \end{vmatrix} = 0 \tag{1.41}$$

This is a cubic equation for the eigenvalue λ, which can be written as follows:

$$-\lambda^3 + I_1\lambda^2 - I_2\lambda + I_3 = 0 \tag{1.42}$$

Moreover the coefficients I_1, I_2, I_3 are algebraic functions of the elements of **D**, that is to say

$$I_1 = D_{xx} + D_{yy} + D_{zz} = \mathrm{tr}\,\mathbf{D} \tag{1.43}$$

$$\begin{aligned} I_2 &= D_{xx}D_{yy} + D_{yy}D_{zz} + D_{zz}D_{xx} - D_{xy}^2 - D_{xz}^2 - D_{yz}^2 \\ &= \frac{1}{2}\left[(\mathrm{tr}\,\mathbf{D})^2 - \mathrm{tr}\,\mathbf{D}^2\right] \end{aligned} \tag{1.44}$$

$$I_3 = \begin{vmatrix} D_{xx} & D_{xy} & D_{xz} \\ D_{xy} & D_{yy} & D_{yz} \\ D_{xz} & D_{yz} & D_{zz} \end{vmatrix} = \det\mathbf{D} \tag{1.45}$$

They are called *principal invariants* of the tensor **D** because their numerical values are independent of the chosen system of coordinates. With a change of the system of coordinates the individual components of the tensor change too, but not the algebraic combinations of the matrix elements given by equations (1.43) to (1.45). The linear invariant I_1 is often called the *trace* of the tensor.

The eigenvalue equation (1.42) always has for symmetrical tensors three real solutions λ_1, λ_2, λ_3, several of which can be coincident. To each unique eigenvalue, say λ_1, there belongs (up to the sign) exactly one unit vector $\mathbf{n}_1$, which satisfies the condition $\mathbf{D}\mathbf{n}_1 = \lambda_1\mathbf{n}_1$. If all the eigenvalues are unique, then the eigenvectors are perpendicular to each other and form the Cartesian system of coordinates in which **D** takes on the principal axis form (1.39) mentioned.

From the fact that one can in particular bring the (symmetrical) rate of strain tensor to the principal axis form by choosing an appropriate local system of coordinates, it follows that the instantaneous change of shape of a fluid particle can always be considered as pure strain in three mutually perpendicular directions. Thus for example in Fig. 1.4 the deformation drawn for an originally rectangular particle in simple shear flow can be considered as elongation and contraction in two directions which form an angle of 45° with the x-direction. The relevant rates of strain are $c/2$ and $-c/2$ respectively. A rotation about the z-axis of angular velocity $-c/2$ is superposed on the deformation of the particle in accordance with equation (1.27).

For a tensor field $\mathbf{D}$ the components D_{xx}, D_{xy}, etc. are generally functions of position. Thus the eigenvectors too, hence the orientation of the principal axis system, depend on x, y, z. Hence if one specifies as a base a system of coordinates in which $\mathbf{D}$ for a given position takes on a diagonal form, then for another position there arise secondary diagonal members. A universal principal axis system only exists if the elements of $\mathbf{D}$, as for example in equation (1.37) are constant with respect to x, y, z.

In order to measure the relative influence of the rotation rates and the rates of strain on a particular flow, one can put in relation the magnitude of the angular velocity $|\boldsymbol{\omega}| = \frac{1}{2}|\operatorname{curl} \mathbf{v}|$ and the magnitude of the rate of strain tensor $|\frac{1}{2}\operatorname{tr} \mathbf{D}^2|^{1/2}$. The local dimensionless *kinematic vorticity number* thus obtained

$$W_k := \frac{|\operatorname{curl} \mathbf{v}|}{|2 \operatorname{tr} \mathbf{D}^2|^{1/2}}$$

of course vanishes for non-rotational movements ($\operatorname{curl} \mathbf{v} = 0$), which one also refers to as potential flows. On the other hand if the liquid rotates without any deformation like a rigid body, then this vorticity number increases over all boundaries, $W_k \to \infty$. One can show that for axial shear flows (steady or unsteady) with straight stream lines $W_k = 1$, and this would apply to a fully developed pipe flow. This flow form is of considerable interest in engineering applications and we shall come across it several times, especially in Sections 3.2, 3.4, 4.4, 5.1, and 5.2.

Starting from the expression (1.17) for the acceleration vector one can also conveniently define a *dynamic vorticity number* as the ratio of the part $|\mathbf{v} \times \text{curl } \mathbf{v}|$ to the rest of the acceleration

$$W_d := \frac{|\mathbf{v} \times \text{curl } \mathbf{v}|}{\left|\frac{\partial \mathbf{v}}{\partial t} + \text{grad } \frac{|\mathbf{v}|^2}{2}\right|}$$

$W_d = 0$ applies again for potential flows; on the other hand $W_d = 2$ applies to a rigid body rotation. If the acceleration of the material point vanishes, as say in a pipe flow, then numerator and denominator naturally are equal, i.e. $W_d = 1$. In a boundary layer W_d and W_k vary in space, but because of the 'proximity' of the boundary layer to an unaccelerated shear flow, they take on a numerical value close to unity.

1.4 Rivlin-Ericksen tensors

In Section 1.3 we obtained the rate of strain tensor by differentiation with respect to time of the distance vector between infinitesimally separated material points. We now wish to examine higher order differentials with respect to time and for this purpose we must define the change of position of the individual points more accurately than was formerly necessary. We first consider only one material point. Its position at time $t + \Delta t$ can be calculated for an error of magnitude $O(\Delta t^3)$ from the position, velocity and acceleration at time t (Fig. 1.5):

$$\mathbf{r}(t + \Delta t) = \mathbf{r}(t) + \mathbf{v}(t)\,\Delta t + \frac{1}{2}\,\mathbf{a}(t)\,\Delta t^2 + O(\Delta t^3) \tag{1.46}$$

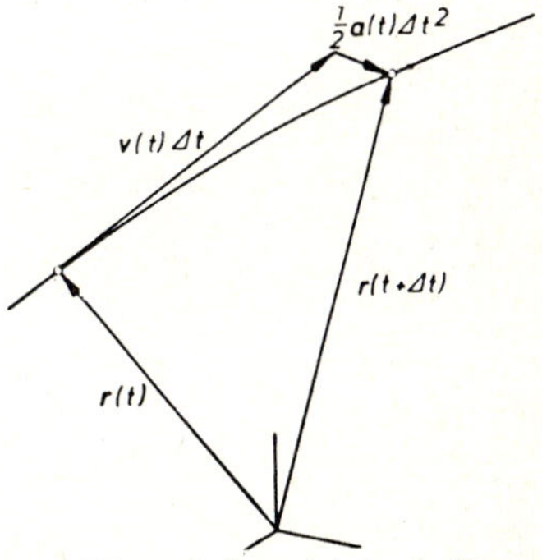

Fig. 1.5 Position change of a material point as regards its acceleration

Hence the expression (1.18) relating to Fig. 1.2, which defines the relative positions of two material points after time Δt, can be more accurately written as:

$$d\mathbf{r}^* = d\mathbf{r} + (\mathbf{v}_2 - \mathbf{v}_1)\,\Delta t + \frac{1}{2}(\mathbf{a}_2 - \mathbf{a}_1)\,\Delta t^2 + O(\Delta t^3) \tag{1.47}$$

Because the two material points considered are infinitesimally separated, $\mathbf{v}_2 - \mathbf{v}_1 = (\text{grad } \mathbf{v})\mathbf{dr}$ (cf. equation (1.20)) applies as before and correspondingly $\mathbf{a}_2 - \mathbf{a}_1 = (\text{grad } \mathbf{a})\mathbf{dr}$. Hence grad **a** defines the acceleration gradient tensor which has the matrix form

$$\text{grad } \mathbf{a} = \begin{bmatrix} \frac{\partial a_x}{\partial x} & \frac{\partial a_x}{\partial y} & \frac{\partial a_x}{\partial z} \\ \frac{\partial a_y}{\partial x} & \frac{\partial a_y}{\partial y} & \frac{\partial a_y}{\partial z} \\ \frac{\partial a_z}{\partial x} & \frac{\partial a_z}{\partial y} & \frac{\partial a_z}{\partial z} \end{bmatrix} \tag{1.48}$$

Thus for the relative position vector of two infinitesimally separated material points after time Δt up to the terms that vanish for $\Delta t \to 0$, at least like Δt^3

$$\mathbf{dr}^* = \mathbf{dr} + \mathbf{L}\,\mathbf{dr}\,\Delta t + \frac{1}{2}(\text{grad } \mathbf{a})\,\mathbf{dr}\,\Delta t^2 + O(\Delta t^3) \tag{1.49}$$

applies.

By scalar multiplication of $\mathbf{dr}^*$ with itself one obtains for the square of the distance vector:

$$\mathbf{dr}^* \cdot \mathbf{dr}^* = \mathbf{dr} \cdot \mathbf{dr} + \mathbf{dr} \cdot \mathbf{L}\mathbf{dr}\,\Delta t + \mathbf{L}\mathbf{dr} \cdot \mathbf{dr}\,\Delta t + \mathbf{dr} \cdot \frac{1}{2}(\text{grad } \mathbf{a})\,\mathbf{dr}\,\Delta t^2$$

$$+ \mathbf{L}\mathbf{dr} \cdot \mathbf{L}\mathbf{dr}\,\Delta t^2 + \frac{1}{2}(\text{grad } \mathbf{a})\,\mathbf{dr} \cdot \mathbf{dr}\,\Delta t^2 + O(\Delta t^3) \tag{1.50}$$

If one collects the terms having the same Δt-exponent and takes into account the previously used identity $\mathbf{T}\mathbf{n} \cdot \mathbf{t} = \mathbf{n} \cdot \mathbf{T}^T\mathbf{t}$, then this formula takes on the following form:

$$\mathbf{dr}^* \cdot \mathbf{dr}^* = \mathbf{dr} \cdot \mathbf{dr} + \Delta t\,\mathbf{dr} \cdot \mathbf{A}_1\mathbf{dr} + \frac{\Delta t^2}{2}\mathbf{dr} \cdot \mathbf{A}_2\mathbf{dr} + O(\Delta t^3) \tag{1.51}$$

For abbreviation

$$\mathbf{A}_1 := \mathbf{L} + \mathbf{L}^T \tag{1.52}$$

$$\mathbf{A}_2 := \text{grad } \mathbf{a} + (\text{grad } \mathbf{a})^T + 2\,\mathbf{L}^T\mathbf{L} \tag{1.53}$$

are introduced.

$\mathbf{A}_1$ is twice the rate of strain tensor; $\mathbf{A}_1 = 2\mathbf{D}$ (equation (1.24)).

Using equation (1.51) the magnitude of the square of a line element at a later time (when $\Delta t > 0$) or its past 'history' (when $\Delta t < 0$) is approximated by the first term of the Taylor expansion for the present time. The coefficients of the expansion, which indicate the nth material derivative with respect to time of the square of the line element, can be written in the form

$$\frac{D^n|\mathbf{dr}|^2}{Dt^n} = \mathbf{dr} \cdot \mathbf{A}_n\mathbf{dr} \tag{1.54}$$

The kinematic quantities defined in this way $\mathbf{A}_n$ (n=1,2,3,...) are called *Rivlin-Ericksen tensors*. For a given velocity field one can find them by differentiation and matrix multiplication all at the same time. The recursion formula (1.55) is especially useful for this, and one can satisfy oneself about its accuracy as follows. Starting from the defining equation (1.54) one obtains

$$\mathbf{dr} \cdot \mathbf{A}_{n+1}\mathbf{dr} = \frac{D^{n+1}|\mathbf{dr}|^2}{Dt^{n+1}} = \frac{D}{Dt}(\mathbf{dr} \cdot \mathbf{A}_n\mathbf{dr})$$

$$= (\mathbf{dr})^{\cdot} \cdot \mathbf{A}_n\mathbf{dr} + \mathbf{dr} \cdot \frac{D\mathbf{A}_n}{Dt}\mathbf{dr} + \mathbf{dr} \cdot \mathbf{A}_n(\mathbf{dr})^{\cdot}$$

$$= \mathbf{dr} \cdot \left[\mathbf{L}^T\mathbf{A}_n + \frac{D\mathbf{A}_n}{Dt} + \mathbf{A}_n\mathbf{L}\right]\mathbf{dr}$$

The last equality follows from $(\mathbf{dr})^{\cdot} = \mathbf{L}\mathbf{dr}$ (equation (1.23)). If one now assumes the symmetry of all the Rivlin-Ericksen tensors one can conclude that

$$\mathbf{A}_{n+1} = \frac{D\mathbf{A}_n}{Dt} + \mathbf{L}^T\mathbf{A}_n + \mathbf{A}_n\mathbf{L} \tag{1.55}$$

The operator which, applied to $\mathbf{A}_n$, yields $\mathbf{A}_{n+1}$ is called the *Oldroyd derivative* with respect to time.

The *Jaumann time derivative* plays a part alongside the Oldroydian in the formulation of constitutive equations for viscoelastic fluids. This is

$$\mathring{\mathbf{A}} := \frac{D\mathbf{A}}{Dt} - \mathbf{W}\mathbf{A} + \mathbf{A}\mathbf{W} \tag{1.56}$$

which we shall denote by a small circle. With this abbreviation the recursion formula (1.55) can also be written in the form

$$\mathbf{A}_{n+1} = \mathring{\mathbf{A}}_n + \mathbf{D}\mathbf{A}_n + \mathbf{A}_n\mathbf{D} \tag{1.57}$$

1.5 Strain tensor

The Rivlin-Ericksen tensors are closely related to the Cauchy-Green strain tensor. Because it is convenient for the formulation of constitutive laws for non-Newtonian fluids to use this and a strain tensor closely related to it, we shall clarify its kinematic significance and derive the above-mentioned connection with the Rivlin-Ericksen tensors.

With the introduction of significant deformation quantities one must remember that the deformation in a continuum is generally dependent upon position. Thus for example a part of the fluid can remain completely undeformed, whilst the remainder is more or less severely deformed. One must therefore define deformation quantities locally for each fluid particle, and in addition take into account their dependence on time. Thus we shall first of all follow back into the past an arbitrarily chosen material point which is located at the position **r** at the time t, hence observing the path which this point has traversed. We must also state its former position **r*** as a function of the previous time t - s or, more appropriately, as a function of the time difference s from the present time. We must think of the path line of the observed material point in the form

$$\begin{aligned} x^* &= x^*(x, y, z, s; t) \\ y^* &= y^*(x, y, z, s; t) \\ z^* &= z^*(x, y, z, s; t) \end{aligned} \tag{1.58}$$

If we interpret here not only s as a variable, but also regard x,y,z as variables, then we obtain all the path lines of the other material points. The motion of the total continuum is therefore described by three functions which are dependent on the four variables x,y,z,s (t is a constant parameter in this respect).

The deformation of a fluid particle referred to the present state now depends on how the corresponding material point has moved in the past relative to its immediate surroundings. We therefore consider in addition to the material point that is located at a position **r** at time t an infinitesimally adjacent material point which at time t is located at **r** + d**r**, and consider the relative position vector d**r*** of the two points at the time t - s. Its components result from the three functions describing the motion of the fluid (1.58) in accordance with

$$\begin{aligned} dx^* &= \frac{\partial x^*}{\partial x} dx + \frac{\partial x^*}{\partial y} dy + \frac{\partial x^*}{\partial z} dz \\ dy^* &= \frac{\partial y^*}{\partial x} dx + \frac{\partial y^*}{\partial y} dy + \frac{\partial y^*}{\partial z} dz \\ dz^* &= \frac{\partial z^*}{\partial x} dx + \frac{\partial z^*}{\partial y} dy + \frac{\partial z^*}{\partial z} dz \end{aligned} \tag{1.59}$$

These formulas can be assembled into the vector relationship

$$d\mathbf{r}^* = \mathbf{F}_t d\mathbf{r} \tag{1.60}$$

We have introduced here for brevity the *relative deformation gradient* $\mathbf{F}_t$ which has the Cartesian components

$$\begin{bmatrix} F_{xx} & F_{xy} & F_{xz} \\ F_{yx} & F_{yy} & F_{yz} \\ F_{zx} & F_{zy} & F_{zz} \end{bmatrix} = \begin{bmatrix} \frac{\partial x^*}{\partial x} & \frac{\partial x^*}{\partial y} & \frac{\partial x^*}{\partial z} \\ \frac{\partial y^*}{\partial x} & \frac{\partial y^*}{\partial y} & \frac{\partial y^*}{\partial z} \\ \frac{\partial z^*}{\partial x} & \frac{\partial z^*}{\partial y} & \frac{\partial z^*}{\partial z} \end{bmatrix} \tag{1.61}$$

According to equation (1.60) and neglecting a parallel displacement, it transfers the line joining infinitesimally adjacent material points from its position at time t to that at time t - s. The subscript t symbolises the relative (based on the position at time t) character of the deformation gradient $\mathbf{F}_t$. Because we consider t to be a fixed moment in time, and s on the other hand as variable in time ($0 \leq s < \infty$), $\mathbf{F}_t$ strictly speaking is the past *history* of the relative deformation gradient. In other respects $\mathbf{F}_t$ of course also generally depends on position, so that we can for convenience write in more detail $\mathbf{F}_t(\mathbf{r},s)$.

The significance of the individual elements of the relative deformation gradient is clear when one specially traces back such line elements which are oriented at time t directly along one of the coordinate directions. From their positions at time t - s one can in a simple way find the quantities F_{xx}, F_{xy}, etc. For example suppose that one starts from a line element of length dx oriented along the x-direction (dy=dz=0), then the relationships (1.59) reduce to $dx^* = F_{xx}dx$, $dy^* = F_{yx}dx$, and $dz^* = F_{zx}dx$. Thus the components of the position vector of this line element at time t - s are directly connected with the elements F_{xx}, F_{yx}, F_{zx} (Fig. 1.6).

When the infinitesimal surroundings of a material point are displaced by translation alone, then the orientation and the length of all line elements remain constant, and the relative deformation gradient reduces to the unit tensor **E**. If $\mathbf{F}_t$ differs from **E**, then this does not mean implicitly that strains occur. For example one can imagine a special type of 'deformation' which only consists of the continuum being rotated as a whole (like a rigid body) round an axis. Hence in general $\mathbf{F}_t$ is not equal to **E**, and no strains occur at any place in the material. The deformation gradient is therefore not a suitable measure of the strain.

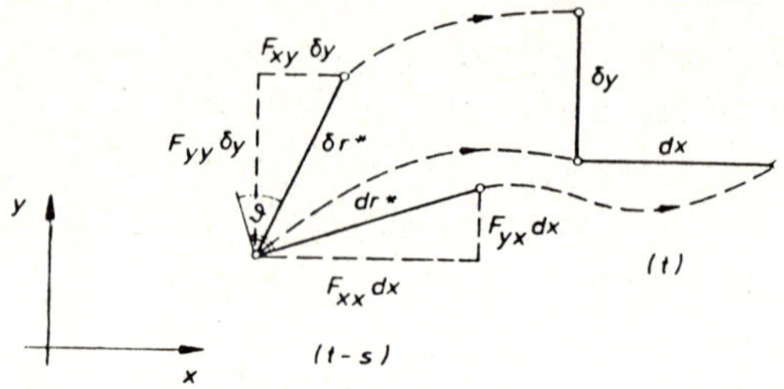

Fig. 1.6 Diagram of the elements of the deformation gradient and of the elements of the strain tensor

It is demonstrated in books about continuum mechanics that one can consider every deformation of a material volume element either as an extension (or a compression) of the element in three mutually perpendicular directions, followed by rotation of the strained element, or as a rotation of the undeformed element followed by an extension (or compression) in three mutually perpendicular directions. Hence it comes about, and we shall understand this in what follows, that the symmetrical tensor

$$\mathbf{C}_t(\mathbf{r}, s) := \mathbf{F}_t^T(\mathbf{r}, s)\, \mathbf{F}_t(\mathbf{r}, s) \tag{1.62}$$

describes the strain in the volume element, but not its rotation, and thus represents a suitable measure of the strain. $\mathbf{C}_t$ is called the (relative) *right-Cauchy-Green tensor*, conventionally also the *Cauchy strain tensor* In order to obtain an illustrative interpretation of its Cartesian components C_{xx}, C_{xy}, etc., we consider in the configuration at time t - s the scalar product of two material line elements which at the actual time are described by the vectors $\mathbf{dr}$ and $\delta\mathbf{r}$:

$$\mathbf{dr}^* \cdot \delta\mathbf{r}^* = \mathbf{F}_t\mathbf{dr} \cdot \mathbf{F}_t\delta\mathbf{r} = \mathbf{dr} \cdot \mathbf{F}_t^T\mathbf{F}_t\delta\mathbf{r} = \mathbf{dr} \cdot \mathbf{C}_t\delta\mathbf{r} \tag{1.63}$$

Let us first of all consider the two line elements as coincident ($\mathbf{dr} = \delta\mathbf{r}$) and choose their position to be parallel to a coordinate direction, so that on the right-hand side of equation (1.63) only a principal diagonal of $\mathbf{C}_t$ occurs. For two adjacent material points for example which at the moment lie at a distance dx on a line parallel to the x-axis, and whose distance at time t - s is defined as $\mathbf{dr}^*$ (Fig. 1.6), equation (1.63) reduces to

$$C_{xx} = \left(\frac{\mathbf{dr}^*}{\mathbf{dx}}\right)^2 \tag{1.64}$$

Corresponding formulas hold for $\mathbf{C}_{yy}$ and $\mathbf{C}_{zz}$. The diagonal elements of the right-Cauchy-Green tensor therefore describe how the line elements which are oriented in the reference state in the coordinate direction have shortened or lengthened. The non-diagonal elements refer to the angles which two line elements, at present oriented in different coordinate directions, have previously enclosed. We choose for example in accordance with Fig. 1.6 $d\mathbf{r} = dx\mathbf{e}_x$ and $\delta\mathbf{r} = \delta y\mathbf{e}_y$, and we designate the angle previously formed by these line elements as $90^\circ - \vartheta$; thus the equation (1.63) gives the relationship $d\mathbf{r}^*\delta\mathbf{r}^*\sin\vartheta = \mathbf{C}_{xy}dx\delta y$. By using the result (1.64) one obtains the following relationship between the elements of the strain tensor and the angle introduced:

$$\mathbf{C}_{xy} = \sin\vartheta\sqrt{\mathbf{C}_{xx}\mathbf{C}_{yy}} \tag{1.65}$$

If one knows the elements of the strain tensor $\mathbf{C}_t$ one can hence specify from which parallelepiped a cuboid material element has originated, the edges of which are at the moment oriented in the coordinate directions. The spatial orientation of the epiped is (significantly) not described by the strain tensor.

It is incidentally found to be troublesome that the strain tensor $\mathbf{C}_t$ (like the deformation gradient $\mathbf{F}_t$) reduces to unity when the continuum remains locally undeformed. Therefore one often prefers to use the *Green strain tensor* $(\mathbf{C}_t - \mathbf{E})/2$, which in the undeformed state vanishes, but otherwise concerning the secondary diagonal elements and the differences of the main diagonal elements is in agreement with the right-Cauchy-Green tensor up to a factor of 1/2.

The connection wanted at the beginning of this discussion between the Rivlin-Ericksen tensors and the history of the right-Cauchy-Green tensor results from the comparison of formulas (1.63) and (1.51). Both define the square of the distance between two material points at time t - s, so long as in equation (1.63) $\delta\mathbf{r} = d\mathbf{r}$ and in equation (1.51) Δt is put as -s. By equating the right-hand sides,

$$d\mathbf{r}\cdot\mathbf{C}_t d\mathbf{r} = d\mathbf{r}\cdot(\mathbf{E} - s\,\mathbf{A}_1 + \frac{s^2}{2}\mathbf{A}_2 \mp \dots)\,d\mathbf{r} \tag{1.66}$$

one can conclude, because $d\mathbf{r}$ is an arbitrary vector and not only the strain tensor but also the Rivlin-Ericksen tensors are symmetrical, that

$$\mathbf{C}_t(\mathbf{r}, s) = \mathbf{E} - s\,\mathbf{A}_1(\mathbf{r}; t) + \frac{s^2}{2}\mathbf{A}_2(\mathbf{r}; t) \mp \dots \tag{1.67}$$

Accordingly the Rivlin-Ericksen tensors are the development coefficients of the past history of the right-Cauchy-Green tensor, and it is true that

$$\mathbf{A}_n(\mathbf{r};t) = (-1)^n \left.\frac{\partial^n \mathbf{C}_t(\mathbf{r},s)}{\partial s^n}\right|_{s=0} \tag{1.68}$$

Because in fluid mechanics it is not so much the motion of the fluid in the form (1.58), but in general the velocity field which is interesting, and the Rivlin-Ericksen tensors can be derived from the velocity field, it is obvious to refer to equation (1.67) for the calculation of the past history of the relative right-Cauchy-Green tensor. This is especially recommended for those flows for which the Rivlin-Ericksen tensors of a certain degree onwards vanish, so that the right-hand side of equation (1.67) degenerates to a sum of a finite number of terms. This is especially so for all steady shear flows, cf. Section 1.6. With flow fields for which the higher order Rivlin-Ericksen tensors do not vanish, one will preferably express the motion (1.58) by means of the velocity field, and determine $\mathbf{C}_t$ (1.62) from the relative deformation gradient. A simple but not trivial example will illustrate the method. We consider a steady flow which has the velocity field

$$u = u(y), \qquad v = -V = \text{const}, \qquad w = 0 \tag{1.69}$$

This is the flow field at a greater distance from the front edge of a plane plate along which there is a flow, and at which the liquid is drawn away homogeneously, i.e. with a spatially constant velocity perpendicular to the wall. We shall study the dynamics of this flow in more detail in Section 8.2.2; only some kinematic aspects will be briefly considered here. Because the velocity component in the y-direction is constant and equal to -V, the y-coordinate of a chosen material point in the time interval s varies about Vs, and because w = 0 the z-coordinate remains constant. For the position of a material point $\mathbf{r}^*$ at time t - s, which at time t lies at position $\mathbf{r}$, therefore the following relationships apply

$$y^* = y + Vs, \qquad z^* = z \tag{1.70}$$

A third relationship follows from the consideration that the material point has traversed in the element of time ds in the x-direction the path $u(y^*)ds$:

$$-dx^* = u(y + Vs)\,ds \tag{1.71}$$

The minus sign is explained by the fact that the time variable s is counted 'backwards', i.e. from the present back into the past. On the right-hand side y^* can be replaced in accordance with equation (1.70) by $y + Vs$. If one integrates with the initial conditions $x^* = x$ at $s = 0$, then one obtains the relationship

$$x^* = x - \frac{1}{V}\int_{y}^{y+Vs} u(\sigma)\,d\sigma \tag{1.72}$$

Using equations (1.70) and (1.72) the motion is now considered in the form of (1.58). As a result of the steadiness of the flow the actual time t does not occur explicitly. By partial differentiation of the three expressions with respect to x,y,z one obtains for the relative deformation gradient (see equation (1.61))

$$F_t(r, s) = \begin{bmatrix} 1 & -\frac{1}{V}[u(y + Vs) - u(y)] & 0 \\ 0 & 1 & 0 \\ 0 & 0 & 1 \end{bmatrix} \tag{1.73}$$

Hence this yields the following expressions for the history of the relative right-Cauchy-Green tensor (according to equation (1.62)):

$$C_t(r, s) = \begin{bmatrix} 1 & -\frac{1}{V}[u(y + Vs) - u(y)] & 0 \\ -\frac{1}{V}[u(y + Vs) - u(y)] & 1 + \frac{1}{V^2}[u(y + Vs) - u(y)]^2 & 0 \\ 0 & 0 & 1 \end{bmatrix} \tag{1.74}$$

It is noteworthy that besides the constant velocity component V normal to the wall and the local velocity component $u(y)$ parallel to the wall, only the velocity component $u(y + Vs)$ enters, which a material point possessed at the time in the past s.

One can easily satisfy oneself from this fact that in the present case the determinant of the tensor C_t always has the value unity. Because the determinant is one of the principal invariants, its numerical value remains constant if one changes over to the main axis system of $\mathbf{C}_t$. The product of the three main diagonal elements then occurring alone is therefore equal to unity. The square roots of these elements can now be thought of from equation (1.64) as extension ratios of three line elements, which make up a cube-shaped material volume element, and from this cube there results a square, because in the main axis system all secondary diagonal elements vanish, and therefore according to equation (1.65) the right angles between the line elements are maintained. If det $\mathbf{C}_t = 1$, then accordingly the products of the three edge lengths of the cube and the square agree, i.e. the flow has the property that the volume of each fluid particle remains constant with respect to time. Therefore this is an isochoric flow.

1.6 Kinematics of steady shear flows

We study in what follows some kinematically very simple aspects of important steady flows which can be described as follows: the flow field consists of a set of material surfaces which during the motion remain, undeformed and untwisted, but which slide over one another. Each shearing surface behaves as a rigid surface, i.e. it is displaced or rotated as a whole. Any shearing surface can therefore be replaced by a correspondingly moved rigid wall. Flows that possess this property are conveniently called *viscometric*. Because they thus possess special kinematic properties, the kinematic tensors take on a specific form. With reference to a suitably chosen orthogonal coordinate system (but not always a Cartesian system), the first Rivlin-Ericksen tensor $\mathbf{A}_1$ and hence the rate of strain tensor $\mathbf{D} = \frac{1}{2}\mathbf{A}_1$ and the second Rivlin-Ericksen tensor $\mathbf{A}_2$ respectively possess only one essential instead of six independent elements. For $\mathbf{A}_2$ it is a main diagonal element; for $\mathbf{A}_1$ two components are affected on the secondary diagonal because of the symmetry of the tensor. In both cases the elements are determined by the shear rate $\dot{\gamma}$ alone, the rate at which an instantaneous right angle changes with time between a line element normal to the shearing surface and one suitably chosen line element in the shearing surface. The shear rate is in general a function of position, in particular cases spatially constant. All higher Rivlin-Ericksen tensors $\mathbf{A}_n (n \geq 3)$ vanish for steady shear flows.

1.6.1 Plane shear flow

The sliding surfaces are parallel planes with y = constant, which all move in the x-direction. Their velocity depends arbitrarily on y:

$$u = u(y), \qquad v = 0, \qquad w = 0 \tag{1.75}$$

The simple shear flow discussed in Section 1.3 is a special case of plane shear flow. The following kinematic tensors obey the velocity field (1.75)

$$\mathbf{L} = \begin{bmatrix} 0 & du/dy & 0 \\ 0 & 0 & 0 \\ 0 & 0 & 0 \end{bmatrix} \tag{1.76}$$

$$\mathbf{A}_1 = 2\,\mathbf{D} = \begin{bmatrix} 0 & du/dy & 0 \\ du/dy & 0 & 0 \\ 0 & 0 & 0 \end{bmatrix} \tag{1.77}$$

Concerning this connection between the velocity derivative du/dy and the shear rate $\dot{\gamma}$, the treatment carried out as in Fig. 1.4 also holds for plane shear flows. The result was

$$\dot{\gamma} = \frac{du}{dy} \tag{1.36}$$

We now calculate the other Rivlin-Ericksen tensors by using equation (1.55). Because time independent laminar flows are *steady in material coordinates* (D/Dt = 0), the first term on the right-hand side of equation (1.55) vanishes. Using (1.76) and (1.77) one then calculates

$$\mathbf{A}_2 = \begin{bmatrix} 0 & 0 & 0 \\ \dot{\gamma} & 0 & 0 \\ 0 & 0 & 0 \end{bmatrix} \begin{bmatrix} 0 & \dot{\gamma} & 0 \\ \dot{\gamma} & 0 & 0 \\ 0 & 0 & 0 \end{bmatrix} + \begin{bmatrix} 0 & \dot{\gamma} & 0 \\ \dot{\gamma} & 0 & 0 \\ 0 & 0 & 0 \end{bmatrix} \begin{bmatrix} 0 & \dot{\gamma} & 0 \\ 0 & 0 & 0 \\ 0 & 0 & 0 \end{bmatrix}$$

$$= \begin{bmatrix} 0 & 0 & 0 \\ 0 & 2\dot{\gamma}^2 & 0 \\ 0 & 0 & 0 \end{bmatrix} \tag{1.78}$$

$$\mathbf{A}_3 = \begin{bmatrix} 0 & 0 & 0 \\ \dot{\gamma} & 0 & 0 \\ 0 & 0 & 0 \end{bmatrix} \begin{bmatrix} 0 & 0 & 0 \\ 0 & 2\dot{\gamma}^2 & 0 \\ 0 & 0 & 0 \end{bmatrix} + \begin{bmatrix} 0 & 0 & 0 \\ 0 & 2\dot{\gamma}^2 & 0 \\ 0 & 0 & 0 \end{bmatrix} \begin{bmatrix} 0 & \dot{\gamma} & 0 \\ 0 & 0 & 0 \\ 0 & 0 & 0 \end{bmatrix}$$

$$= \begin{bmatrix} 0 & 0 & 0 \\ 0 & 0 & 0 \\ 0 & 0 & 0 \end{bmatrix} \tag{1.79}$$

After $\mathbf{A}_3$ all higher Rivlin-Ericksen tensors vanish,

$$\mathbf{A}_n = 0 \text{ for } n \geq 3. \tag{1.80}$$

This property is characteristic of all steady shear flows. The two principal invariants $\mathbf{A}_1$ and $\mathbf{A}_2$ are zero in both cases. The non-vanishing invariants are related in a simple way to $\dot{\gamma}^2$: $I_2(\mathbf{A}_1) = -\dot{\gamma}^2$, $I_1(\mathbf{A}_2) = 2\dot{\gamma}^2$.

It is expedient for later use to make the following clear: for a steady plane shear flow an observer travelling on an arbitrarily chosen material point sees in the immediate vicinity of the point a simple shear flow which always remains constant. Therefore he is always looking back on the same 'deformation history', and the deformation state can be described by a single scalar quantity, the shear rate $\dot{\gamma}$. The memory of the past therefore consists of a single bit of information, the quantity $\dot{\gamma}$ which is constant for each fluid particle, so viscometric flows are special *flows with a constant stretch history.*

1.6.2 Poiseuille flow

The shearing surfaces in this flow are coaxial circular section cylindrical surfaces which move in the axial direction. The velocity of the individual shearing surfaces depends only on the distance r from the axis of the cylinder. It is expedient here to use cylindrical polar coordinates r, φ, z, instead of Cartesian coordinates, so that the common axis of the shearing surfaces is identical with the z-axis (Fig. 1.7).

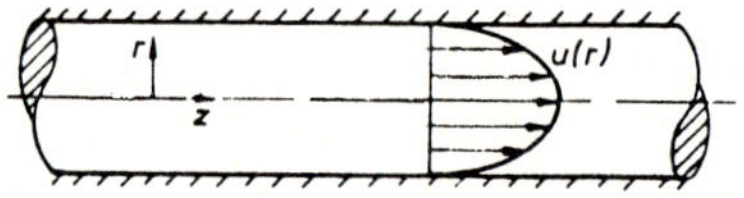

Fig. 1.7 Poiseuille flow in a cylindrical pipe with circular cross-section

The velocity field is then $v_r = 0, \quad v_\varphi = 0, \quad v_z = u(r)$ (1.81)

For the shear rate as the rate of change of the angle between a line element in the z-direction and another in the r-direction the relationship

$$\dot{\gamma} = \frac{du}{dr} \tag{1.82}$$

applies, as for a plane shear flow.

1.6.3 Couette flow

A *Couette flow* is one in which the shearing surfaces are coaxial cylinders which rotate with individual angular velocities $\omega(r)$ round the common z-axis. Using cylindrical polar coordinates the velocity field is described as shown in Fig. 1.8 by

$$v_r = 0, \qquad v_\varphi = r\,\omega(r), \qquad v_z = 0 \tag{1.83}$$

The relationship between the shear rate $\dot{\gamma}$ and the angular velocity ω becomes clear when one considers two material line elements which are oriented directly along the r-direction and φ-direction respectively. The rate of change of the angle is obtained from the displacement of the three points in the time interval Δt as

$$\dot{\gamma} = r\,\frac{d\omega}{dr} \tag{1.84}$$

as shown in Fig. 1.9.

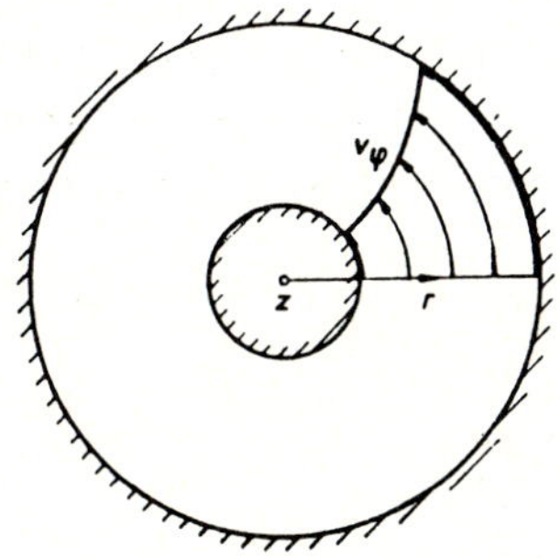

Fig. 1.8 Couette flow between rotating coaxial cylinders

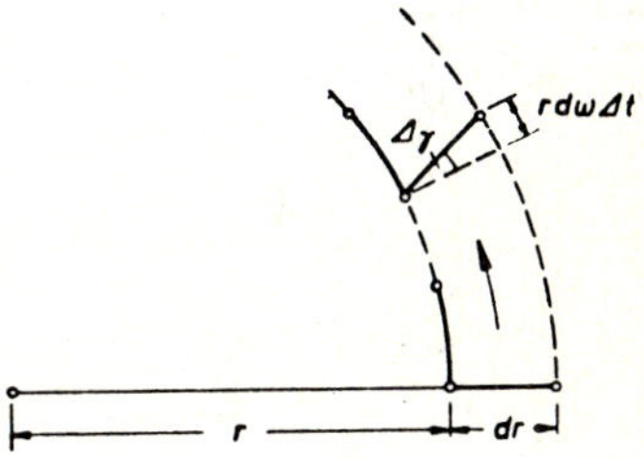

Fig. 1.9 The derivation of equation (1.84)

Flows of this type occur in a gap filled with a viscous liquid between two cylindrical walls (radii r_1 and $r_1 + \Delta r$), when these rotate at constant angular velocities ω_1 and $\omega_1 + \Delta\omega$. For a narrow gap, that is if $\Delta r \ll r_1$, the factor r can be approximately replaced by r_1, and $d\omega/dr$ by $\Delta\omega/\Delta r$ in the formula (1.84). For a narrow gap the shear rate in the field is therefore approaching a constant value

$$\dot{\gamma} \simeq r_1 \frac{\Delta\omega}{\Delta r} \tag{1.85}$$

and the flow is locally equal to a simple shear flow.

1.6.4 Helical flow

The superposition of the velocity fields (1.81) and (1.83) leads to a shear flow in which the circular cylindrical shearing surfaces not only rotate but also move axially

$$v_r = 0, \qquad v_\varphi = r\,\omega(r), \qquad v_z = u\,(r) \tag{1.86}$$

Each fluid particle thus maintains its distance r from the axis of rotation and moves along a helical path with the helix angle $u(r)/r\omega(r)$.

Note that with respect to the base used here consisting of the unit vectors $\mathbf{e}_r$, $\mathbf{e}_\varphi$, $\mathbf{e}_z$ the strain rate tensor has two independent elements and the second Rivlin-Ericksen tensor has only one rr-element. Use the formulas given in the Appendix to calculate the tensors in cylindrical coordinates.

$$\mathbf{A}_1 = 2\mathbf{D} = \begin{bmatrix} 0 & r\dfrac{d\omega}{dr} & \dfrac{du}{dr} \\ r\dfrac{d\omega}{dr} & 0 & 0 \\ \dfrac{du}{dr} & 0 & 0 \end{bmatrix}, \qquad \mathbf{A}_2 = \begin{bmatrix} 2\left(\dfrac{du}{dr}\right)^2 + 2r^2\left(\dfrac{d\omega}{dr}\right)^2 & 0 & 0 \\ 0 & 0 & 0 \\ 0 & 0 & 0 \end{bmatrix} \tag{1.87}$$

The higher Rivlin-Ericksen tensors vanish.

By introducing a suitably chosen base one can bring the matrices of $\mathbf{A}_1$ and $\mathbf{A}_2$ to that form which describes simple shear flows (equations (1.77) and (1.78)).

$$\mathbf{A}_1 = 2\mathbf{D} = \begin{bmatrix} 0 & \dot{\gamma} & 0 \\ \dot{\gamma} & 0 & 0 \\ 0 & 0 & 0 \end{bmatrix}, \qquad \mathbf{A}_2 = \begin{bmatrix} 0 & 0 & 0 \\ 0 & 2\dot{\gamma}^2 & 0 \\ 0 & 0 & 0 \end{bmatrix} \tag{1.88}$$

One learns from this that the deformation of the fluid particles, apart from a rotation, consists only of a simple shear. The relation between the shear rate $\dot{\gamma}$, the axial velocity field $u(r)$ and the azimuthal rotation velocity field $\omega(r)$ results from considering that the invariants of the matrices (1.87) and (1.88) for $\mathbf{A}_1$ and $\mathbf{A}_2$ must be equal.

$$\dot{\gamma} = \sqrt{\left(\frac{du}{dr}\right)^2 + r^2 \left(\frac{d\omega}{dr}\right)^2} \tag{1.89}$$

The unit vectors $\mathbf{e}_1$, $\mathbf{e}_2$, $\mathbf{e}_3$ of the *natural base* on which the matrices in (1.88) are based, are connected with the base $\mathbf{e}_r$, $\mathbf{e}_\varphi$, $\mathbf{e}_z$:

$$\begin{aligned} \mathbf{e}_1 &= \frac{1}{\dot{\gamma}} \left[r \frac{d\omega}{dr} \mathbf{e}_\varphi + \frac{du}{dr} \mathbf{e}_z \right] \\ \mathbf{e}_2 &= \mathbf{e}_r \\ \mathbf{e}_3 &= \frac{1}{\dot{\gamma}} \left[\frac{du}{dr} \mathbf{e}_\varphi - r \frac{d\omega}{dr} \mathbf{e}_z \right] \end{aligned} \tag{1.90}$$

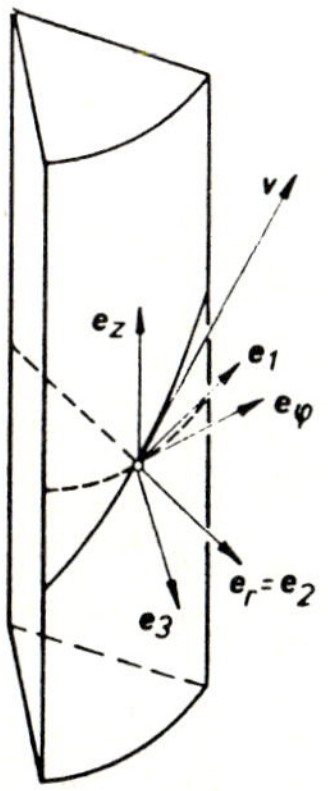

Fig. 1.10 Natural base for helical flow

These relationships are found by comparing the expressions (1.87) and (1.88). It follows directly from the two matrices for $\mathbf{A}_2$ that $\mathbf{e}_2$ is equivalent to $\mathbf{e}_r$. The first Rivlin-Ericksen tensor can according to (1.87) be represented on the one hand by the form $r \frac{d\omega}{dr} (\mathbf{e}_r \otimes \mathbf{e}_\varphi + \mathbf{e}_\varphi \otimes \mathbf{e}_r) + \frac{du}{dr} (\mathbf{e}_r \otimes \mathbf{e}_z + \mathbf{e}_z \otimes \mathbf{e}_r)$, and on the other hand as $\dot{\gamma} (\mathbf{e}_2 \otimes \mathbf{e}_1 + \mathbf{e}_1 \otimes \mathbf{e}_2)$ according to (1.88). Here $\otimes$ signifies the dyadic product of two vectors. The first relationship (1.90) follows from the equality of the two expressions because $\mathbf{e}_2 = \mathbf{e}_r$. The third relationship follows from the requirement that the unit vector $\mathbf{e}_3$ is normal to $\mathbf{e}_1$ and $\mathbf{e}_2$, and with these forms a rectangular system.

As in a simple shear flow $\mathbf{e}_2$ specifies the direction normal to the shearing surfaces. The vector $\mathbf{e}_1$ generally does not lie in the direction of motion given by the velocity vector (cf. Fig. 1.10). It is tangential to the helical path line only if $\omega(r)/u(r)$ is constant.

1.6.5 Torsional flow

The shearing surfaces are parallel to the planes z = constant, which rotate with individual angular velocities round the z-axis. The velocity field is then given by

$$v_r = 0, \qquad v_\varphi = r\,\omega(z), \qquad v_z = 0 \tag{1.91}$$

Similarly, as has been previously explained, one can derive the shear rate

$$\dot{\gamma} = r\frac{d\omega}{dz} \tag{1.92}$$

A flow of this type would take place in an imaginary inertialess, viscous fluid between two parallel planes rotating round the same axis (Fig. 1.11). In this case the term $d\omega/dz$ in equation (1.92) would be spatially constant and equal to $\Delta\omega/h$.

1.6.6 Cone-and-plate flow

The shearing surfaces are a disc and a cone with a common point of contact, which rotate with individual angular velocities round the common axis. It is convenient to use spherical polar coordinates R, ϑ, φ (Fig. 1.12) to describe the flow field. In this case there is only one velocity component in the φ-direction, and this depends simply on the angular velocity ω of the cone-shaped shearing surfaces:

$$v_R = 0, \qquad v_\vartheta = 0, \qquad v_\varphi = R\sin\vartheta \cdot \omega(\vartheta) \tag{1.93}$$

The relative velocity of infinitesimally adjacent cones is $R\sin\vartheta \cdot d\omega$ and their distance apart is $Rd\vartheta$. Thus one obtains the rate of shear

$$\dot{\gamma} = \sin\vartheta\,\frac{d\omega}{d\vartheta} \tag{1.94}$$

The described flow form can be approximately realised in a viscous liquid between a rotating cone ($\vartheta_K = 90^\circ - \beta$) and a rotating plate ($\vartheta_P = 90^\circ$) which are in contact. One of the two surfaces may be at rest. Of particular interest here is the case where the cone is very obtuse, i.e. $\beta \ll 1$. In this case the terms $\sin\vartheta$ and $d\omega/d\vartheta$ in formula (1.94) can to a good approximation be replaced by 1 and $\Delta\omega/\beta$ respectively, in which $\Delta\omega$ represents the difference between the angular velocities of plate and cone:

$$\dot{\gamma} \simeq \frac{\Delta\omega}{\beta} \tag{1.95}$$

Under these conditions, as in the case of a simple shear flow, the shear rate assumes the same value everywhere in the flow field.

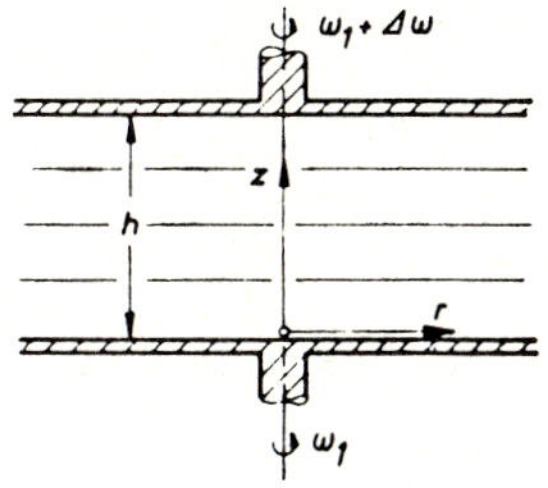

Fig. 1.11 Creation of a torsional flow

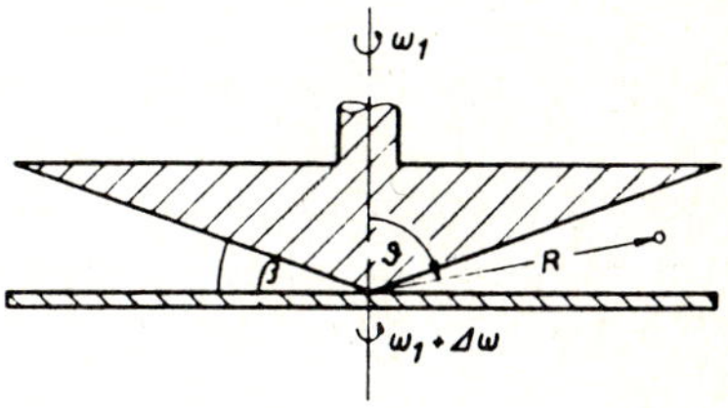

Fig. 1.12 Cone-and-plate flow

1.7 Continuity equation

So far we have been concerned with the kinematics of continua in motion, and for that we have only needed the *velocity field* $\mathbf{v}(x,y,z,t)$. We now introduce the *density field* $\rho(x,y,z,t)$, which describes the mass distribution of the continuum per unit volume as a function of position and time. Further, we consider a spatially constant control volume of arbitrary size V, which is completely filled by the flowing material. The mass contained in V is obtained by integration over the density field $\iiint_V \rho(x,y,z,t)\,dx\,dy\,dz$, which we shall write more briefly as $\iiint_V \rho dV$. This quantity can only increase with time when more liquid flows out across the surface A of the control volume. Now the mass flux through a surface element dA results from the product of the local density, the local velocity component normal to the surface element, and the size of the surface element itself. Therefore the total mass flux emerging from the control volume is given by the integral $\iint_A \rho\mathbf{v}\cdot\mathbf{n}dA$. The quantity $\mathbf{n}$ always designates here and in what follows the unit vector of the external normals to the surface area of the control volume (Fig. 1.13). Hence the following integral form of the law of conservation of mass applies:

$$\iiint_V \frac{\partial \rho}{\partial t}\, dV = -\iint_A \rho\, \mathbf{v} \cdot \mathbf{n}\, dA \qquad (1.96)$$

The surface integral can be transformed by the Gaussian theorem

$$\iint_A \mathbf{b} \cdot \mathbf{n}\, dA = \iiint_V \operatorname{div} \mathbf{b}\, dV \qquad (1.97)$$

into a volume integral. Thus the integral over V is obtained with the integrand $\partial\rho/\partial t + \operatorname{div}(\rho\,\mathbf{v})$. Because of the identity $\operatorname{div}(\rho\,\mathbf{v}) = \mathbf{v} \cdot \operatorname{grad} \rho + \rho \operatorname{div} \mathbf{v}$, and by use of the material time derivative given in equation (1.12) this can also be written in the form $D\rho/Dt + \rho \operatorname{div} \mathbf{v}$. Because the domain of integration V is arbitrary, the integrand must vanish. Therefore one obtains the *continuity equation* as the differential form of the mass balance

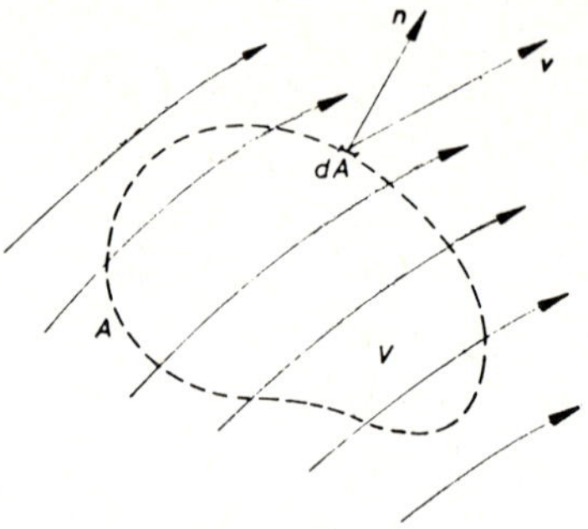

Fig. 1.13 Flow through a control volume

$$\frac{D\rho}{Dt} + \rho \operatorname{div} \mathbf{v} = 0 \qquad (1.98)$$

For reasons of conservation of mass a connection exists between the velocity field and the density field.

We shall later consider predominantly liquid fields, whose density depends only insignificantly on pressure and temperature. With such 'incompressible', or more appropriately *constant density* fluids the density and therefore the volume of each fluid particle remain constant with time ($D\rho/Dt = 0$), therefore only isochoric motions are possible. In the simplest and at the same time most important special case the density ρ is a material constant, which one can determine when the fluid is in a state of rest. For constant density fluids the continuity equation simplifies to

$$\operatorname{div} \mathbf{v} = 0 \qquad (1.99)$$

By use of Cartesian coordinates this relationship is written out in detail thus

$$\frac{\partial u}{\partial x} + \frac{\partial v}{\partial y} + \frac{\partial w}{\partial z} = 0 \qquad (1.100)$$

In the expression on the left-hand side one recognises the first invariant (the trace) of the strain rate tensor (equations (1.24) and (1.43)).

In constant density fluids accordingly only those flows are realisable for which the divergence of the velocity field and therefore the trace of the strain rate tensor and hence of the first Rivlin-Ericksen tensor vanishes.

A flow is called plane if the velocity component in one fixed direction, e.g. in the z-direction everywhere and at all times vanishes, and the velocity field does not depend on z, hence if

$$u = u(x, y, t), \qquad v = v(x, y, t), \qquad w = 0 \tag{1.101}$$

In the case of plane flows of constant density fluids the continuity equation reduces to

$$\frac{\partial u}{\partial x} + \frac{\partial v}{\partial y} = 0 \tag{1.102}$$

This differential equation can be integrated by the introduction of a *stream function* $\Psi(x,y,t)$. If one puts

$$u = \frac{\partial \Psi}{\partial y}, \qquad v = -\frac{\partial \Psi}{\partial x} \tag{1.103}$$

which by the way means $\mathbf{v} = \text{curl}(\Psi\,\mathbf{e}_z)$, then the continuity equation (1.102) is obviously satisfied. According to these relations the vectorial velocity field with two components is reduced to a single scalar field. Furthermore the stream function Ψ has an obvious meaning. The curves $\Psi(x,y,t)$ = constant at a certain time are of course the stream lines, because if one proceeds along such a curve

$$0 = d\Psi = \frac{\partial \Psi}{\partial x}dx + \frac{\partial \Psi}{\partial y}dy = -\,v dx + u dy \tag{1.104}$$

applies.

The relationship $dy/dx = v/u$ which follows from that is the differential equation of the stream lines (equation (1.7)).

For an illustration we shall consider once more a plane elongational flow having the velocity field (1.3). This obviously satisfies the requirements of (1.102), so that a stream function exists. To determine the stream function one inserts the given velocity components into the left-hand side of equation (1.103) and integrates. The first relationship yields $\Psi = a(t)xy + C(x,t)$; the second excludes the constant of integration C from being dependent on x. Now a time dependent term added to the stream function is insignificant because it disappears when constructing the velocity field according to equation (1.103). It can therefore be made equal to zero without loss of generality.

Thus the expression $\Psi = a(t)xy$ describes the function of the plane elongational flow. If one makes Ψ constant then one obtains the stream lines by taking into account equation (1.104). In the above cases therefore there is agreement with the findings obtained earlier concerning rectangular hyperbolas, xy = constant (Fig. 1.1).

1.8 Stress and volume force

Concerning the forces that act on a chosen part of a continuum, one has to distinguish between *volume forces* and *surface forces*. Volume forces act on the interior of the body; surface forces act on the boundary between the material volume considered and its surroundings. One uses the volume force density $\mathbf{f}$, which is the force per unit volume, and the stress vector $\mathbf{t}$, the force per unit surface area (Fig. 1.14). Therefore

$$\iint_A \mathbf{t}\, dA + \iiint_V \mathbf{f}\, dV \tag{1.105}$$

is the total force acting on a material volume V which has a surface area A.

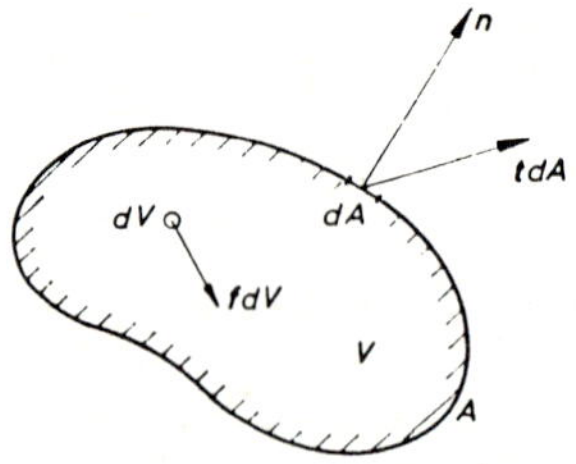

Fig. 1.14 Stress vector and volume force density

The volume force in most applications is only the gravitational force. In this case $\mathbf{f} = -\rho g \mathbf{e}_z$, provided that the z-axis is perpendicularly upwards (g is the acceleration caused by the Earth's gravity).

One designates as *normal stress* that component of the stress vector $\mathbf{t}$ which belongs to the direction of the external normal $\mathbf{n}$, hence the scalar product $\mathbf{t}\cdot\mathbf{n}$. The other components oriented tangential to the surface are designated as *shear stresses*. With the help of these terms in what follows the only interesting fluids can be separated from solids.

A deformable substance is called a fluid if the shear forces acting on it, no matter how small, continuously deform the substance, and within an unlimited period cause unlimited large deformations. From this it follows that a fluid in contrast to a solid body can only be at rest when it is not subjected to any shear forces. Normal stresses alone are present on the surface of an arbitrarily chosen part of a fluid at rest. An important concept of elementary fluid mechanics states that the normal stress in a fluid at rest is independent of the orientation of the surface elements concerned. Therefore a scalar field function suffices to describe the state of stress of a fluid at rest, and this is dependent on position and possibly also on time, and is called the *hydrostatic pressure* p. Because this pressure force is oriented along the external normals, the equation $\mathbf{t} = -p\mathbf{n}$ describes the stress vector in a fluid at rest.

Shear stresses are generally also present in a moving fluid. The normal stress generally depends on the orientation of the surface elements and is therefore of different magnitude in various directions. It is usual to introduce special symbols for the components of those three stress vectors which belong to the normals $\mathbf{e}_x$, $\mathbf{e}_y$, and $\mathbf{e}_z$:

$$\mathbf{t}(\mathbf{e}_x) = \begin{bmatrix} -p + \tau_{xx} \\ \tau_{yx} \\ \tau_{zx} \end{bmatrix}; \quad \mathbf{t}(\mathbf{e}_y) = \begin{bmatrix} \tau_{xy} \\ -p + \tau_{yy} \\ \tau_{zy} \end{bmatrix}; \quad \mathbf{t}(\mathbf{e}_z) = \begin{bmatrix} \tau_{xz} \\ \tau_{yz} \\ -p + \tau_{zz} \end{bmatrix} \tag{1.106}$$

For a cube-shaped element whose edges are oriented parallel to the coordinate axes, the components $-p + \tau_{xx}$, $-p + \tau_{yy}$ and $-p + \tau_{zz}$ appear as normal stresses, τ_{xy}, τ_{xz}, etc. as shear stresses (Fig. 1.15).

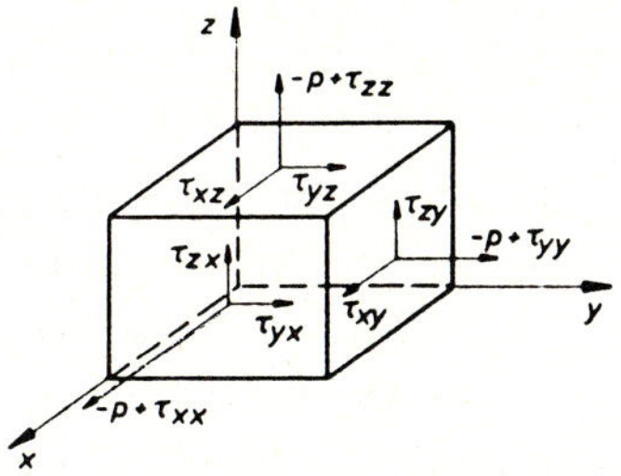

Fig. 1.15 Stress tensor elements

In the textbooks about mechanics it is demonstrated that the stress vector belonging to an arbitrarily oriented surface element is composed of those nine elements and the components n_x, n_y, n_z of the normal vector according to

$$\begin{bmatrix} t_x \\ t_y \\ t_z \end{bmatrix} = \begin{bmatrix} -p + \tau_{xx} & \tau_{xy} & \tau_{xz} \\ \tau_{yx} & -p + \tau_{yy} & \tau_{yz} \\ \tau_{zx} & \tau_{zy} & -p + \tau_{zz} \end{bmatrix} \begin{bmatrix} n_x \\ n_y \\ n_z \end{bmatrix} \tag{1.107}$$

These *Cauchy stress formulas* can be written in the short form

$$\mathbf{t} = \mathbf{S}\,\mathbf{n} = -\,p\,\mathbf{n} + \mathbf{T}\,\mathbf{n} \tag{1.108}$$

In this $\mathbf{S}$ denotes the *stress tensor*, p the *pressure*, $\mathbf{T}$ with the components τ_{xx}, τ_{xy}, etc. the *tensor of the viscous or extra-stresses*. The decomposition

$$\mathbf{S} = -\,p\,\mathbf{E} + \mathbf{T} \tag{1.109}$$

indicated in equation (1.108), where $\mathbf{E}$ is the unit tensor, corresponds to the fact that in a fluid at rest the state of stress is spherical symmetric, and in a flowing fluid additional stresses arise, caused by the internal friction. The separation of the pressure from the total stresses for incompressible fluids is of course arbitrary. In actual applications it is expedient to fix p over the boundary conditions for the stresses. For compressible fluids on the other hand p signifies the thermodynamic pressure.

We interpret as being empirically established the fact that for all interesting engineering substances the stress tensor is symmetrical ($\mathbf{S} = \mathbf{S}^T$ and hence $\mathbf{T} = \mathbf{T}^T$), i.e.

$$\tau_{yx} = \tau_{xv}, \qquad \tau_{zy} = \tau_{yz}, \qquad \tau_{xz} = \tau_{zx} \tag{1.110}$$

One therefore requires six scalar quantities for the description of the stress condition in a continuum, and these represent the components of a symmetrical tensor. Except for simple situations they of course depend just as the components of the velocity vector on position and possibly also on time: $\tau_{xy} = \tau_{xy}(x,y,z,t)$, etc.

1.9 Equations of motion

The Cauchy stress formulas allow the first terms in (1.105) to be represented in the form $\int\!\!\int_A \mathbf{S}\,\mathbf{n}\,dA$, that is the resultant surface force acting on an arbitrarily chosen part of the moving fluid. We first of all consider only the x-components of this force and transform the surface integral (assuming a differentiable stress field) with the help of Gauss' theorem into a volume integral:

$$\iint_A [(-p+\tau_{xx})n_x + \tau_{xy}n_y + \tau_{xz}n_z]\, dA = \iiint_V \left[\frac{\partial(-p+\tau_{xx})}{\partial x} + \frac{\partial \tau_{xy}}{\partial y} + \frac{\partial \tau_{xz}}{\partial z}\right] dV \quad (1.111)$$

If one adds the corresponding components of the resulting volume force, namely $\iiint_V f_x\, dV$, then one obtains the total effective force acting in the x-direction on the material volume considered in the form of a single integral over the volume. If one chooses as the domain of integration a sufficiently small volume Δv, then one finds that

$$\left[-\frac{\partial p}{\partial x} + \frac{\partial \tau_{xx}}{\partial x} + \frac{\partial \tau_{xy}}{\partial y} + \frac{\partial \tau_{xz}}{\partial z} + f_x\right] \Delta V \quad (1.112)$$

represents the resultant force acting on this volume element in the x-direction. According to Newton's laws this force is given by the product of the mass Δm of the element and its acceleration a_x. Because $\Delta m/\Delta v$ is the density ρ of the volume element considered, this statement is represented by the formula

$$\rho a_x = -\frac{\partial p}{\partial x} + \frac{\partial \tau_{xx}}{\partial x} + \frac{\partial \tau_{xy}}{\partial y} + \frac{\partial \tau_{xz}}{\partial z} + f_x \quad (1.113)$$

Similar reasoning applied to the forces acting in the y-direction and the z-direction respectively leads to the two relationships

$$\rho a_y = -\frac{\partial p}{\partial y} + \frac{\partial \tau_{yx}}{\partial x} + \frac{\partial \tau_{yy}}{\partial y} + \frac{\partial \tau_{yz}}{\partial z} + f_y \quad (1.114)$$

$$\rho a_z = -\frac{\partial p}{\partial z} + \frac{\partial \tau_{zx}}{\partial x} + \frac{\partial \tau_{zy}}{\partial y} + \frac{\partial \tau_{zz}}{\partial z} + f_z \quad (1.115)$$

It is sometimes expedient to bring these three *equations of motion* into the vector form

$$\rho \mathbf{a} = -\operatorname{grad} p + \operatorname{div} \mathbf{T} + \mathbf{f} \quad (1.116)$$

in which div **T** represents a vector whose components arise from equations (1.113) to (1.115). Because the volume element considered in the derivation was arbitrary, the equations of motion are valid at every place and at all times in the fluid. Therefore only those motions are possible for which the relationship (1.116) applies between the fields of density, acceleration, pressure, extra-stresses and volume force. In order to be able to solve actual flow problems constitutive equations are required in addition to the equations of motion, particularly relationships between the stresses and the deformation state in the moving fluid. This will be considered in detail later.

The equations of motion also apply to fluids at rest. The viscous forces in a fluid in a state of rest vanish by definition ($\mathbf{T} = 0$), and all particles are obviously not accelerated ($\mathbf{a} = 0$). Hence equation (1.116) reduces to the relationship grad $p = \mathbf{f}$. From this it follows that a fluid, in contrast to a rigid body, can be at rest only when acted on by a conservative volume force field with the property curl $\mathbf{f}=0$.

1.10 Energy equation for fluid flow

The *energy equation* is a statement about the change of the total energy of a material volume V with time. It is made up of the *kinetic energy* $\iiint\limits_V \frac{1}{2}|\mathbf{v}|^2 \rho dV$ and the internal energy, which one can represent by the scalar field e of the *specific* (i.e. based on the unit of mass) *internal energy* in the form $\iiint\limits_V e \rho\, dV$. The equation indicates that the change with respect to time of the total energy of the material volume is equal to the rate of work of all the forces acting on the volume and the incoming energy per unit time. One can describe the latter by the vector of the energy flux density $\mathbf{q}$ in such a way that $-\mathbf{q}\cdot\mathbf{n}\, dA$ gives the inflowing energy per unit time through an element of surface with external normal $\mathbf{n}$ and size dA. If one disregards the radiation and diffusion energy, $\mathbf{q}$ is the vector of the *heat flux density*. The equation is therefore

$$\frac{D}{Dt}\iiint\limits_V \left(e + \frac{1}{2}|\mathbf{v}|^2\right)\rho dV = \iint\limits_A \mathbf{v}\cdot\mathbf{t}\, dA + \iiint\limits_V \mathbf{v}\cdot\mathbf{f}\, dV - \iint\limits_A \mathbf{q}\cdot\mathbf{n}\, dA \quad (1.117)$$

The scalar product $\mathbf{v}\cdot\mathbf{t}$ can be replaced by using the Cauchy stress formulas (1.108) by $-p\mathbf{v}\cdot\mathbf{n} + \mathbf{v}\cdot\mathbf{T}\mathbf{n}$ and after recasting of the second term finally by $-p\mathbf{v}\cdot\mathbf{n} + (\mathbf{T}^T\mathbf{v})\cdot\mathbf{n}$. Thus $\mathbf{T}^T\mathbf{v}$ is a vector formed from the transposed extra-stress tensor and the velocity vector which has the components

$$\begin{aligned}
(\mathbf{T}^T\mathbf{v})_x &= u\,\tau_{xx} + v\,\tau_{yx} + w\,\tau_{zx} \\
(\mathbf{T}^T\mathbf{v})_y &= u\,\tau_{xy} + v\,\tau_{yy} + w\,\tau_{zy} \\
(\mathbf{T}^T\mathbf{v})_z &= u\,\tau_{xz} + v\,\tau_{yz} + w\,\tau_{zz}
\end{aligned} \quad (1.118)$$

Hence Gauss' theorem can be applied to both surface integrals in equation (1.117):

$$\iint_A (\mathbf{v} \cdot \mathbf{t} - \mathbf{q} \cdot \mathbf{n})\, dA = \iiint_V \operatorname{div}(-p\,\mathbf{v} + \mathbf{T}^T \mathbf{v} - \mathbf{q})\, dV \tag{1.119}$$

On the left-hand side of equation (1.117) one can interchange the material derivative with respect to time and the integration according to the following rule:

$$\frac{D}{Dt} \iiint_V \left(e + \frac{1}{2} |\mathbf{v}|^2 \right) \rho\, dV = \iiint_V \rho \frac{D}{Dt} \left(e + \frac{1}{2} |\mathbf{v}|^2 \right) dV \tag{1.120}$$

Note that the factor ρ in the integrand remains unaffected. In order to understand this peculiarity, one must remember that a material volume deforms with respect to time, and hence the range of integration in equation (1.120) varies with time. The time derivative on the left-hand side therefore concerns not only the integrand but also the domain of integration. If however one transforms the integral in a way that the integration remains constant with time, then the time derivative is naturally only concerned with the integrand. Such a transformation is met for example in the conversion from the three dimensional space of observation where the flow occurs to a space in which each element of mass occupies its fixed position. A material volume in this space has the same shape at all times. If therefore one replaces $\iiint_V \varphi \rho \, dV$ by $\iiint_M \varphi \, dm$ and notes that now the integration domain M remains constant with respect to time, one can see that the material derivative acts only on φ, in equation (1.120) therefore on the factor $(e + \frac{1}{2}|\mathbf{v}|^2)$ in the integrand.

The energy equation (1.117) can be written by use of the relationships (1.119) and (1.120) in such a way that only volume integrals arise. Because the integration domain V is arbitrary there must be equality between the integrands on both sides. One thus obtains the following differential form of the energy equation:

$$\rho \left(\frac{De}{Dt} + u \frac{Du}{Dt} + v \frac{Dv}{Dt} + w \frac{Dw}{Dt} \right) = \operatorname{div}(-p\,\mathbf{v} + \mathbf{T}^T \mathbf{v} - \mathbf{q}) + \mathbf{v} \cdot \mathbf{f} \tag{1.121}$$

The material derivatives with respect to time Du/Dt, Dv/Dt and Dw/Dt are the components of the acceleration vector **a**. If one eliminates them by using the equations of motion (1.113) - (1.115), writes the divergence as the sum of partial derivatives of x, y and z, and inserts the expression (1.118) for the components of the vector $\mathbf{T}^T\mathbf{v}$, then several terms cancel out in pairs and there remains

$$\rho \frac{De}{Dt} = -p \operatorname{div} \mathbf{v} + \Phi - \operatorname{div} \mathbf{q} \tag{1.122}$$

The *dissipation function* Φ introduced here is composed of the elements of the (symmetric) extra-stress tensor $\mathbf{T}$ and the strain rate tensor $\mathbf{D}$:

$$\Phi = \tau_{xx} \frac{\partial u}{\partial x} + \tau_{yy} \frac{\partial v}{\partial y} + \tau_{zz} \frac{\partial w}{\partial z} + \tau_{xy} \left(\frac{\partial u}{\partial y} + \frac{\partial v}{\partial x} \right) + \tau_{yz} \left(\frac{\partial v}{\partial z} + \frac{\partial w}{\partial y} \right) + \tau_{zx} \left(\frac{\partial w}{\partial x} + \frac{\partial u}{\partial z} \right) \tag{1.123}$$

This describes the mechanical energy dissipated irreversibly per unit time and unit volume by internal friction. The expression (1.123) can be interpreted as the first invariant (the 'trace') of the tensor product $\mathbf{T}\,\mathbf{D}$:

$$\Phi = \operatorname{tr}(\mathbf{T}\,\mathbf{D}) \tag{1.124}$$

This expression has the advantage of being valid in arbitrary coordinates, whereas equation (1.123) is written in terms of Cartesian coordinates.

We shall now consider especially flows of liquids for which because of continuity $\operatorname{div} \mathbf{v} = 0$ (Section 1.7). In this case we can take as the energy equation the following shortened form as compared with equation (1.122):

$$\rho \frac{De}{Dt} = \Phi - \operatorname{div} \mathbf{q} \tag{1.125}$$

Hence the specific internal energy of a liquid particle changes with time in two ways, namely because of dissipation and because of thermal conductivity, unless the dissipated energy is completely conducted away in the form of heat ($\operatorname{div} \mathbf{q} = \Phi$)

Regarding constitutive laws for the newly introduced quantities e and $\mathbf{q}$ in the energy equation, we generally have sufficient classical assumptions, according to which the specific internal energy of a liquid depends only on the *absolute temperature* Θ, $e = e(\Theta)$, and the heat flux is proportional to the temperature gradient, $\mathbf{q} = -\lambda \operatorname{grad} \Theta$. Hence equation (1.125) takes on the special form

$$\rho c \frac{D\Theta}{Dt} = \Phi + \operatorname{div}(\lambda \operatorname{grad} \Theta) \tag{1.126}$$

Thus the *specific heat capacity* $c := de/d\Theta$ and the *thermal conductivity* λ of the liquid can in general be regarded as given functions of temperature, and of temperature and pressure respectively, and in the simplest case as constant material quantities.

Problems

1.1 The velocity field having the components

$$u = -U \sin \omega t \sin kx \cosh ky$$

$$v = U \sin \omega t \cos kx \sinh ky$$

$$w = 0$$

$(-\infty < x < \infty, 0 \leqslant y \lesssim H)$ describes a plane standing gravity wave in a liquid layer located above a horizontal bottom (for which $y = 0$). U is a constant reference velocity. The wave number k, the angular velocity ω and the water depth H are related to each other by the dispersion law, which is however not needed here if one takes k and H to be given quantities.

Calculate and sketch the stream lines and the path lines, and discuss how an arbitrarily chosen material point moves along its path lines. Further, determine the stream function $\Psi(x,y,t)$ of the plane motion.

1.2 Show that the rate of strain tensors for the two velocity fields (1.3) and (1.34) have the same eigenvalues in the case $c = 2a$. Liquid particles are deformed in the same way in both flows because of this. Hence what is the difference?

1.3 Investigate the kinematic properties of the following plane steady flows which are characterised by a velocity gradient tensor $\mathbf{L}$, which is independent of position and for which the relationship $\mathbf{v} = \mathbf{Lr}$ applies:

$$\mathbf{L} = \begin{bmatrix} 0 & -c \\ c & 0 \end{bmatrix}, \quad \mathbf{L} \begin{bmatrix} 0 & c \\ c & 0 \end{bmatrix}, \quad \mathbf{L} = \begin{bmatrix} c & 0 \\ 0 & c \end{bmatrix}$$

Determine in particular the stream lines in the x-y plane, the rate of strain tensor $\mathbf{D}$, its eigenvalues and principal axis, and the vorticity tensor $\mathbf{W}$ with the associated vector curl $\mathbf{v}/2$, and using this information discuss how a chosen fluid particle of suitably chosen shape deforms with time.

1.4 For a steady torsional flow between a fixed disc and one rotating parallel to it all the material points move along circular paths with an angular velocity $\Delta\omega z/h$ (Fig. 1.11). Then the velocity has the Cartesian coordinates

$$u = -\frac{\Delta\omega}{h} zy, \qquad v = \frac{\Delta\omega}{h} zx, \qquad w = 0$$

Calculate for this motion the first three Rivlin-Ericksen tensors with their principal invariants. What can be said about $\mathbf{A}_n (n \geqq 3)$?

1.5 The velocity field

$$u = \dot{\epsilon}_1 x, \qquad v = \dot{\epsilon}_2 y, \qquad w = \dot{\epsilon}_3 z$$

describes a steady simple elongational flow when the elongation rates $\dot{\varepsilon}_1$, $\dot{\varepsilon}_2$, and $\dot{\varepsilon}_3$ are constant with respect to space and time (homogeneous deformation).

Show that for such an elongational flow the higher Rivlin-Ericksen tensors can be reduced to the first order Rivlin-Ericksen tensor according to $\mathbf{A}_n = \mathbf{A}_1^n$ and determine all the Jaumann time derivatives of $\mathbf{A}_1$ (hence $\overset{\circ}{\mathbf{A}}_1$, $\overset{\circ\circ}{\mathbf{A}}_1$, etc.).

1.6 Determine for a steady plane elongational flow which has the velocity field (1.3) in which a is a constant, the Cartesian coordinates of the relative right-Cauchy-Green tensor $\mathbf{C}_t(x,y,z,s)$ and explain clearly the significance of the individual elements by means of Fig. 1.1.

1.7 With a Maxwell-Chartoff rheometer one tries to create a velocity field of the form

$$u = -\omega y + \omega\psi z, \qquad v = \omega x, \qquad w = 0$$

in which ω and ψ are constants. Is this a plane flow? Determine the path line of that particle which at time t is located at x,y,z and then find the history of the relative right-Cauchy-Green tensor. Is this flow viscometric?

1.8 Can a liquid of constant density remain at rest when acted upon by a volume force field which has the components

$$f_r = \frac{a}{r^2}\sin\varphi, \qquad f_\varphi = -\frac{a}{r^2}\cos\varphi, \qquad f_z = -\rho g$$

(r, φ, z are cylindrical coordinates; a is a constant)? If the answer is "yes", derive the relation between the hydrostatic pressure and the volume force, from which the pressure field could be calculated.

2 MATERIAL PROPERTIES OCCURRING IN STEADY SHEAR FLOWS

The friction conditions must be known before the evaluation of the equations for actual flow situations. These are defined by a constitutive equation which connects the tensor of the friction stresses with the velocity field. We wish first of all to explain how this relationship appears in the applications to important classes of steady shear flow, and how the material properties thereby occurring operate. Hence we shall investigate at the same time the possibilities of the experimental determination of these properties. Moreover the caloric state equation of the fluid and an equation for the heat flow will be necessary for the evaluation of the energy equation. We have already mentioned this in Section 1.10.

For many fluids occurring in nature and engineering, say air and other gases, water, motor oils, alcohols, simple hydrocarbon compounds and others, the actual friction behaviour under the conditions that are interesting in this context is described with sufficient accuracy by a linear relation between the extra-stresses and the rate of strain. Fluids that have these properties are called *Newtonian* fluids. We shall become acquainted with the general constitutive equation of the type described in Section 8.1. Concerning the interesting engineering applications it is sufficient for us to consider first of all kinematically severely constricted motion, namely shear flows. Because in a shear flow the strain rate tensor has only one important parameter, namely the shear rate $\dot{\gamma}$, the law of friction of a Newtonian fluid in this case reduces to a scalar statement

$$\tau = \eta \dot{\gamma} \tag{2.1}$$

according to which the shear stress τ between the layers increases linearly with $\dot{\gamma}$. The (always positive) coefficient of proportionality η is called the *shear viscosity*, or briefly viscosity, and it is a characteristic material quantity for the fluid concerned, which in general depends on temperature and pressure, but not on the amount of the rate of shear.

Understanding of the following reasoning will be easier if we for the moment examine an unsteady shear flow. For the sake of simplicity consider a motion which has the velocity field $\mathbf{v} = u(y,t)\mathbf{e}_x$, for which each plane layer y = constant moves with a time dependent velocity u(y,t) in the x-direction (Fig. 2.1).

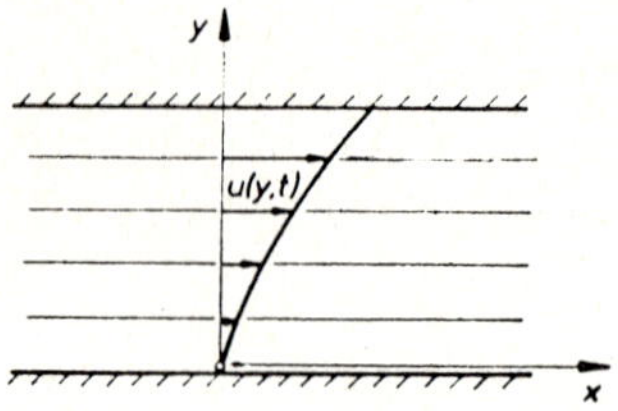

Fig. 2.1 Unsteady plane shear flow

Such a flow occurs for example in a viscous flow in a channel between two parallel plates when these are moved with different velocities in a tangential direction. It is clear that under these conditions the shear rate, $\dot{\gamma} := \partial u/\partial y$, and the shear stress depend on position and time. We shall therefore write equation (2.1) in more detail in the form

$$\tau(y,t) = \eta\,\dot{\gamma}(y,t) \qquad (2.2)$$

This relationship, valid for Newtonian fluids, is very simple from three aspects.

It states firstly that the shear stress acting at the actual time on an arbitrarily chosen fluid particle depends only on the momentary state of motion of the particle. Its motion at other times does not affect the shear stress occurring at the time (*instantaneous reaction*).

It states secondly that the shear stress at a certain point depends only on the shear rate and therefore on the velocity gradient at this point. Higher order derivatives of the velocity field, or even the motion of spatially distant layers have no effect on the state of stress of a chosen fluid particle (*local action*).

It states thirdly that the local shear stress is directly proportional to the shear rate. Doubling of the shear rate doubles the stress (*linearity*).

In recent years liquids have been appearing in increasing numbers which do not have Newtonian properties. Molten plastics, polymer solutions, dyes, varnishes, suspensions, and natural liquids like blood for example behave in certain circumstances in a way different from what one would expect for a Newtonian fluid. We shall analyse in this book some typical phenomena of such *non-Newtonian fluids*. It turns out that all hitherto observed non-Newtonian effects can be explained by the following hypotheses concerning the friction behaviour of the fluid.

The extra-stresses acting on a fluid particle at a particular time depend, besides on the momentary state of motion, possibly also on the motion of the fluid in the past; one can say that the fluid has a *memory*. The future movement however does not influence the state of affairs at the present time (*determinism*).

The extra-stresses acting on the fluid particle depend only on the deformation history of the particle that is being observed; and more accurately on the history of its relative deformation gradient. The motion of spatially distant particles continues to have no effect on the state of stress of the chosen liquid particle (*local action*).

Materials that have these properties are called rheologically *simple fluids*. Newtonian fluids, which have no memory and behave according to a linear constitutive law are included as a special case in the class of simple fluids.

We now apply these statements about the friction behaviour of simple fluids to an unsteady plane shear flow (Fig. 2.1). Because an arbitrarily chosen point therefore moves at all times in the x-direction, it is always located on a sliding surface y = constant (here y has the significance of a *material coordinate*). Therefore the deformation history of a fluid particle consists of a shear with the local shear rate $\dot{\gamma}(y,t-s) := \partial u(y,t-s)/\partial y$ at the present time ($s = 0$), and in the past ($s > 0$). According to the previously formulated material hypotheses this kinematic quantity determines for a plane shear flow the extra-stresses of a simple fluid at the actual time concerned. Hence follows the general relationship for the shear stress $\tau(y,t)$ between the layers.

$$\tau(y,t) = \mathop{\mathrm{F}}_{s=0}^{\infty} [\dot{\gamma}(y,t-s)] \tag{2.3}$$

A comparison with the relationship (2.2) valid for Newtonian fluids shows that instead of the constant viscosity η there appears here a largely arbitrary and in general non-linear functional of the history of the shear rate. It may for example be an integral or an algebraic expression, which consists of time derivatives of $\dot{\gamma}(y,t)$. In the latter case the fluid would have an only infinitesimally short memory. Such laws will play a role in Section 5 (equations (5.6) and (5.118)). It would also be conceivable that the right-hand side of equation (2.3) reduces to $\eta \cdot \dot{\gamma}\ (y,\ t - \lambda)$. For the present shear stress therefore the shear rate at the time going back by λ would be responsible. The constant λ would therefore describe the range of the memory and would have the significance of a dead time.

These few examples show how universal the constitutive law in the form of (2.3) really is. Difficulties are involved in determining the general functional of a given liquid. In this respect one would have to record the stress behaviour of the infinite multiplicity of all possible unsteady shear flows, and to make conclusions about the form of the functional from all the observations. Because this is quite impossible the general law of real fluids in the form given by (2.3) remains unknown and then equation (2.3) is of little practical use. On the other hand it is clear that the observation of any motion, or a certain class of motion, provides some partial information about the law. But as long as one is only interested in the class of motion considered, this incomplete knowledge about the law is sufficient, and the general form (2.3) is not needed at all. The practical way is to limit oneself to such motions for which the right-hand side of equation (2.3) reduces to expressions that can be easily manipulated. In Section 5 we shall consider motions of small amplitude and besides these some slow and slowly varying processes. Before that we shall look at applications of the important class of steady shear flows in detail.

If the plane shear flow sketched in Fig. 2.1 is steady, then the velocity field and hence the rate of strain of each layer is constant with time. Therefore every fluid particle experiences, apart from a rotation, at all times a constant shear rate $\dot{\gamma}(y)$. Hence it looks back to a peculiar monotonous deformation history, in which there were no variations at all, so that the memory of the past only contains one bit of information, namely the quantity $\dot{\gamma}(y)$, which remains constant for the particle concerned. One therefore also conveniently speaks of a *motion with constant stretch history*. Whatever memory properties the fluid may also have, the shear stress $\tau(y)$ in a simple fluid can only depend on $\dot{\gamma}(y)$ under these conditions. Equation (2.3) accordingly reduces for the steady state to a generally non-linear relationship between the local shear stress and the local shear rate

$$\tau = F(\dot{\gamma}) \qquad (2.4)$$

We omit here and in what follows the coordinate y, because thus the equation (2.4) holds not only for plane flows, but also for others, particularly those referred to in Section 1.6 as *steady shear flows* which one can also designate as *viscometric flows*.

Because in equation (2.4) only local quantities occur, there is no memory which might otherwise occur. For viscometric flows therefore liquids with finite memory are not to be differentiated from those which have no memory.

For the following it is important to know that for a viscometric flow two of three independent shear stress components will vanish. We explain this for a plane shear flow by introducing a reference system of the type that brings an arbitrarily chosen point to rest. We select in the surroundings of the stationary point a cube-shaped volume element with its edges parallel to the axes, and write the shear stresses on the surface which a first observer sees with reference to his coordinate system (left-hand part of Fig. 2.2), and which a second observer sees with reference to his coordinate system rotated by 180° round the z-axis (right-hand part of Fig. 2.2). Because both observers see exactly similar flows, they actually record the same stresses[1]. Figure 2.2 shows that they would come to different conclusions about the shear stresses τ_{xz} and τ_{yz} if these two quantities were present. It follows from this reasoning that for a plane shear flow $\tau_{xz} = \tau_{yz} = 0$ holds, i.e. only the shear stress τ_{xy} (and with it the component $\tau_{yx} = \tau_{xy}$) occur. Correspondingly for other viscometric flows only one important shear stress component occurs: for a Poiseuille flow it is τ_{zr}, for a Couette flow $\tau_{\varphi r}$, for a torsional flow $\tau_{\varphi z}$, and for a cone-and-plate flow $\tau_{\varphi\vartheta}$ (Section 1.6). Because the other two independent shear stress components vanish and therefore no confusion is possible we shall in what follows omit the subscripts and simply write τ.

---

1 Behind this statement lies the '*principle of material objectivity*' according to which different observers of one and the same motion record the same state of stress.

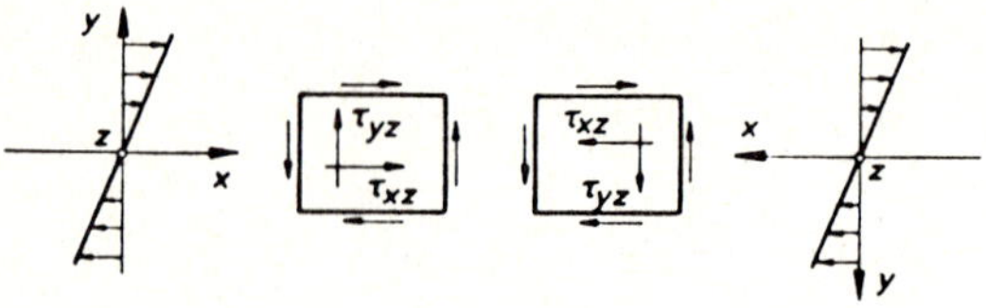

Fig. 2.2 Shear stresses in a rectangular element

2.1 The flow function

According to the above statements, for steady shear flows of simple fluids at an arbitrarily chosen position, a shear stress τ occurs between the layers, which as regards the kinematic field quantities, only depends on the local shear rate $\dot{\gamma}$. We have first written this *flow law* in the form $\tau = F(\dot{\gamma})$ (equation (2.4)). There $F(\dot{\gamma})$ is a characteristic function for the liquid considered, which describes the flow properties occurring in a viscometric flow. For Newtonian fluids $F(\dot{\gamma}) = \eta\dot{\gamma}$ applies, in which the apparent viscosity η represents a material quantity independent of the shear rate. In analogy to this relationship for Newtonian fluids, equation (2.4) is often written in the form

$$\tau = \eta(\dot{\gamma}) \cdot \dot{\gamma} \qquad (2.5)$$

Hence one introduces the quotient made up from shear stress and shear rate, designates it again as apparent viscosity, and gives it the symbol η. One must remember that it can be a quantity which depends on the shear rate. The function $\eta(\dot{\gamma})$ is just as suitable for characterising the flow properties of real fluids in steady flows as the function $F(\dot{\gamma})$, because both are related to each other in a simple way by $F(\dot{\gamma}) = \eta(\dot{\gamma})\cdot\dot{\gamma}$. Fig. 2.3 shows the dependence of the viscosity of a polymer melt on the shear rate and temperature. One learns from this that at fixed temperatures and at relatively small shear rates the melt has a finite value of the viscosity, which we designate as the *lower Newtonian limiting viscosity*, or more briefly as *zero-shear viscosity*, and give it the symbol η_0.

$$\eta_0 := \lim_{\dot{\gamma}\to 0} \eta(\dot{\gamma}) \qquad (2.6)$$

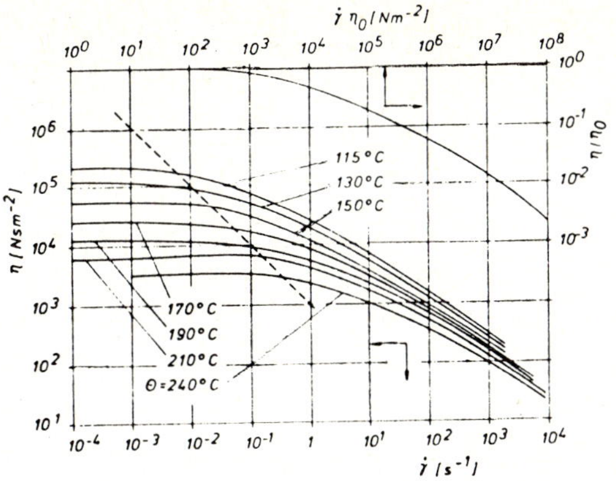

Fig. 2.3 Viscosity of a polymer melt (low density polyethylene) as a function of the shear rate and temperature (after Meissner) and its universal graphical illustration

The viscosity decreases rapidly with increasing shear rate. Liquids like those for which η decreases as $\dot{\gamma}$ increases are called *pseudoplastic*, or *shear thinning*. Liquids for which in contrast the viscosity increases with increasing shear rate are called *dilatant* or *shear thickening*.

Concerning the dependence of viscosity on temperature, one can see in Fig. 2.3 a characteristic property of liquids over a wide range: the higher the temperature the lower is the viscosity for a given shear rate. Polymer solutions exhibit liquid properties similar to those of molten plastics, especially marked pseudoplastic behaviour. The important difference is that the absolute values of the viscosity are distinctly lower than for a melt.

It is an interesting fact that for one and the same material the viscosity curves for various temperatures have the same shape. For a log-log scale as in Fig. 2.3 all the curves converge if one shifts them along a straight line of slope -1. By suitable representation they can be made to coincide. One obtains this by shifting the individual viscosity curves corresponding to the appropriate zero-shear viscosity $\eta_0(\Theta)$ by the distance $\log\eta_0(\Theta)$ to the right and at the same time downwards. This displacement is made in order to display the reduced viscosity η/η_0 against the quantity $\eta_0\dot{\gamma}$. In Fig. 2.3 the 'master curve' obtained in this way is drawn in above. Because the effect of temperature is no longer noticeable, we speak of a *universal representation*, and all flow curves can consequently be represented by a single function characteristic of the material.

$$\frac{\eta(\dot{\gamma};\Theta)}{\eta_0(\Theta)} = H\left(\frac{\eta_0(\Theta)\,\dot{\gamma}}{\tau_*}\right) \tag{2.7}$$

In order to make the product $\eta_0\dot{\gamma}$ dimensionless, we make use of a constant (independent of temperature) reference stress τ_*, which one can arbitrarily define say at $\tau_* = 1\ N/m^2$, or chose in another suitable way. The significance of equation (2.7) consists of the fact that the dependence of the viscosity on the shear rate and temperature is factorised when the temperature influence is taken into account via the zero-shear viscosity (*law of similarity* for material values).

The zero-shear rate of a fluid besides depending on the temperature is also weakly dependent on the pressure p, and in the case of polymer melts significantly on the average molecular weight $\bar{M}$. It turns out that the viscosity curves for different molecular weights are similar, and the above described proposed reduction with the zero-shear rate results in a single universal curve which is characteristic of the substance, if the molecular weight exceeds a certain limiting value (Fig. 2.4). This reduction assumes in each case a similar molecular weight distribution for the melts relative to each other. The average molecular weight, just like the temperature, affects only the zero-shear rate, but not the shape of the viscosity curve. The same applies to the dependence on pressure. Equation (2.7) can therefore be generally written in the form

$$\frac{\eta(\dot{\gamma};\Theta,p,\bar{M})}{\eta_0(\Theta,p,\bar{M})} = H\left(\frac{\eta_0(\Theta,p,\bar{M})\,\dot{\gamma}}{\tau_*}\right) \tag{2.8}$$

From the data represented in Fig. 2.4 one also obtains a constant of proportionality for the zero-shear rate at $\bar{M}^{3.5}$.

The concentration c is an important parameter for solutions. Fig. 2.5 shows the viscosity of polymer solutions as a function of the shear rate with concentration as a parameter. Here also the reduction with the zero-shear rate will lead to a universal representation of the viscosity curves, supposing that the concentration exceeds a certain threshold value. It is clear that very dilute solutions do not lend themselves to the reduction of the viscosity curve reduced with η_0, because for $c \to 0$ only the viscosity η_L of the Newtonian solvent substance can make itself perceptible. This difficulty does not arise if one represents the difference $\eta - \eta_L$ instead of η, and makes it dimensionless with $\eta_0 - \eta_L$ instead of with η_0. For polymer solutions it would therefore be necessary to modify equation (2.8), replacing η by $\eta - \eta_L$ and correspondingly η_0 by $\eta_0 - \eta_L$. This modification would however be almost insignificant for the properties represented in Fig. 2.5 because of the low viscosity of water as the solvent.

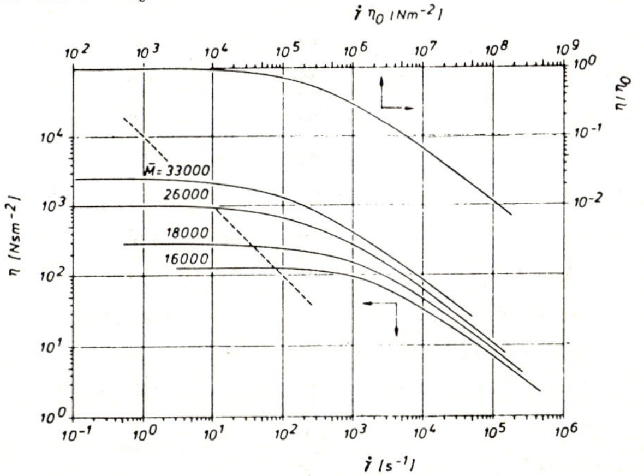

Fig. 2.4 Viscosity function of polymer melts (polyamide 6) at different average molecular weights, Θ = 250°C (after Laun)

Fig. 2.5 shows that the viscosity of the polymer solution for increase in shear rate does not drop without limit, but finally reaches a value independent of the shear rate. Actual liquids therefore have Newtonian flow properties not only at relatively low but also at very high shear rates. The *upper Newtonian limiting viscosity* is correspondingly defined in the same way as the zero-shear viscosity and accordingly is designated by η_∞ (cf. equation (2.6)).

$$\eta_\infty := \lim_{\dot{\gamma}\to\infty} \eta(\dot{\gamma}) \tag{2.9}$$

For every actual flow the shear rates occurring in the fluid are limited upwards. Only the left-hand part of the flow curve then plays a part up to a maximum shear rate. Hence the deflection of the viscosity curve at the upper limiting value can often be entirely neglected

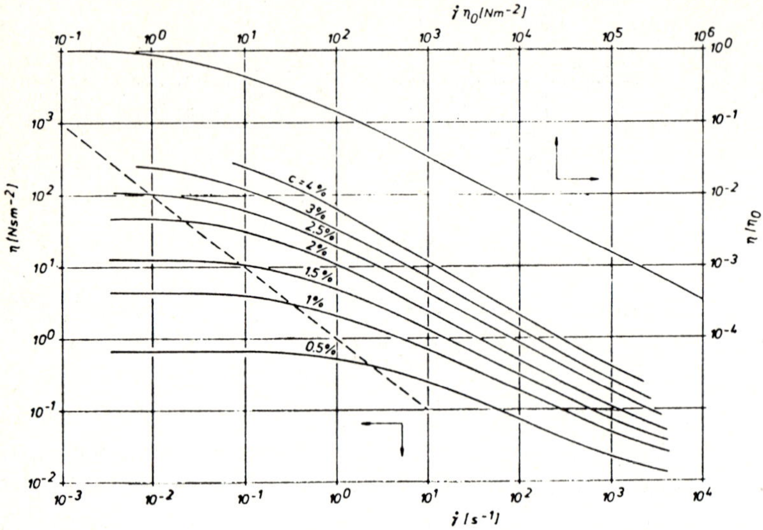

Fig. 2.5 Viscosity function of polymer solutions at different concentrations (polyacrylamide in water), Θ = 20°C

and the actual procedure for the mathematical treatment of the flow can be replaced by a curve modified in the upper region, which decreases without limit. It is therefore often possible to regard a pronounced pseudoplastic liquid with a finite upper Newtonian limiting viscosity as a substance for which η_∞ vanishes.

The *differential viscosity*

$$\hat{\eta} := \frac{d\tau}{d\dot{\gamma}} \tag{2.10}$$

plays a part in some applications in addition to the usual viscosity defined by equation (2.5).

The difference in meaning between viscosity and differential viscosity will be made clear in what follows. We start from a viscometric flow which has the shear rate $\dot{\gamma}$ and the shear stress τ associated with it, and consider the change $\delta\tau$ of the shear stress when the shear rate is changed by $\delta\dot{\gamma}$. We shall thus allow for the fact that it also changes the direction in which the shear stress acts on a sliding surface.

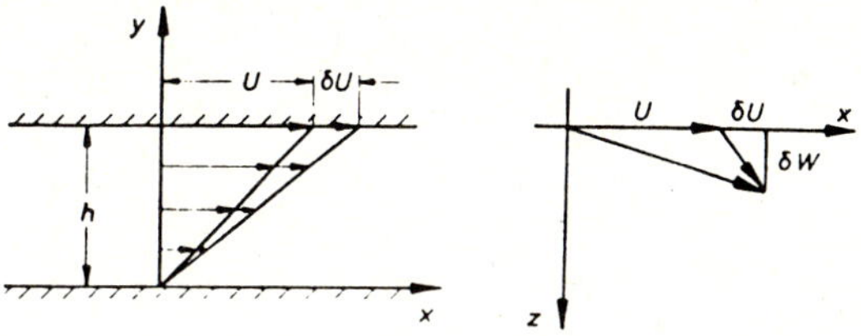

Fig. 2.6 Simple shear changed by change in wall velocity

For example think of a simple shear flow between a rigid plate and a plate moved parallel to it at a distance h with a velocity U in the x-direction (Fig. 2.6). The shear rate obviously has the value U/h, and the shear stress acts on the sliding surfaces y = constant in the x-direction. If one now changes the velocity of the moving plate by a small amount δU in the x-direction and by δW in the z-direction, then on the one hand the shear stress along the x-direction changes and on the other hand a shear stress arises along the z-direction. The increase in stress parallel to the original direction is designated below as $\delta\tau_{\parallel}$, and that in the normal direction as $\delta\tau_{\perp}$. Because in our treatment of this the direction of the shear stress acting on a sliding surface also changes, we can regard the shear stress as a vector quantity. We indicate its amount as before as τ. Accordingly we interpret the shear rate temporarily as a vector $\dot{\boldsymbol{\gamma}}$ parallel to $\boldsymbol{\tau}$, whose amount will henceforth be called $\dot{\gamma}$. The relationship

$$\boldsymbol{\tau} = \frac{\tau}{\dot{\gamma}} \dot{\boldsymbol{\gamma}} \tag{2.11}$$

therefore applies not only to the steady shear flow which we started from, but also to the changed shear flow. If one derives the complete differential form of equation (2.11) and notes that because $\dot{\gamma}^2 = \dot{\boldsymbol{\gamma}} \cdot \dot{\boldsymbol{\gamma}}$ the quantity $\dot{\gamma}\delta\dot{\gamma}$ can be replaced by the scalar product $\dot{\boldsymbol{\gamma}} \cdot \delta\dot{\boldsymbol{\gamma}}$, then one obtains the following relationship between the vector quantities $\delta\boldsymbol{\tau}$ and $\delta\dot{\boldsymbol{\gamma}}$.

$$\delta\boldsymbol{\tau} = \frac{1}{\dot{\gamma}^2}\left(\frac{d\tau}{d\dot{\gamma}} - \frac{\tau}{\dot{\gamma}}\right)\dot{\boldsymbol{\gamma}}\left(\dot{\boldsymbol{\gamma}} \cdot \delta\dot{\boldsymbol{\gamma}}\right) + \frac{\tau}{\dot{\gamma}}\delta\dot{\boldsymbol{\gamma}} \tag{2.12}$$

If one introduces here the usual viscosity η and the differential viscosity $\hat{\eta}$ (cf. equations (2.5) and (2.10), and the vector $\dot{\boldsymbol{\gamma}}\,(\dot{\boldsymbol{\gamma}}\cdot\delta\dot{\boldsymbol{\gamma}})$ as the products of the dyads $\dot{\boldsymbol{\gamma}} \otimes \dot{\boldsymbol{\gamma}}$ with the vector $\delta\dot{\boldsymbol{\gamma}}$, then equation (2.12) takes on the following form:

$$\delta\boldsymbol{\tau} = \left[(\hat{\eta} - \eta)\frac{\dot{\boldsymbol{\gamma}} \otimes \dot{\boldsymbol{\gamma}}}{\dot{\gamma}^2} + \eta\,\mathbf{E}\right]\delta\dot{\boldsymbol{\gamma}} \tag{2.13}$$

Thus the relationship between $\delta\boldsymbol{\tau}$ and $\delta\dot{\boldsymbol{\gamma}}$ is given by a 'viscosity tensor', contained within square brackets. If one now resolves $\delta\dot{\boldsymbol{\gamma}}$ into a component $\delta\dot{\boldsymbol{\gamma}}_{\parallel}$ parallel to the direction $\dot{\boldsymbol{\gamma}}/\dot{\gamma}$ characterising the base flow, and another component $\delta\dot{\boldsymbol{\gamma}}_{\perp}$ normal to it, then equation (2.13) resolves into the two relationships $\delta\boldsymbol{\tau}_{\parallel} = \hat{\eta}\,\delta\dot{\boldsymbol{\gamma}}_{\parallel}$ and $\delta\boldsymbol{\tau}_{\perp} = \eta\delta\dot{\boldsymbol{\gamma}}_{\perp}$, in accordance with which the changes $\delta\dot{\boldsymbol{\gamma}}_{\parallel}$ and $\delta\dot{\boldsymbol{\gamma}}_{\perp}$ give rise to corresponding stress increments, where different viscosity values control the amounts of these. Because the subscripts $_{\parallel}$ (parallel to) and $_{\perp}$ (normal to the original shear direction) give information about the direction, it suffices to write down the scalar form of both relationships:

$$\delta\tau_{\parallel} = \hat{\eta}\,\delta\dot{\gamma}_{\parallel} \tag{2.14}$$

$$\delta\tau_{\perp} = \eta\,\delta\dot{\gamma}_{\perp} \tag{2.15}$$

Therefore the differential viscosity $\hat{\eta}$ determines the change of shear stress caused by a change of the shear in the tangential direction. For a change of shear in the normal direction on the other hand the viscosity η occurs as a constant of proportionality. In the example illustrated in Fig. 2.6 $\delta\dot{\gamma}_{\parallel} = \delta U/h$ and $\delta\dot{\gamma}_{\perp} = \delta W/h$; $\delta\tau_{\parallel}$ would be identifiable with $\delta\tau_{xy}$ and $\delta\tau_{\perp}$ with τ_{zy}.

With regard to the applications to be considered in Section 3 it is expedient to rearrange the relationship (2.4) and write it in the dimensionless form

$$\frac{\eta_*}{\tau_*}\dot{\gamma} = \mathrm{f}\left(\frac{\tau}{\tau_*}\right) \tag{2.16}$$

In this η_* and τ_* are constant positive reference quantities for the viscosity and the shear stress. We shall in future identify η_* with the zero-shear viscosity η_0 so far as the fluid considered has such a lower viscosity limit, and τ_* is chosen so that the numerical parameters occurring in the analytical form of the dimensionless flow function take on the simplest possible values. According to the previous statements about the universal representation of flow curves, a dependence on external parameters like temperature, pressure, molecular weight or the concentration only makes itself noticeable in the reference quantity η_*, but not in τ_* (cf. equation (2.8)). Moreover the dimensionless flow function is related as follows to the reduced viscosity

$$\frac{\eta}{\eta_*} = \frac{\tau/\tau_*}{f(\tau/\tau_*)} \tag{2.17}$$

Hence Newtonian flow properties make themselves felt by a linear increase in the flow function, pseudoplasticity by an initial convex upwards increase, and dilatancy by a concave upwards increase (Fig. 2.7). According to equation (2.17) the reduced viscosity can be indicated in a flow diagram like Fig. 2.7 as a reciprocal value of the slope of the secant. The slope of the tangent to the flow curve is connected with the differential viscosity (cf. equations (2.10) and (2.16)) by

$$\frac{\hat{\eta}}{\eta_*} = \frac{1}{f'(\tau/\tau_*)} \tag{2.18}$$

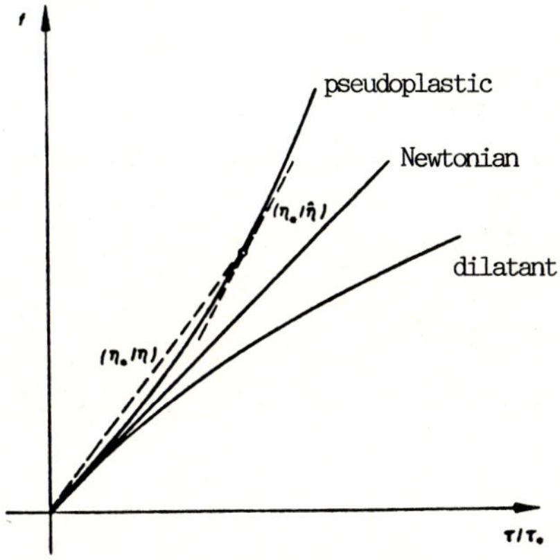

Fig. 2.7 The dimensionless flow functions of a pseudoplastic, of a Newtonian and of a dilatant fluid. The slopes of chords and tangents determine the viscosity and the differential viscosity

In equation (2.18) f' is the derivative of the flow function with respect to its argument τ/τ_*. For pseudoplastic fluids $\hat{\eta} < \eta$ applies, for dilatant fluids $\hat{\eta} > \eta$.

If the actual viscosity curve $\eta(\dot{\gamma})$ is known, the flow function $f(\tau/\tau_*)$ can be determined as follows. Using the relationship $\tau = \eta\dot{\gamma}$ one represents the viscosity as a function of the shear stress and the reduced viscosity η/η_* as a function of the reduced shear stress τ/τ_* by first of all putting an arbitrary value for τ_*, e.g. $\tau_* = 1N/m^2$. From equation (2.17) this immediately gives the function $f(\tau/\tau_*)$. By changing the definition of τ_* the analytical expression for $f(\tau/\tau_*)$ is simplified.

It is expedient to define the flow function $f(\tau/\tau_*)$ not only for positive but also for negative arguments. Because a change of sign of the shear stress ($\tau \to -\tau$) brings about a reversal of the flow direction ($\dot{\gamma} \to -\dot{\gamma}$), the flow function is an odd function of its argument.

$$f(-\sigma) = -f(\sigma). \tag{2.19}$$

Moreover, for reasons of stability the value of the differential $d\tau(\dot{\gamma})/d\dot{\gamma}$ for real fluids is always positive, i.e. the flow function increases monotonically

$$\frac{df(\sigma)}{d\sigma} > 0 \tag{2.20}$$

The first property follows from $f(0) = 0$, the second from $f(\sigma) > 0$ for $\sigma > 0$.

There have been numerous attempts to approximate real fluid properties by simple analytical functions. Some special flow laws used in the literature are collected in Table 2.1. They include in each case free parameters, which in the actual case can be determined by comparison with the experimentally obtained flow properties. Note that the reference quantity τ_* is also one such parameter which has to be suitably chosen.

Table 2.1 Flow functions of some fluid models

Fluid model	Dimensionless flow function	Type of flow behaviour
Newton	$f(\sigma) = \sigma$	Newtonian
Rabinowitsch	$f(\sigma) = \sigma + \sigma^3$	pseudoplastic
Prandtl-Eyring	$f(\sigma) = \sinh\sigma$	pseudoplastic
Ellis	$f(\sigma) = \sigma + \lvert\sigma\rvert^m\sigma \ (m>0)$	pseudoplastic
Reiner-Philippoff	$f(\sigma) = \dfrac{\sigma + \sigma^3}{1 + \dfrac{\eta_\infty}{\eta_*}\sigma^2}$	pseudoplastic for $\eta_\infty<\eta_*$ dilatant for $\eta_\infty > \eta_*$
Reiner	$f(\sigma) = \dfrac{\eta_*}{\eta_\infty}\sigma - \left(\dfrac{\eta_*}{\eta_\infty} - 1\right)\sigma\cdot e^{-\sigma^2}$	pseudoplastic for $\eta_\infty<\eta_*$ dilatant for $\eta_\infty>\eta_*$
Ostwald-de Waele	$f(\sigma) = \lvert\sigma\rvert^m\sigma$	pseudoplastic for $m>0$ dilatant for $m<0$
Bingham	$f(\sigma) = \begin{cases} \sigma - \operatorname{sgn}\sigma & \text{for } \lvert\sigma\rvert > 1 \\ 0 & \text{for } \lvert\sigma\rvert < 1 \end{cases}$	above the flow limit pseudoplastic

The models put forward by *Ostwald-de Waele* and *Bingham* occupy a special place since the reference viscosity there, η_*, used for obtaining the dimensionless form, does not designate the zero-shear viscosity. In the Bingham model, which applies to fluids that have a yield point, a zero-shear viscosity naturally does not exist at all. There τ_* indicates the yield stress, and η_* describes the differential viscosity in the flow range above the yield stress (Fig. 2.10). In the Ostwald-de Waele model the zero-shear viscosity either vanishes or is infinitely large. One should therefore use this fluid model only when the shear stress nowhere vanishes in the flow field. If we describe processes like the combined pressure-drag flow in a channel, for which the shear stress takes on partly positive, partly negative values, we omit the Ostwald-de Waele model because of its unrealistic behaviour for $\sigma \to 0$. All the other models of liquids given in Table 2.1 have a finite zero-shear viscosity, and the *Reiner* and *Philippoff* models also have an upper non-zero limiting viscosity, and they all are suited for the illustration of phenomena that have non-Newtonian fluid properties.

Care is needed in the application of the Reiner-Philippoff and the Reiner models to the description of dilatant fluid properties. In both models the stability requirement (2.20) is infringed if the viscosity ratio η_∞/η_* exceeds a certain value. In the first case the critical value is 9; in the second case it is as low as 3.24.

Because we wanted to identify the reference quantity η_* with the lower Newtonian viscosity, in the case where it exists, the flow function $f(\sigma)$ for all models, except the Ostwald-de Waele and the Bingham models, is normalised so that $df/d\sigma = 1$ for $\sigma = 0$.

2.2 The normal stress functions

Having in the previous section become acquainted with the fluid properties dependent on the shear rate, we now turn our attention to another characteristic of non-Newtonian fluids. This is the fact that with viscometric flows in general different normal stresses arise in three different directions. In a Newtonian fluid for a shear flow all three normal stresses are equal. One single quantity there suffices to describe them, the pressure p. But in a non-Newtonian fluid for a plane shear flow (Fig. 2.1) the normal stresses $-p + \tau_{xx}$ in the flow direction, $-p + \tau_{yy}$ in the shear direction, and $-p + \tau_{zz}$ in the indifferent direction are not equal to each other. It suffices for many applications to limit oneself to constant density fluids (liquids). As has already been mentioned elsewhere the separation of the pressure p from the normal stresses is arbitrary. In consequence not the three extra-stresses τ_{xx}, τ_{yy}, τ_{zz} themselves, but their differences $\tau_{xx} - \tau_{yy}$ and $\tau_{yy} - \tau_{zz}$ are determined by the motion of the liquid. Taking the differences the arbitrary reference level for the pressure drops out, so that $\tau_{xx} - \tau_{yy}$ and $\tau_{yy} - \tau_{zz}$ are friction stresses in the strict sense. According to the statements introduced in Section 2 those stresses for a simple fluid in a steady shear flow can only be dependent on the local shear rate $\dot{\gamma}$. The following expressions in particular therefore hold for a steady plane shear flow

$$\tau_{xx} - \tau_{yy} = N_1(\dot{\gamma}) \tag{2.21}$$

$$\tau_{yy} - \tau_{zz} = N_2(\dot{\gamma}) \tag{2.22}$$

Here $N_1(\dot{\gamma})$ and $N_2(\dot{\gamma})$ are two new characteristic material functions for a non-Newtonian fluid. These *normal stress functions* combined with the flow function $F(\dot{\gamma})$ (equation (2.4)) define the friction behaviour in a steady plane shear flow.

With the same functions $N_1(\dot{\gamma})$ and $N_2(\dot{\gamma})$ the expressions (2.21) and (2.22) are also valid for other viscometric flows. Thereby one has to note that the normal stress functions are defined by means of a special basis, namely that in which the first two Rivlin-Ericksen tensors have the form

$$\mathbf{A}_1 = \begin{bmatrix} 0 & \dot{\gamma} & 0 \\ \dot{\gamma} & 0 & 0 \\ 0 & 0 & 0 \end{bmatrix}, \quad \mathbf{A}_2 = \begin{bmatrix} 0 & 0 & 0 \\ 0 & 2\dot{\gamma}^2 & 0 \\ 0 & 0 & 0 \end{bmatrix} \tag{2.23}$$

All higher Rivlin-Ericksen tensors vanish, as is well known, for steady shear flows (Section 1.6). If we indicate the directions of the unit vectors of this so-called *natural basis* with the subscript digits 1, 2 and 3, the first normal stress function corresponds to the difference $\tau_{11} - \tau_{22}$, and the second normal stress function to the difference $\tau_{22} - \tau_{33}$:

$$\tau_{11} - \tau_{22} = \mathrm{N}_1(\dot{\gamma}) \tag{2.24}$$

$$\tau_{22} - \tau_{33} = \mathrm{N}_2(\dot{\gamma}) \tag{2.25}$$

For very simple flows, say the plane shear flow, to which equations (2.21) and (2.22) relate, the first natural direction agrees with the flow direction, and the second gives the direction normal to the sliding surfaces. This also occurs for example in a pipe flow with the velocity field (1.81), so that there τ_{11} would be identified with the normal stress in the axial direction, τ_{22} with that in the radial direction, and τ_{33} with that in the neutral azimuthal direction. The example of a helical flow described in Section 1.6.4 shows however that the first natural direction is not obliged to coincide with the flow direction. Equation (1.90) defines the spatial orientation of the natural basis in this case (cf. also Fig. 1.10).

Because with a flow reversal, hence when $\dot{\gamma}$ changes its sign, the normal stress differences remain unchanged, N_1 and N_2 are even functions of its argument.

$$N_1(-\dot{\gamma}) = N_1(\dot{\gamma}), \qquad N_2(-\dot{\gamma}) = N_2(\dot{\gamma}) \tag{2.26}$$

In a fluid in the state of rest ($\dot{\gamma} = 0$) no normal stress differences exist, as is well known. The two normal stress functions therefore have the property

$$N_1(0) = 0, \qquad N_2(0) = 0 \tag{2.27}$$

For that very reason a factor $\dot{\gamma}^2$ can be separated, which is an advantage for many purposes

$$N_1(\dot{\gamma}) = \nu_1(\dot{\gamma}) \cdot \dot{\gamma}^2 \tag{2.28}$$

$$N_2(\dot{\gamma}) = \nu_2(\dot{\gamma}) \cdot \dot{\gamma}^2 \tag{2.29}$$

Instead of the normal stress function $N_1(\dot{\gamma})$ and $N_2(\dot{\gamma})$ we shall therefore conveniently use the *normal stress coefficients* $\nu_1(\dot{\gamma})$ and $\nu_2(\dot{\gamma})$ which have been defined here. Note that N_1 and N_2 have the dimensions of pressure $[N/m^2]$. The coefficients ν_1 and ν_2 on the other hand have the dimensions pressure $\cdot$ (time)^2 $[Ns^2/m^2]$. The ratio of a characteristic value for the normal stress coefficients and a characteristic viscosity $[Ns/m^2]$ therefore represents a characteristic relaxation time of the fluid.

Possible ways of measuring the two normal stress functions will be considered in Chapter 4. Here we shall first consider the results of such measurements. Fig. 2.8 shows, and this corresponds generally to observations on polymer systems, that the first normal stress function is positive; the second normal stress function on the other hand is always negative. This signifies for example that for a plane shear flow, compared with the normal stress in the neutral direction, because of the shear a tensile force arises in the flow direction and a compressive force in the shear direction. One can furthermore read off from Fig. 2.8 that the second normal stress function in absolute values is considerably smaller than the first. It is therefore more difficult to measure, and often its values are affected by some degree of uncertainty. In addition Fig. 2.8 shows that the normal stress functions for relatively low shear rates increase first with $\dot{\gamma}^2$, but with increasing shear rate more and more slowly. We can therefore conclude from this that the normal stress coefficients ν_1 and ν_2 of real fluids have finite lower limiting values, which we shall denote with ν_{10} and

ν_{20}:

$$\nu_{10} := \lim_{\dot\gamma \to 0} \nu_1(\dot\gamma), \qquad \nu_{20} := \lim_{\dot\gamma \to 0} \nu_2(\dot\gamma) \tag{2.30}$$

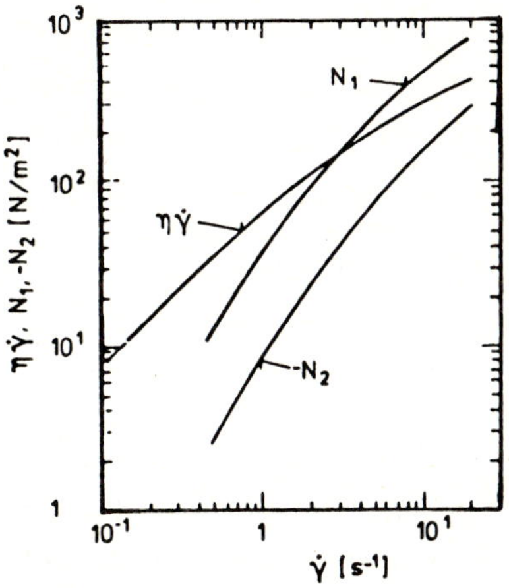

Fig. 2.8 Shear stress and normal stress differences of a polymer solution (10 1/2% polyisobutylene in decalin) as a function of the shear rate at 25°C (after Ginn and Metzner)

In other respects the coefficient functions $\nu_1(\dot\gamma)$ and $\nu_2(\dot\gamma)$ for polymer systems have similar properties as the viscosity function $\eta(\dot\gamma)$ because all three quantities decrease with increasing shear rate monotonically.

By means of Figs. 2.3 to 2.5 it has been found that external parameters like temperature, pressure, molecular weight, or concentration only affect the flow curve of non-Newtonian materials to the extent that they alter the zero-shear viscosity. By representing the flow curve reduced with the zero-shear viscosity the dependence on these parameters can be totally eliminated. One may thus presume that a universal representation of the other two viscometric functions $N_1(\dot\gamma)$ and $N_2(\dot\gamma)$ is also possible, and arises from the fact that the quantities $\dot\gamma$ and N_1, N_2 are made dimensionless by using the zero-shear viscosity η_0 and a constant reference stress τ_*. If the dependence of the normal stress functions on the temperature, etc. is given by the reference quantity $\eta_0(\Theta,\ldots)$, then the laws of similarity

$$\frac{N_{1,2}(\dot{\gamma};\Theta,p,\bar{M})}{\tau_*} = \tilde{N}_{1,2}\left(\frac{\eta_0(\Theta,p,\bar{M})\,\dot{\gamma}}{\tau_*}\right) \tag{2.31}$$

apply, so that $\tilde{N}_1$ and $\tilde{N}_2$, which are characteristic of the material, would be functions independent of Θ, p, $\bar{M}$. Accordingly the normal stress properties determined at various temperatures, etc. would have to coincide on a single curve if one plots N_1 and N_2 against $\eta_0\dot{\gamma}$. To date there has been almost no systematic research into the dependence of the normal stress functions of real fluids on external parameters like temperature, pressure, molecular weight, or concentration. The small amount of data which is available however support the above stated hypothesis. Thus Fig. 2.9 shows a log-log graph of the first normal stress function $N_1(\dot{\gamma})$ for similar types of polymer melts which differ in average molecular weight. All the curves have the same shape and coincide if one moves them to the right along the distance $\log\eta_0(\bar{M})$, i.e. if one plots N_1 against $\eta_0\dot{\gamma}$. The individual values of the zero-shear viscosity η_0 are found from Fig. 2.4, which shows the flow properties of these polymer melts. Hence there are definite reasons for the assumption that the normal stress parameters can also be represented in universal form in accordance with equation (2.31).

According to Fig. 2.8 the first normal stress difference for sufficiently large shear rates exceeds the value of the shear stress. The intersection point of the two graphs ($N_1 = \eta\dot{\gamma}$) defines a stress value τ_c characteristic of the material and a certain shear rate $\dot{\gamma}_c$, which will be designated as the 'natural frequency' of the material. The way by which the fluid properties and normal stress properties can be represented in universal form implies that the natural frequency so defined depends on external parameters like temperature, molecular weight etc. in the same way as the reciprocal value of the zero-shear viscosity, i.e. $\dot{\gamma}_c(\Theta, p, \bar{M}) \sim 1/\eta_0(\Theta, p, \bar{M})$.

Moreover it follows that the stress value τ_c does not depend at all on the external parameters, hence it represents a true material constant. It suggests that this 'natural stress' τ_c can be used as a reference stress, hence $\tau_* = \tau_c$. But one can obviously identify τ_* just as well with the product $\eta_0\dot{\gamma}_c$. Both values however do not differ very much from each other. In the latter case the argument of the universal functions on the right-hand side of equation (2.31) and of equation (2.8) would be $\dot{\gamma}/\dot{\gamma}_c$; i.e. a universal representation of the viscometric properties also results when one makes the shear rate dimensionless with the natural frequency $\dot{\gamma}_c$ and stresses with $\eta_0\dot{\gamma}_c$.

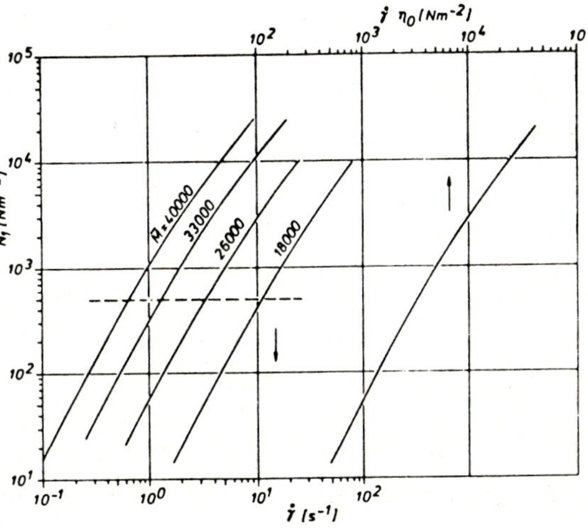

Fig. 2.9 First normal stress function of polymer melts (polyamide 6) at different molecular weights, Θ = 250°C (after Laun)

If one restricts equation (2.31) to that range of shear rates where the normal stress functions increase like $\dot{\gamma}^2$, then one finds that the lower limiting values of the normal stress coefficients as defined in equation (2.30) depend like the square of the zero-shear viscosity on temperature, pressure, etc.:

$$\nu_{10}(\Theta, p, \bar{M}) \sim \eta_0^2(\Theta, p, \bar{M}) \tag{2.32}$$

$$\nu_{20}(\Theta, p, \bar{M}) \sim \eta_0^2(\Theta, p, \bar{M}) \tag{2.33}$$

The expression (2.32) is proved for polymer melts with the properties shown in Figs. 2.4 and 2.9 by the statement that the coefficient ν_{10}

is proportional to $\overline{M}^7$, whereas for the zero-shear viscosity $\eta_0 \sim \overline{M}^{3.5}$ applies.

We conclude the statements about the material properties arising in steady shear flows by assembling the scalar equations (2.5), (2.24), and (2.25) into a coordinate invariant tensor relationship. We have already seen that for an incompressible fluid there is some degree of freedom when specifying the 'pressure' p. One often designates the arithmetic mean of the three normal stresses as -p. But we can use the symbol p just as well for the negative normal stress in the neutral direction. Following this definition, for a viscometric flow of an incompressible non-Newtonian fluid with reference to a suitably chosen basis in accordance with equations (2.5), (2.24), and (2.25) the following stress components occur

$$\mathbf{S} = \begin{bmatrix} -p + N_1(\dot\gamma) + N_2(\dot\gamma) & \eta(\dot\gamma)\dot\gamma & 0 \\ \eta(\dot\gamma)\dot\gamma & -p + N_2(\dot\gamma) & 0 \\ 0 & 0 & -p \end{bmatrix} \tag{2.34}$$

As explained above the representation of the stress tensor **S** is based on the natural basis, with reference to which the two non-vanishing Rivlin-Ericksen tensors are given by equation (2.23). For the square of $\mathbf{A}_1$ the following obviously applies

$$\mathbf{A}_1^2 = \begin{bmatrix} \dot\gamma^2 & 0 & 0 \\ 0 & \dot\gamma^2 & 0 \\ 0 & 0 & 0 \end{bmatrix} \tag{2.35}$$

A comparison of the right-hand side of equation (2.34) with the expressions in equations (2.23) and (2.35) shows that one can represent the stress tensor **S** as the sum of a spherically symmetric term and three proportional terms $\mathbf{A}_1$, $\mathbf{A}_2$, and $\mathbf{A}_1^2$. By use of the normal stress coefficients ν_1 and ν_2 (cf. equations (2.28) and (2.29) equation (2.34) can be written briefly in the form

$$\mathbf{S} = -p\mathbf{E} + \eta(\dot\gamma)\mathbf{A}_1 + [\nu_1(\dot\gamma) + \nu_2(\dot\gamma)]\mathbf{A}_1^2 - \frac{1}{2}\nu_1(\dot\gamma)\mathbf{A}_2 \tag{2.36}$$

In equation (2.36) there is a coordinate invariant form of the general *constitutive equation for viscometric flows* of incompressible fluids.

The quantity $2\dot{\gamma}^2$ denotes the linear principal invariant of the tensor $\mathbf{A}_1^2$; cf. equation (2.35).

Problems

2.1 Attempt to make an approximation (cf. Table 2.1) of the flow properties of a polymer solution represented in Fig. 2.5 by using a Reiner-Philippoff model. Specify to this end using the best possible fit of the model to the actual viscosity curve the two parameters τ_* and η_∞/η_*, and discuss the quality of the approximation by means of the deviations remaining.

2.2 There are materials which do not flow until the shear stress exceeds a critical value, the yield stress τ_*. As a model for such pseudoplastic substances we consider a hypothetical material which has the flow properties shown in Fig. 2.10 (Bingham plastic) which has the distinguishing feature of a constant differential viscosity η_*. Write the constitutive law in the form $\eta_*\dot{\gamma}/\tau_* = f(\tau/\tau_*)$, and determine the dimensionless flow function $f(\sigma)$.

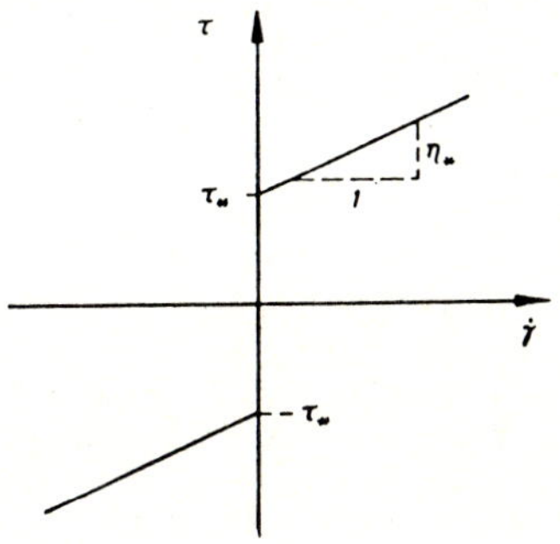

Fig. 2.10 Bingham material flow properties

2.3 Write down the usual viscosity and differential viscosity for a Prandtl-Eyring fluid and for a Bingham fluid as functions of $\dot{\gamma}$ and the two reference quantities η_* and τ_*.

2.4 Determine by means of equation (2.36) the components τ_{rr}, $\tau_{r\varphi}$, τ_{rz}, etc. of the extra-stress tensor $\mathbf{S} + p\mathbf{E}$ for a helical flow which has the velocity field (1.86) as a function of the velocity field $u(r)$ and $\omega(r)$ as well as the three viscometric functions.

3 PROCESSES THAT ARE CONTROLLED BY THE FLOW FUNCTION

3.1 Rotational viscometer

We shall first consider the matter of how one can experimentally determine the flow curve of a non-Newtonian fluid. For this one always has to create a steady shear flow, because only this kind of motion can be described by the flow function. Therefore some types of apparatus with which the viscosity of Newtonian fluids can be determined are not suitable for measuring the viscosity of non-Newtonian fluids. It is for example not valid to make conclusions about the viscosity of a test fluid from the time of descent of a sphere in the fluid. Because the flow round a sphere is not a motion which has a constant stretch history, complex properties manifest themselves, particularly memory effects, and it is not certain which property is mainly being measured. For the viscometry of non-Newtonian fluids therefore it is essential to have apparatus which creates steady shear flows. Hence it is obvious that one often also designates these kinematically strongly restricted motions as viscometric flows.

Under certain conditions the shear rate $\dot{\gamma}$ for a certain viscometric flow can be spatially constant over the domain. Because the shear stress τ only depends on $\dot{\gamma}$, it has the same value everywhere in the liquid and particularly at the surface of the bounding walls. As was established in Sections 1.6.3 and 1.6.6, this is true particularly for a Couette flow between two cylinders, and approximately so for a cone-and-plate flow if the gap between the cylinders is narrow, and the cone is very blunt respectively. The simplest viscometers are accordingly devices in which the test fluid passes between two coaxial cylinders with the smallest possible gap between them, or the gap between a plate and a rotating cone located on it.

In the first case we shall denote the radius of the inner cylinder as r_0, its wetted length as b, the width of the gap as h, and the angular velocity of the rotating cylinder as ω. One of the two cylinders is usually stationary. The shear rate of the liquid then has the value $\dot{\gamma} = r_0\omega/h$ (cf. equation (1.85)), hence is determined up to the numerical factor r_0/h by the angular velocity. A torque is needed in order to maintain the rotation of the cylinder, and this depends in a simple way on the shear stress on the cylinder which is acting in the direction of rotation. Because the shear stress τ is constant there arises a torque on a wetted cylinder surface which is $2\pi r_0 b$ at a distance r_0 from the axis of rotation.

$$M = 2\pi r_0^2 b\,\tau \tag{3.1}$$

Hence up to the instrument constant $2\pi r_0^2 b$ the torque M corresponds to the shear stress τ in the liquid, and in fact this is independent of the angular velocity of the rotating cylinder. For the viscosity η the ratio of τ and $\dot{\gamma}$ (equation (2.5)) therefore yields the relationship

$$\eta = \frac{h}{2\pi r_0^3 b} \cdot \frac{M}{\omega} \tag{3.2}$$

Hence one can determine the viscosity curve $\eta(\dot{\gamma})$ of a non-Newtonian fluid step by step by varying the rotation rate, measuring the torque acting on one of the cylinders, then calculating the value of the right-hand side of equation (3.2), and entering the value as an ordinate in a diagram which shows $r_0\omega/h$ along the abscissa. On the other hand if one represents the torque M as a function of the angular velocity ω, then this graph gives the relationship between τ and $\dot{\gamma}$, hence the flow function $F(\dot{\gamma})$ as in equation (2.4), because M is proportional to τ and ω to $\dot{\gamma}$.

We now consider a device in which the fluid is located between a flat plate and a cone placed on it (Fig. 1.12). We denote the radius of the circular top of the cone by r_0. If the angle between the cone and the plate is sufficiently small, the overall shear rate in the liquid has the same value, $\dot{\gamma} = \omega/\beta$, in which ω is the angular velocity of the plate relative to the stationary cone. Thus the shear stress in the field is also constant. Its amount is found from the torque on the cone.

$$M = \int_0^{r_0} 2\pi r^2 \tau\, dr = \frac{2\pi}{3} r_0^3 \tau \tag{3.3}$$

One obtains an analogous formula for the viscosity of the fluid as for Couette viscometers:

$$\eta = \frac{3\beta}{2\pi r_0^3} \cdot \frac{M}{\omega} \tag{3.4}$$

This again makes it possible, as described above, to determine the viscosity at given rotation speeds by measuring the torque which the motion creates.

For practical applications equation (3.2) needs a correction to take into account the *end effect*. Every Couette viscometer naturally has a finite length. The cylindrical housing at the bottom end is blanked off by a plate, in the vicinity of which in general no shear flow occurs. This unwanted end effect at the base can be completely counteracted if one makes the inner cylinder conical below and places its tip on the plate (Fig. 3.1). Hence the Couette arrangement is combined with the cone-and-plate arrangement, and hence a shear flow arises not only between the cylindrical container and the cylinder, but also at the base, which has a constant shear rate. It is advantageous to choose the cone angle so that the shear rate and therefore the shear stresses in both regions match each other: $\beta = h/r_0$. Note that the individual contributions (3.1) and (3.3) add to the total torque, so that one obtains instead of equation (3.2) the modified expression

$$\eta = \frac{h}{2\pi r_0^3 b\left(1 + \frac{r_0}{3b}\right)} \cdot \frac{M}{\omega} \qquad (3.5)$$

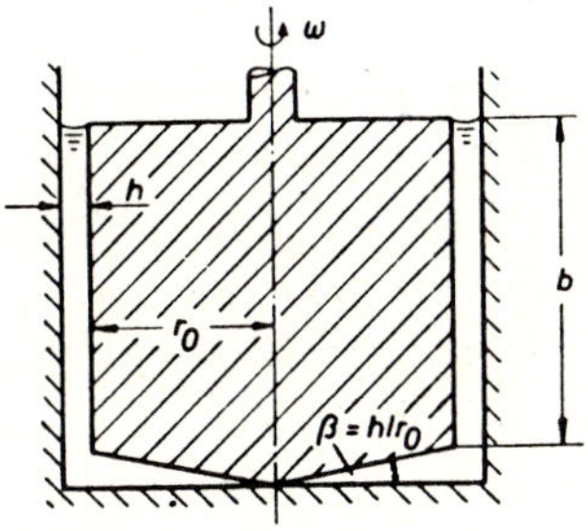

Fig. 3.1 Modified Couette viscometer with controlled end effect at the base

Comparison with equation (3.2) shows that the end effect at the base expresses itself in a constant correction factor, namely the term in brackets in the denominator in equation (3.5). Note however that for another choice of cone angle ($\beta \neq h/r_0$) different stresses arise in the two regions and the measured torque corresponds to the weighted average of the two stresses. The use of an expression of the type $\eta \sim M/\omega$ is then questionable because with fixed rotation speed in fact two viscosity values are controlling, whilst the formula only gives one weighted average. In addition it is not certain which shear rate must be correlated with this calculated average viscosity value.

In a rotational viscometer the test fluid is in general not totally enclosed between fixed walls, but it has a partly free surface (Fig. 3.1). For that reason the viscosity measurement is also subject to error. Simple shear no longer applies in the vicinity of the top surface. The surface deforms because of the movement (described in detail in Section 4.4), and there arises a complicated superposed 'secondary flow' on the shear, which extends down to a depth of the same order of magnitude as the width of the spacing, so that viscometric properties are not the only ones operating. In order to be able to find the viscosity in spite of this the Couette arrangement must be of such a type that the end effect at the surface affects the measured torque only relatively little, i.e. the cylinder length b must be large compared with the gap width h. Devices which do not satisfy this requirement cannot truly be called viscometers.

3.1.1 Effect of dissipation

So far it has been assumed implicitly that the test fluid has a constant temperature. However this is not so in practice. Because of the shear the mechanical energy in the interior of the fluid is dissipated as heat, which must flow out to the surroundings. As a result a temperature gradient is created in the fluid, i.e. the temperature varies with position and is not constant. At moderate shear rates the temperature differences as a result of dissipation are of course so low that one can regard the fluid as an isothermal continuum to a close approximation. At increasing shear rates however correspondingly greater temperature differences arise, so that the assumption of isothermal conditions becomes more and more questionable. The following considerations are intended to develop a theory with reference to the dissipation, which opens up a possibility of also determining the flow function with the help of a rotational viscometer, when isothermal conditions no longer apply.

We first of all model the gap between two cylinders with insignificantly different radii, and between a plate and a blunt cone by a plane channel, thus replacing the weakly curved sliding surfaces by plane layers. Let the stationary wall be at y = 0, the moving wall at y = h (in the case of a cone-and-plate viscometer h would be a parameter dependent on z). The velocity of the moving wall is denoted by u_h. A peculiarity of the rotational viscometer is that all state variables depend only on the coordinate perpendicular to the wall, and there is no pressure drop in the flow direction. We are therefore considering a pure drag flow in the x-direction with initially unknown velocity field u(y) and unknown temperature field $\Theta(y)$. Because this is a case of unaccelerated motion and the pressure is constant in the x-direction, the extra-stresses acting on a volume element maintain the equilibrium. Therefore the equation of motion for the x-direction reduces to $d\tau/dy = 0$ (cf. equation (1.113)). The shear stress τ therefore, independent of the thermal conditions, has a constant value across the channel depth which would be derived for the Couette arrangement from the measured torque using equation (3.1). The shear rate is on the contrary in general a function of position because the reference stress τ_* and the right-hand side are constant in the equation (2.16), but the reference viscosity η_* depends on the temperature and hence on y.

After the relevant equation of motion has been evaluated we turn to the energy equation (1.126). In accordance with the previously made statements about the velocity field and the temperature field $D\Theta/Dt = 0$ applies, i.e. the temperature of a material point remains constant with time, which can be explained by the fact that the path lines (y=constant) coincide with the isotherms. The energy balance therefore reduces to the statement that the energy Φ dissipated per unit time and unit volume is completely carried away by heat conduction: $\Phi + \mathrm{div}(\lambda\, \mathrm{grad}\Theta) = 0$. Note that the heat flux is in the y-direction, and that for a shear flow $\Phi = \tau\, du/dy$ applies (cf. equation (1.123)), from which one obtains

$$\frac{d}{dy}\left(\lambda \frac{d\Theta}{dy}\right) = -\tau \frac{du}{dy} \tag{3.6}$$

Because the shear stress is constant the expression can be integrated once:

$$\lambda \frac{d\Theta}{dy} = \tau\,(\alpha\, u_h - u) \tag{3.7}$$

The constant of integration has been represented by an as yet not determined numerical parameter α in the form $\alpha \tau u_h$. As regards the thermal boundary conditions, we assume that the stationary wall has a known temperature Θ_0 impressed from the outside. Concerning the conditions at the moving wall we shall later consider the special cases with likewise given temperature Θ_0 there, or with vanishing heat flux, when the wall is saturated with heat. We choose Θ_0 as the reference temperature, denote the values of the reference viscosity η_* and the thermal conductivity λ belonging to Θ_0 with η_{*0} and λ_0 respectively, and introduce the following dimensionless expressions:

$$\sigma := \frac{\tau}{\tau_*}, \qquad \zeta := \frac{y}{h}$$
$$v(\zeta) := \frac{u(y)}{u_h}, \qquad \vartheta(\zeta) := \frac{\Theta(y) - \Theta_0}{\Theta_0} \tag{3.8}$$
$$S := \frac{\eta_{*0} u_h}{\tau_* h}, \qquad Br := \frac{\eta_{*0} u_h^2}{\lambda_0 \Theta_0}$$

We shall henceforth regard the dimensionless velocity field v and the dimensionless temperature field ϑ as functions of the dimensionless position variable ζ $(0 \leq \zeta \leq 1)$. The 'drag velocity parameter' S can be regarded as the ratio of an 'internal' time η_{*0}/τ_* determined by the behaviour of the material, and an 'external' time h/u_h impressed by the flow.

The ratio of an internal time characteristic of the material and an external time characterising the flow process is often designated as the *Weissenberg number* (elasticity parameter). We employ the term Weissenberg number only when the internal time is composed of the normal stress coefficients (and the viscosity). So far as the internal time is derived from the flow properties alone we make use of a notation like the drag velocity parameter.

The *Brinkman number* Br is a dimensionless quantity for the total dissipated energy per unit time in the channel. As a reference quantity we choose a fictive heat flux formed from the reference temperature

Θ_0 and the channel height. In the following the parameters S and Br are considered as given quantities. We now relate the zero-shear viscosity $\eta_*(\Theta)$ and the thermal conductivity $\lambda(\Theta)$ to the values η_{*0} and λ_0 respectively, appropriately to the reference temperature, therefore forming the dimensionless expressions

$$H_*(\vartheta) := \frac{\eta_*(\Theta)}{\eta_{*0}}, \qquad \Lambda(\vartheta) := \frac{\lambda(\Theta)}{\lambda_0} \tag{3.9}$$

and obtain the temperature dependent quantities H_* and Λ as defined here as known functions of ϑ. Moreover the following holds

$$H_*(0) = \Lambda(0) = 1 \tag{3.10}$$

With these abbreviations the energy balance (3.7) takes on the following form:

$$\Lambda(\vartheta)\frac{d\vartheta}{d\zeta} = \frac{\sigma}{S}\,\mathrm{Br}\,(\alpha - v) \tag{3.11}$$

The flow law (2.16) is added to this as the second equation for the determination of the fields $v(\zeta)$ and $\vartheta(\zeta)$. Because for a plane shear flow $\dot{\gamma}$ is replaced by $du/dy = (u_h/h)\,dv/d\zeta$ equation (2.16) changes to

$$H_*(\vartheta)\frac{dv}{d\zeta} = \frac{f(\sigma)}{S} \tag{3.12}$$

In order to integrate the system of equations (3.11), (3.12) one first expediently divides both relationships by each other, because by doing so the coordinate ζ cancels out:

$$\frac{\Lambda(\vartheta)}{H_*(\vartheta)}\frac{d\vartheta}{dv} = \frac{\sigma\,\mathrm{Br}}{f(\sigma)}(\alpha - v) \tag{3.13}$$

This relationship can be integrated by separating the variables:

$$\int_0^{\vartheta} \frac{\Lambda(\bar{\vartheta})}{H_*(\bar{\vartheta})}\,d\bar{\vartheta} = \frac{\sigma\,\mathrm{Br}}{f(\sigma)}\left(\alpha v - \frac{v^2}{2}\right) \tag{3.14}$$

The new constants of integration thereby arising could be put as zero because at the stationary wall (for $\zeta = 0$) the flow velocity and the temperature difference vanish:

$$v(0) = 0, \qquad \vartheta(0) = 0 \tag{3.15}$$

Equation (3.14) represents a relationship between the fields v and ϑ. The events in the gap can therefore be represented for known material properties and a given Brinkman number in a v-ϑ diagram when the constants of integration σ and α have been determined. We consider equation (3.14) solved for ϑ and write the relationship symbolically in the form

$$\vartheta = g\left(v; \frac{\sigma\,Br}{f(\sigma)}\right) \tag{3.16}$$

By entering this result on the left-hand side of equation (3.12) and integrating once, one obtains

$$G\left(v; \frac{\sigma\,Br}{f(\sigma)}\right) := \int_0^v H_*\left(g\left(\bar{v}; \frac{\sigma\,Br}{f(\sigma)}\right)\right) d\bar{v} = \frac{f(\sigma)}{S}\zeta \tag{3.17}$$

Because of the confining conditions at the wall (equation (3.15)) there is once more no additional constant of integration. By inverting this relationship one obtains the velocity profile $v(\zeta)$. The temperature field $\vartheta(\zeta)$ then follows from equation (3.16), or from equation (3.12) by entering the value for v. In the following example the use of the equation (3.12) proves to be the simpler method.

So far we have only formulated and worked with the boundary conditions at the stationary wall. At the moving wall one has first to note that the fluid is dragged along at the same velocity, i.e. $v(1) = 1$. As regards the thermal constraints we consider two special cases. In one we consider the possibility that the moving wall likewise has the temperature Θ_0, from which the condition $\vartheta(1) = 0$ would follow. In addition we are interested in the case where the wall is thermally insulated from the surroundings and hence cannot lose any heat to them. In the steady thermal state it cannot under these conditions absorb the energy dissipated in the fluid. In other words, the heat flux and with it the temperature gradient in the fluid vanish at the wall; $d\vartheta/d\zeta = 0$ for $\zeta = 1$. In the first case it follows from equation (3.14) that the hitherto still undetermined constant of integration α has the value 1/2.

$$v(1) = 1, \qquad \vartheta(1) = 0 \;\rightarrow\; \alpha = \frac{1}{2} \tag{3.18}$$

In the other case the value 1 is obtained for α from equation (3.11):

$$v(1) = 1, \qquad \left.\frac{d\vartheta}{d\zeta}\right|_{\zeta=1} = 0 \;\rightarrow\; \alpha = 1 \tag{3.19}$$

Thus with given numerical values for S and Br and the known fluid properties only one constant is unknown, the dimensionless value σ of the shear stress in the gap. The equation which determines it is obtained from the condition v(1)=1 for the velocity profile (cf. equation (3.17):

$$G\left(1;\frac{\sigma\,\mathrm{Br}}{f(\sigma)}\right)=\frac{f(\sigma)}{S} \tag{3.20}$$

This is an implicit formula for σ which one may think of solving in the concrete case by means of a suitable iterative method. Hence all the constants that affect the velocity and temperature profile in the gap are determined.

So far we have supposed that the flow function $f(\sigma)$ of the fluid is known and the velocity field of the pure drag flow and the temperature field associated with it will be calculated. But we can also use (3.20) as the conditional equation for the flow function f, thus returning to the question of how one can experimentally obtain the flow curve of non-Newtonian fluids. In this case one would have to obtain the value of the parameter σ experimentally. For a Couette device this comes out in the measurement of the torque because equation (3.1) also remains valid when dissipation has to be considered. By the use of this measured value and the numerical values for the drag velocity parameter S and the Brinkman number Br, the value of the flow function f appropriate to σ may be calculated. This procedure of course assumes that the temperature dependence of the zero-shear viscosity and the thermal conductivity are known. With a change in the drag velocity (caused by a change in rotation speed) there is a change in the coefficients S and Br (equation (3.8)), the measured quantity σ , and hence the calculated value f. In this way it is in principle also possible to obtain the reduced flow curve when marked temperature differences arise by dissipation in the gap.

The train of thought described above is possibly clearer if we use concrete expressions for the thermal properties $H_*(\vartheta)$ and $\Lambda(\vartheta)$. Because with actual fluids the thermal conductivity changes only slightly compared with the viscosity, its temperature dependence can often be disregarded for practical purposes. The temperature dependence of the viscosity can often be very well approximated by a decreasing exponential function. Considering the conditions (3.10) we therefore write

$$\Lambda(\vartheta)=1, \qquad H_*(\vartheta)=e^{-\kappa\vartheta} \tag{3.21}$$

Note that for each application only a finite temperature interval is of interest, and the assumption (3.21) must naturally only be in agreement with the actual material properties within that interval. The constant κ, which occurs in the expression for the viscosity ratio H_* thus depends on the position of the temperature interval, particularly on the reference temperature Θ_0. The integral in equation (3.14) can be obtained explicitly by using the expressions in (3.21). Finally if one solves for ϑ one obtains

$$\vartheta = \frac{1}{\kappa} \ln\left[1 + \kappa \frac{\sigma \, \mathrm{Br}}{f(\sigma)}\left(\alpha v - \frac{v^2}{2}\right)\right] \tag{3.22}$$

This is the equation (3.16) adapted to the special circumstances.

Thus the integral occurring in equation (3.17) takes on the following form:

$$G\left(v; \frac{\sigma \, \mathrm{Br}}{f(\sigma)}\right) = \int_0^v \frac{d\bar{v}}{1 + \kappa \dfrac{\sigma \, \mathrm{Br}}{f(\sigma)}\left(\alpha \bar{v} - \dfrac{1}{2}\bar{v}^2\right)} \tag{3.23}$$

This can be solved by elementary methods. Using the abbreviation

$$\beta := \sqrt{\frac{2 f(\sigma)}{\sigma \kappa \, \mathrm{Br}} + \alpha^2} \tag{3.24}$$

the result of the integration is given by:

$$G\left(v; \frac{\sigma \, \mathrm{Br}}{f(\sigma)}\right) = \frac{2 f(\sigma)}{\sigma \kappa \, \mathrm{Br} \, \beta}\left[\operatorname{artanh} \frac{v - \alpha}{\beta} + \operatorname{artanh} \frac{\alpha}{\beta}\right] \tag{3.25}$$

If one equates this expression in accordance with (3.17) to the expression $f(\sigma)\zeta/S$, then one obtains the explicit formula for the velocity

$$v(\zeta) = \alpha + \beta \tanh\left[\frac{\sigma \kappa \, \mathrm{Br} \, \beta}{2S} \zeta - \operatorname{artanh} \frac{\alpha}{\beta}\right] \tag{3.26}$$

Equation (3.20) can be used to determine the shear stress parameter σ for known flow properties and to determine the reduced flow velocity f for measured values of σ respectively. The transcendental equation resulting from that can be given in both special cases which we wished to consider ($\alpha = 1/2$ and $\alpha = 1$), uniformly in the form

$$\operatorname{artanh} \frac{\alpha}{\beta} = \frac{\alpha \sigma \kappa \, \mathrm{Br} \, \beta}{2S} \tag{3.27}$$

Therefore the analytical expression for the velocity profile can be further simplified:

$$v(\zeta) = \alpha + \beta \tanh\left[\frac{\sigma\kappa\,\mathrm{Br}\,\beta}{2S}(\zeta - \alpha)\right] \tag{3.28}$$

It is expedient for determining the temperature profile in the gap to return to equation (3.12) and to form the derivative $dv/d\zeta$ in it with the help of the result just obtained. In this way one obtains

$$\kappa\,\vartheta(\zeta) = \ln\left\{\frac{\sigma\kappa\,\mathrm{Br}\,\beta^2}{2f(\sigma)}\cdot\frac{1}{\cosh^2\left[\frac{\sigma\kappa\,\mathrm{Br}\,\beta}{2S}(\zeta - \alpha)\right]}\right\} \tag{3.29}$$

Figs. 3.2 and 3.3 illustrate the analytical results. For this a fluid model with Prandtl-Eyring flow properties has been taken as an example ($f(\sigma) = \sinh\sigma$). The relevant values of the shear stress parameter σ must then be calculated from the transcendental equation (3.27) while taking into account equation (3.24) for the specified values of the drag velocity S and the product $\kappa\cdot\mathrm{Br}$. After that the velocity and temperature profiles according to equations (3.28) and (3.29) can be immediately represented. In order to take account of the effect of increasing drag velocity u_h, the parameters S and $\kappa\,\mathrm{Br}$ have been simultaneously changed in accordance with the relationship $\kappa\,\mathrm{Br} = 4S^2$, hence taking into account that $S \sim u_h$ and $\mathrm{Br} \sim u_h^2$ (equation (3.8)).

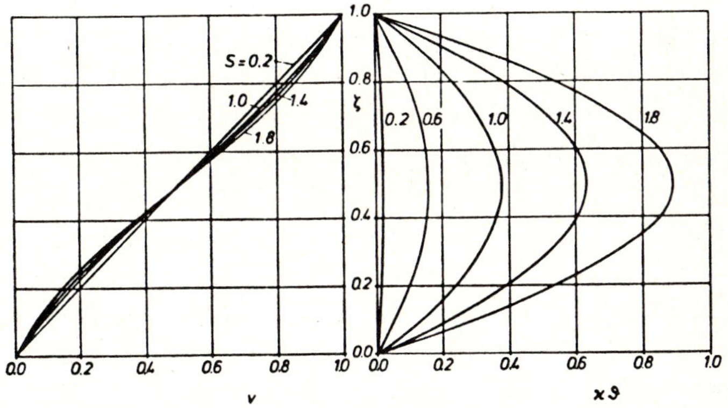

Fig. 3.2 Velocity and temperature profiles in accordance with equations (3.28) and (3.29) for the case of two isothermal walls ($\alpha = 1/2$) for various drag velocity parameters S and $\kappa\,\mathrm{Br} = 4S^2$; Prandtl-Eyring model

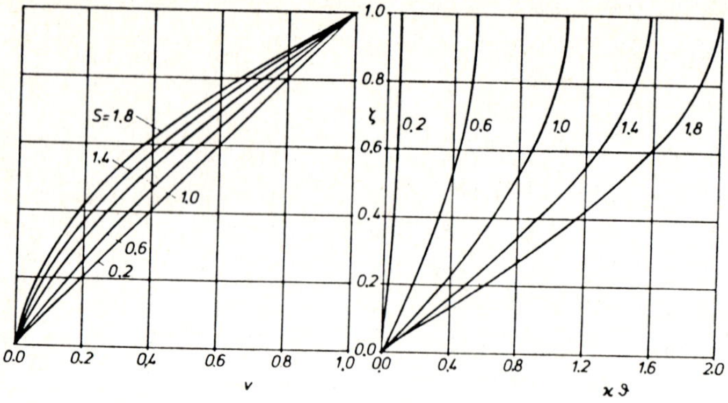

Fig. 3.3 Velocity and temperature profiles for the case of an adiabatic wall (α = 1) for various drag velocity parameters S and $\kappa Br = 4S^2$; Prandtl-Eyring model

In the case of two isothermal walls (Fig. 3.2) the highest temperature occurs in the centre of the channel, and the temperature profile is symmetric about the centre. For relatively low drag velocities with correspondingly small effect of dissipation the flow velocity increases practically linearly with ζ, which is characteristic of a totally isothermal pure drag flow. With increasing drag velocity departures from linearity become apparent, so that the velocity profile in the centre of the channel has a turning point and with reference to this position is point symmetrical. These symmetrical properties are independent of the flow law, therefore they are also valid in particular for Newtonian fluids and can of course be obtained from equations (3.28) and (3.29). In the case of an adiabatic wall the curvature of the velocity profile in the channel has the same sign overall, and the temperature increases monotonically with ζ, whereby the temperature increase diminishes upwards. Finally it is clearly evident from the illustration that in this case the heat dissipated in the fluid can only flow outwards through the bottom wall. The heat flux vector is therefore directed from above downwards at each position, i.e. the temperature decreases from the adiabatic to the isothermal wall. Because the energy amounts dissipated in the individual layers add up, the heat flux and with it the temperature drop from above downwards increase monotonically.

3.2 Pressure-drag flow in a straight channel

The function of a *friction pump* is to deliver a viscous fluid against rising pressure. This is effected by moving, mostly rotating, walls which drag the fluid in the desired direction. By choice of suitable dimensions and the correct drag velocity the drag effect overcomes the counteracting pressure effect, and a specified amount of fluid per unit time flows from the lower to the higher pressure side. The simplest use of such a combined *pressure-drag flow* consists of a simple shear flow between two parallel plane walls moving tangentially to each other. One of the walls can be stationary, so that the pressure in the direction of motion rises or falls. This type of flow is for example the approximate basis for the roller pump commonly used in plastics technology, the helical pump, and melt extruders where the flow channel is curved, but the channel width is small compared with the radius of curvature. Study of pressure-drag flow in straight channels will not only enable us to account for pump and extrusion phenomena, but at the same time will reveal an approach to the theory of rolling and of hydrodynamic lubrication with many applications. We therefore consider the steady plane shear flow of a viscous fluid between two parallel walls (at y = 0 and y = h) which move at the velocities u_0 and u_h in the x-direction (Fig. 3.4). We shall disregard temperature effects and hence assume isothermal conditions. Because the fluid particles move without acceleration the pressure and friction forces are in equilibrium in the rectangular element shown in Fig. 3.4.

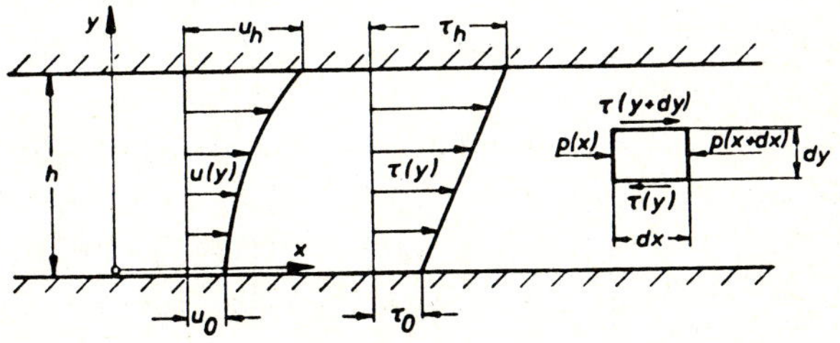

Fig. 3.4 Combined pressure-drag flow in a straight channel

We can combine with the pressure any potential of conservative volume force fields, particularly the force of gravity. Concerning the balance of forces in the direction of movement, this leads to the equation $[\tau(y + dy) - \tau(y)]bdx - [p(x + dx) - p(x)]bdy = 0$ (b is the width of the rectangular element perpendicular to the plane of the drawing). After dividing by the volume $b \cdot dx \cdot dy$ a simple relationship between shear stress τ and pressure p results:

$$\frac{d\tau}{dy} = \frac{dp}{dx} \tag{3.30}$$

On one side the shear stress is only dependent on the coordinate y normal to the wall, which has already been established in connection with the discussion of the flow law. On the other side the pressure, or more accurately the negative normal stress in the y-direction, is only dependent on x. The last follows directly from a balance of forces in the direction normal to the wall. The left-hand side of equation (3.30) can accordingly be a function only of y; the right-hand side a function only of x. Equality of the two expressions can only exist under these conditions if both sides are constant. We can therefore state that the pressure gradient dp/dx is spatially constant, and therefore the pressure itself increases or decreases linearly with x. The shear stress correspondingly varies linearly with the coordinate y.

$$\tau(y) = \frac{dp}{dx} y + \tau_0 \tag{3.31}$$

The constant τ_0 indicates the shear stress at the bottom wall. We accordingly introduce the symbol τ_h for the shear stress at the upper wall (Fig. 3.4). Both stress values depend on each other in accordance with equation (3.31) with the relationship

$$\tau_h = \frac{dp}{dx} h + \tau_0 \tag{3.32}$$

For a pressure increase ($dp/dx > 0$) τ_h is greater than τ_0; on the other hand it is smaller for a pressure drop. We now bring in the flow law in the form (2.16), whereby for a plane shear flow $\dot{\gamma}$ is equal to du/dy:

$$\frac{du}{dy} = \frac{\tau_*}{\eta_*} f\left(\frac{\tau}{\tau_*}\right) \tag{3.33}$$

Because τ has already been determined as a function of y, the *velocity profile* in the channel can be found by integrating. By use of equation (3.31) and noting the boundary condition $u(0) = u_0$ one obtains

$$u(y) - u_0 = \frac{\tau_*^2}{\eta_* \frac{dp}{dx}} \int_{\tau_0/\tau_*}^{\tau(y)/\tau_*} f(\sigma)\, d\sigma \tag{3.34}$$

The still unknown quantity τ_0/τ_* in this equation is found from the boundary conditions $u(h) = u_h$:

$$\frac{\eta_*(u_h - u_0)}{\tau_* h} = \frac{\tau_*}{h \frac{dp}{dx}} \int_{\tau_0/\tau_*}^{\tau_h/\tau_*} f(\sigma)\, d\sigma \tag{3.35}$$

Hence the velocity profile in the channel naturally depends on the flow properties of the material, but only on that part of the flow function $f(\sigma)$ which represents the shear stresses (between τ_0 and τ_h) which prevail in the channel.

Integration over the cross-section gives the total volume $\dot{V}$ passing through the channel per unit time:

$$\dot{V} = b \int_0^h u(y)\, dy = b\,(u_0 + u_h)\,\frac{h}{2} - b \int_0^h \left(y - \frac{h}{2}\right) \frac{du}{dy}\, dy \tag{3.36}$$

In this expression b is the channel width normal to the plane of the flow. The second equality follows after partial integration.

By eliminating du/dy in the second term by use of the constitutive law (3.33) and (y - h/2) by using equations (3.31) and (3.32) one obtains the relationship

$$\frac{\dot{V}}{b} = (u_0 + u_h)\frac{h}{2} - \frac{\tau_*^3}{\eta_* \left(\frac{dp}{dx}\right)^2} \int_{\tau_0/\tau_*}^{\tau_h/\tau_*} \left(\sigma - \frac{\tau_0 + \tau_h}{2\tau_*}\right) f(\sigma)\, d\sigma \tag{3.37}$$

The term without an integral describes that volume flux which occurs with a pure drag flow. If there is no pressure gradient (dp/dx = 0) the velocity profile is of course linear, and this holds independently of the flow properties, so that the flow rate is equal to the product of the mean velocity $(u_0 + u_h)/2$ and the cross-section area bh. From that it follows, and can of course be proved, that the second term in the limiting case $dp/dx \to 0$ vanishes. In equation (3.37) it is therefore evident that the volume flux for a combined pressure-drag flow splits up into the part $bh(u_0 + u_h)/2$ for a pure drag flow, and another part which is added for the pressure drop. In this form the formula for the flow rate reminds one of the relationship for Newtonian fluids (subscript N)

$$\frac{\dot{V}_N}{b} = (u_0 + u_h)\frac{h}{2} - \frac{h^3}{12\eta_*}\frac{dp}{dx} \tag{3.38}$$

It is included as a special case in equation (3.37). In the derivation $f(\sigma)$ would be made equal to σ and the expression (3.32) relating the wall shear stresses would be taken into account. Here the wall velocities and pressure rise appear separately, so that the volumes delivered by the drag effect (first term) and by the pressure effect (second term) add or subtract. In other words: for a Newtonian fluid the pure drag effect and the pure pressure effect are additively superposed. This also occurs for the velocity profile, because for $f(\sigma) = \sigma$ equation (3.34) gives by use of (3.31) and (3.35):

$$u_N(y) = u_0 + (u_h - u_0)\frac{y}{h} - \frac{1}{2\eta_*}\frac{dp}{dx}(yh - y^2) \tag{3.39}$$

This shows that the characteristic linear profile for a flow without a pressure gradient, and the characteristic parabolic profile which is symmetrical about the middle of the channel for a pure pressure flow (at stationary walls) add in a combined pressure-drag flow. This property of additive superposition of the pure parts without interaction does not apply to non-Newtonian fluids. The second term in equation (3.37) includes, if one imagines τ_0 and τ_h to be eliminated through (3.32) and (3.35), not only the pressure gradient dp/dx, but also the velocity difference $u_h - u_0$. In addition the influence of the pressure rise is no longer linear, but depending on the flow properties is more or less complex. Hence the shape of the velocity profile varies when the pressure gradient is changed.

We explain this for the sake of simplicity for a pure pressure flow with a pressure drop in the x-direction ($dp/dx < 0$ and $u_0 = u_h = 0$). One can in this case conclude from equation (3.35), because the left-hand side vanishes and the flow function $f(\sigma)$ is odd, that the shear stresses at both walls except for the sign are equal, $\tau_h = -\tau_0$. This is also immediately apparent because with stationary walls a flow symmetrical about the centre of the channel is set up. Hence equation (3.32) gives for the value of the wall shear stress $\tau_0 = -(h/2)dp/dx$. If one takes this into account in equation (3.34) one learns that the shape of the velocity profile depends on the value of the numerical parameter $-(h/2\tau_*)\cdot dp/dx$, which we shall abbreviate by σ_0, because it is the same as the wall shear stress related to τ_*. It is expedient to normalise the local velocity u(y) with the mean velocity $\bar{u} := \dot{V}/bh$. One obtains after a short calculation using equations (3.34) and (3.37)

$$\frac{u(y)}{\bar{u}} = \frac{\sigma_0}{\int_0^{\sigma_0} \sigma f(\sigma)d\sigma} \cdot \int_{\sigma_0\left(1-2\frac{y}{h}\right)}^{\sigma_0} f(\sigma)d\sigma \tag{3.40}$$

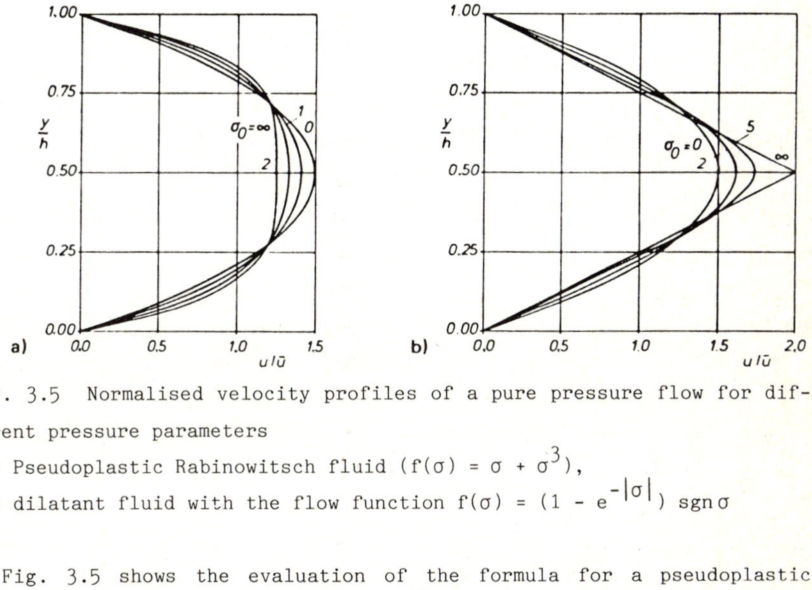

Fig. 3.5 Normalised velocity profiles of a pure pressure flow for different pressure parameters
(a) Pseudoplastic Rabinowitsch fluid ($f(\sigma) = \sigma + \sigma^3$),
(b) dilatant fluid with the flow function $f(\sigma) = (1 - e^{-|\sigma|})\,\mathrm{sgn}\,\sigma$

Fig. 3.5 shows the evaluation of the formula for a pseudoplastic Rabinowitsch fluid (with $f(\sigma) = \sigma + \sigma^3$), and a dilatant fluid with the flow function $f(\sigma) = (1 - e^{-|\sigma|})\,\mathrm{sgn}\,\sigma$. Hence the integrals can be evaluated elementarily, and the right-hand side reduces to a simple function of position with one variable. For a small pressure drop ($\sigma_0 \ll 1$) the fluid flows so slowly that only the low-shear Newtonian properties appear. The characteristic parabolic velocity profile for Newtonian fluids is correspondingly associated with the parameter $\sigma_0 = 0$. With an increasing pressure parameter σ_0 we see blunter velocity profiles for the Rabinowitsch fluid (Fig. 3.5a). This property is typical of pseudoplastic fluids. Conversely for dilatant fluids the velocity profile becomes more and more pointed with increasing pressure drop, so that the relative deviation of the maximum velocity in the middle of the channel becomes increasingly greater than the mean velocity (Fig. 3.5b).

3.2.1 Flow characteristics, pumping efficiency

We return to the pump situation mentioned at the start, in which the fluid is pumped against increasing pressure as the result of the drag effect of the walls. Because for friction pumps in general one of the two walls which make up the flow channel is stationary, we put $u_0 = 0$ in what follows. The volume $\dot{V}$ pumped per unit time depends critically on the drag velocity u_h and on the amount of the pressure rise dp/dx. The total of the combined amounts of pressure increase and volume flux associated with it for constant drag velocity yields a curve in the diagram of pressure against flow rates, which is known as the *flow* or *pump characteristic*. If the drag velocity varies, then one obtains a set of flow characteristics, hence collectively they represent the characteristics of the pump. Equations (3.35) and (3.37) provide the basis for the calculation of these characteristics. For a given flow function equation (3.35) is a relationship between three dimensionless quantities, the pressure parameter

$$K := \frac{h}{\tau_*} \frac{dp}{dx} \tag{3.41}$$

the drag parameter

$$S := \frac{\eta_* u_h}{\tau_* h} \tag{3.42}$$

and the ratio $\sigma_0 := \tau_0/\tau_*$ of the shear stress at the stationary wall and the reference value τ_*. With these abbreviations equation (3.35) takes the following form:

$$S = \frac{1}{K} \int_{\sigma_0}^{K+\sigma_0} f(\sigma) d\sigma \tag{3.43}$$

We use the quantity

$$Q := \frac{\eta_* \dot{V}}{\tau_* bh^2} \tag{3.44}$$

for the dimensionless measure of the volume flux. Hence by elimination of the drag velocity u_h by use of equation (3.37), equation (3.35) transforms to

$$Q = \frac{1}{K^2} \int_{\sigma_0}^{K+\sigma_0} (K + \sigma_0 - \sigma) f(\sigma) d\sigma \tag{3.45}$$

Thus the dimensionless volume flux is expressed explicitly by the pressure parameter K and the dimensionless quantity σ_0 for the shear stress at the stationary wall. In conjunction with equation (3.43) there is a parametral expression for Q as a function of K and S, in which σ_0 is the parameter. In this connection only the fluid properties of the material enter, represented by the dimensionless flow function $f(\sigma)$.

As an example we consider a Prandtl-Eyring model in which $f(\sigma)$ is replaced by $\sinh\sigma$. The two fundamental relationships (3.43) and (3.45) then reduce to

$$S = \frac{1}{K} [\cosh (K + \sigma_0) - \cosh \sigma_0] \tag{3.46}$$

$$Q = \frac{1}{K^2} [\sinh (K + \sigma_0) - \sinh \sigma_0 - K \cosh \sigma_0] \tag{3.47}$$

In this case the parameter σ_0 can be analytically eliminated with the result

$$Q = \frac{S}{2} - \sqrt{\frac{S^2}{4} + \left(\frac{\sinh (K/2)}{K}\right)^2} \left(\coth \frac{K}{2} - \frac{2}{K}\right) \tag{3.48}$$

This relationship is shown in Fig. 3.6. For a Newtonian fluid this would give a set of parallel straight lines which have the slope -1/12. But here the flow characteristics are curved in consequence of the non-linear flow behaviour and no longer occur as separate parallel lines. The significance of a diagram of the flow characteristics

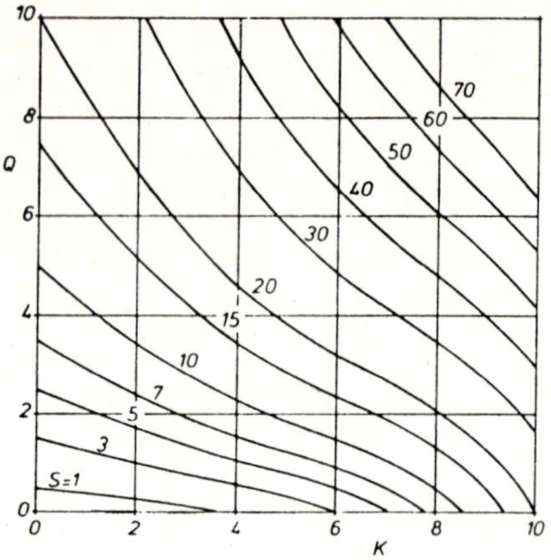

Fig. 3.6 Flow characteristics for a Prandtl-Eyring fluid

like that in Fig. 3.6 lies in the fact that it gives information about the friction pump performance, i.e. what (dimensionless) volume of flow may be expected for given pressure and velocity parameters. For a relatively low pressure rise the characteristics can be approximated by straight lines, the slope of which depends on the drag velocity parameter S. The series expansion of the right-hand side of equation (3.48) in terms of powers of K begins with the following terms:

$$Q = \frac{S}{2} - \sqrt{S^2+1} \cdot \frac{K}{12} + O(K^3)$$

This possibility of approximation is not limited to the fluid model considered. One can easily show that the function Q(K;S), which describes the flow characteristics, has the following properties for a fluid with the general flow function $f(\sigma)$:

$(\partial Q/\partial K)_S = -f'(\sigma_0)/12 = -f'(f^{-1}(S))/12$ for K = 0, and

(independent of the flow law)

$(\partial^2 Q/\partial K^2)_S = 0$ for K = 0

Hence the expansion in terms of K starts thus

$$Q = \frac{S}{2} - f'(f^{-1}(S)) \cdot \frac{K}{12} + O(K^3) \tag{3.49}$$

(f^{-1} indicates the inverse of f)

Because the deviation from a linear relationship does not start until the third order term, the linear approximation generally gives satisfactory results for K-values of order one (Fig. 3.6). Equation (3.49) is similar to the relationship (3.38) valid for a Newtonian fluid, and its dimensionless form $Q_N = S/2 - K/12$, but instead of the constant viscosity η_* for the Newtonian fluid, the differential viscosity $\hat{\eta}(S) = \eta_*/f'(f^{-1}(S))$ associated with the velocity parameter S appears here. This can be understood when one remembers the fact that with small pressure rise the shear stress in the channel departs only slightly from that value which would be characteristic of a pure drag flow. Therefore only a small section of the flow curve in the vicinity of this reference point plays a part, which can be approximated by the tangent at the position and hence by the differential viscosity. The result corresponds to the information obtained in Section 2.1, in which the differential viscosity is critical for the change of shear stress which is caused by a change of shear in the tangential direction.

The important friction pump characteristics are the power P to be maintained in keeping the flow in motion, and the *efficiency* $\bar{\eta}$, i.e. the ratio of the useful work done to the energy expended. The useful work done by a pump is the product of the volume flux $\dot{V}$ and the pressure rise across the pump $\ell dp/dx$, where ℓ denotes the length of the flow channel. The work done results from the velocity of the moving wall and the shear stress acting on it, $P = \tau_h b\ell u_h$. By using the dimensionless parameter for the work done

$$\Pi := \frac{\eta_* P}{\tau_*^2 b\ell h} \tag{3.50}$$

the just introduced parameters initially in the form $\Pi = (K + \sigma_0)S$ and $\bar{\eta} = QK/\Pi$, can be finally represented with the help of equations (3.43) and (3.45) as functions of K and σ_0:

$$\Pi = \frac{K + \sigma_0}{K} \int_{\sigma_0}^{K+\sigma_0} f(\sigma)d\sigma \tag{3.51}$$

$$\bar{\eta} = 1 - \frac{\int_{\sigma_0}^{K+\sigma_0} \sigma f(\sigma)d\sigma}{(K + \sigma_0) \int_{\sigma_0}^{K+\sigma_0} f(\sigma)d\sigma} \tag{3.52}$$

One can think of the pair of equations (3.45) and (3.51) as a parametral representation of the surface $Q(K;\Pi)$ or, what is the same thing, as the parametral form of the characteristics of constant power in a pressure-rate of flow diagram. Thus σ_0, a dimensionless quantity for the shear stress at the stationary wall, is once more the parameter. Similarly in equations (3.45), (3.52) there is a parametral representation of the characteristics of constant efficiency. For known flow properties of the non-Newtonian fluid the four fundamental relationships (3.43), (3.45), (3.51) and (3.52) can always be illustrated by the explicitly occurring quantities S, Q, Π and $\bar{\eta}$ as functions of K with σ_0 as a parameter. The necessary integrations can be done numerically as required. With restriction to real pumping situations only the regions $K>0$, $S>0$, $Q>0, \Pi>0, \bar{\eta}>0$ are of interest. The four characteristic fields completely describe the properties of the pump, so that in principle one only needs this information. In case of need one can produce new plots from any two of these characterising fields by eliminating the parameter σ_0, particularly the characteristics in the K-Q diagram, which are distinguished by having constant values of the parameter S (as in Fig. 3.6), Π or $\bar{\eta}$.

Figs. 3.7 and 3.8 illustrate this by means of a pseudoplastic Reiner-Philippoff fluid (Table 2.1). Thus the integral in the fundamental relationships (3.43), (3.45), (3.51), and (3.52) can be evaluated analytically, so that the right-hand sides reduce to elementary expressions, in which besides K and σ_0 the numerical parameter η_∞/η_* enters. The evaluation of formulas (3.45) and (3.52) leads to the characteristics shown in Fig. 3.7. The functions $Q(K;\sigma_0)$, $S(K;\sigma_0)$ and $\Pi(K;\sigma_0)$ have noteworthy monotonic properties, and this holds independently of the special form of the flow function $f(\sigma)$. One can easily show that for fluids with monotonically increasing flow curves (equation (2.20)) the quantities $\partial S/\partial K$, $\partial S/\partial\sigma_0$, $\partial Q/\partial K$ and $\partial Q/\partial\sigma_0$ for chosen values of K and σ_0 are always positive. Accordingly not only S but also Q increase monotonically with respect to K (for σ_0 = constant), and with respect to σ_0 (for K = constant). The characteristic field $S(K;\sigma_0)$ therefore appears similar to that field $Q(K;\sigma_0)$ shown in Fig. 3.7. The same holds for the function $\Pi(K;\sigma_0)$, but only for actual pump conditions ($K>0$, $S>0$, $Q>0$), which here however signifies no limitation. There are no monotonic statements of that kind for the function $\bar{\eta}\,(K;\sigma_0)$.

A complex non-monotonic behaviour appears in the actual case (Fig. 3.7). In the regions for which $\sigma_0>0$ there are velocity profiles over the channel cross-section which have positive values throughout. For $\sigma_0<0$ back flow occurs in the vicinity of the stationary wall.

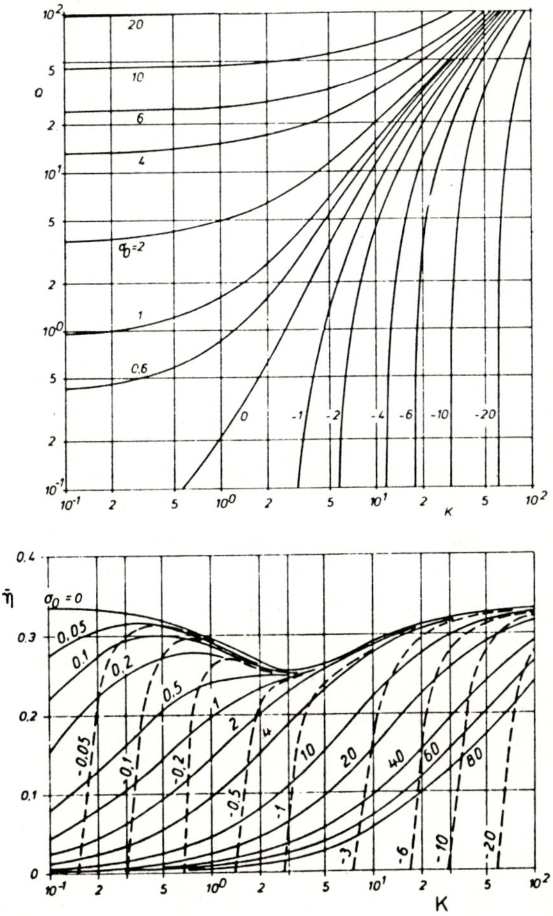

Fig. 3.7 Characteristic fields $Q(K;\sigma_0)$ and $\bar{\eta}(K;\sigma_0)$ for a pseudoplastic Reiner-Philippoff fluid with $\eta_\infty/\eta_* = 0.1$

If one eliminates the parameter σ_0 from two of the described characteristic fields by a graphical method or numerically from the relevant formulas, then one obtains in particular the characteristics of constant required power and constant efficiency given in Fig. 3.8. The lines Π = constant are arranged so that for a constant pressure rise (K = constant) the power increases with increasing volume flux. On the other hand the lines $\bar{\eta}$ = constant have complex properties, which are connected with the non-monotonic behaviour of the curve $\bar{\eta}$ $(K;\ \sigma_0 = 0)$ in Fig. 3.7. There is a singular point corresponding to the minimum of this curve, which has the shape of a saddle, seen in Fig. 3.8.

As can be seen in Fig. 3.7 the pump efficiency during the delivery of a pseudoplastic fluid at best reaches the value 1/3. During delivery of a dilatant fluid on the other hand greater efficiencies are possible (Fig. 3.9).

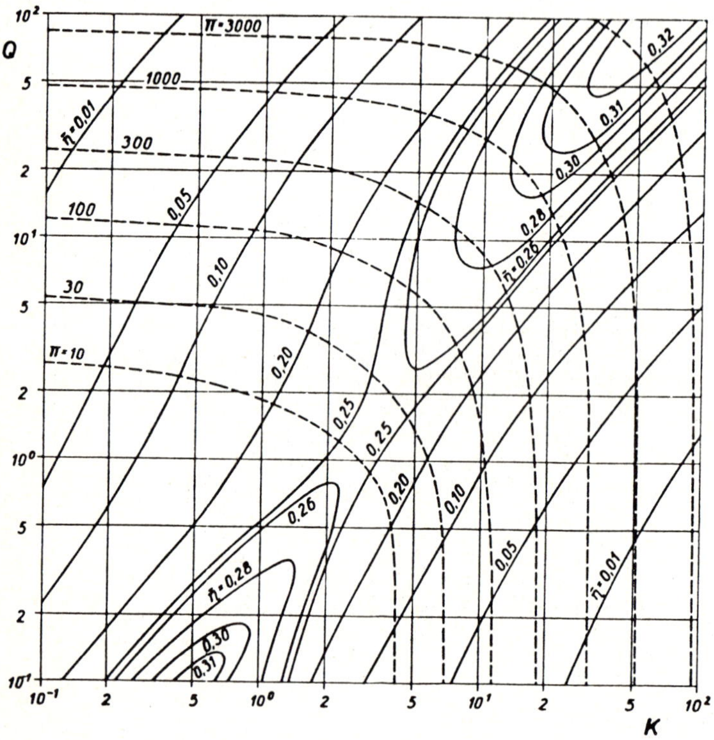

Fig. 3.8 Characteristic fields $Q(K;\Pi)$ and $Q(K;\bar{\eta})$ for a pseudoplastic Reiner-Philippoff fluid with $\eta_\infty/\eta_* = 0.1$ (after Böhme and Nonn)

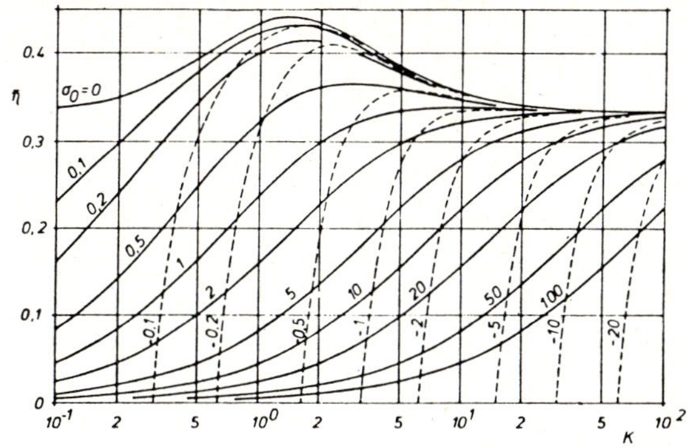

Fig. 3.9 Pumping efficiency as a function of the pressure parameter and of the dimensionless shear stress at the stationary wall; dilatant Reiner-Philippoff fluid with $\eta_\infty/\eta_* = 8$

The greatest possible efficiency is thus (other than for a Newtonian fluid) assumed for a quite definite value of the pressure parameter K. This maximum corresponds in the fluid of the characteristics of constant efficiency to a singular vortex point (Fig. 3.10) Not only for pseudoplastic but also for dilatant fluids the set of curves of constant efficiency therefore have non-interchangeable properties.

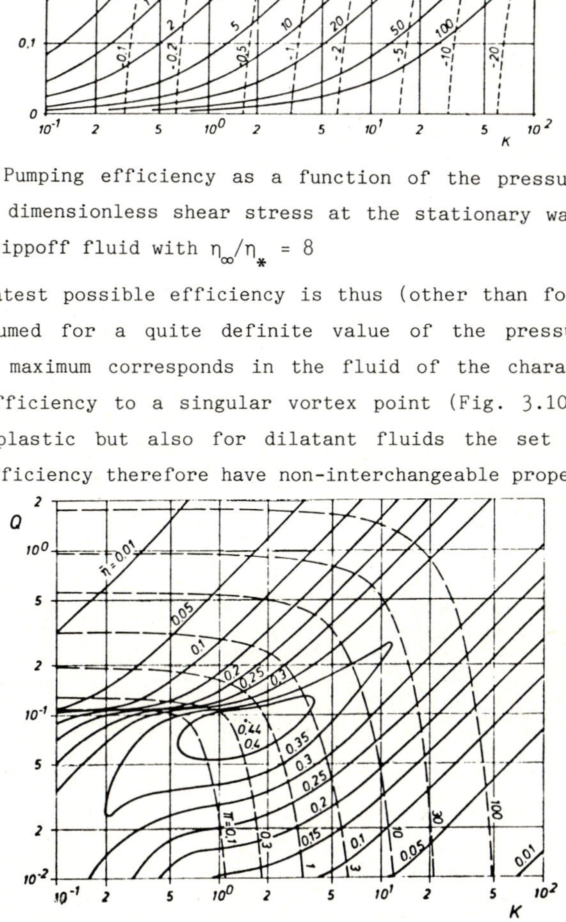

Fig. 3.10 Characteristic fields $Q(K;\Pi)$ and $Q(K;\bar{\eta})$ for a dilatant Reiner-Philippoff fluid with $\eta_\infty/\eta_* = 8$

A diagram of the characteristics permits for instance the determination of the gap width h and the drag velocity u_h by the criterion of optimum efficiency of a pump of given width for given pressure rise and given flow rate. The product K^2Q is determined numerically with the given values b, dp/dx and $\dot{V}$, and with known constants η_*, τ_* (equations (3.41) and (3.44)). The results are plotted on log-log paper and give a straight line K-Q diagram. That point on it where one of the curves for $\bar{\eta}$ = constant in Fig. 3.8 and Fig. 3.10 is touched designates the maximum possible efficiency. Definite values for σ_0 and S are obtained from Fig. 3.7 using the appropriate values for K and Q, and from the characteristics based on equation (3.43). Hence the required gap width h can be calculated from equation (3.41) and the required wall velocity u_h from equation (3.42).

Another important exercise consists of determining the volume flux and the required power and the drag velocity for given dimensions h, b and ℓ of the pump, given the material constants η_*, τ_*, and given the pressure rise dp/dx so that the efficiency is maximum. With the given values the numerical value is first of all determined for K. A definite value of Q for this is found from Fig. 3.8, corresponding to the requirement for maximum efficiency. Fig. 3.7 and equations (3.43) and (3.51) give the numerical values for σ_0, S, and Π. One thus obtains from equations (3.42), (3.44) and (3.50) the required drag velocity, volume flux and the required power.

3.2.2 Extrusion flow

Extruders as drag flow pumps are used particularly in the plastics industry for the continuous supply of polymer melts. The upper diagram in Fig. 3.11 shows the principle of a *screw extruder*. The fluid moves in the gap between a cylindrical housing and a rotating screw round which there is a helical web. For a stationary observer the process is unsteady, but an observer rotating with the screw sees a steady flow. If the pressures in front of and behind the screw are equal, the flow is brought about only by the drag effect of the tube wall. Without the web on the inner screw the fluid in this case would of course follow a circular motion in the direction of rotation. Because the fluid cannot get past the web, there is a component of motion with an axial velocity superposed on the motion in the direction of rotation, so that the desired transporting effect in the axial direction results. In general the screw extruder moves the melt against rising pressure. It is clear that in this way the rate of flow is diminished with increasing counterpressure. The increased pressure behind the screw can in practice serve to press the melt through a moulding die. For a given die and a preselected volume flux the pressure required is fixed and the main task is to design the extruder size suitably. This is achieved by study of the flow processes in the inside and their quantitative description.

We denote the internal radius of the housing by r, the gap width by h, the pitch of the screw by β, and the angular velocity of the screw by ω. Besides that there is the length of the extruder in the axial direction ℓ; also the pressure difference behind and in front of the extruder Δp. In order to be able to calculate for a steady flow it is essential to view the process from the standpoint of an observer rotating with the screw, for whom the housing appears to rotate with the velocity $r\omega$. The theory to be developed is based on several simplifying assumptions. Thus, assuming that $h \ll r$, the curvature of the channel walls can first of all be left out of the calculation. This permits 'unrolling' of the flow channel so that henceforth the bounding walls can be considered as plane. The length of this channel is of course $\ell/\sin\beta$; its width b is given by the consideration that its surface area $b \cdot \ell/\sin\beta$, neglecting the wall thickness of the web must agree with the internal area of the cylinder $2\pi r \ell$: $b = 2\pi r \sin\beta$. If the cross-section of the channel is narrow ($h \ll b$), the velocity component v perpendicular to the housing (in the y-direction, lower diagram in Fig. 3.11) exists only in the immediate vicinity of the flanks of the webs. In these conditions one can entirely disregard v and neglect the dependence of the two other velocity components on z, which likewise becomes noticeable only in the vicinity of the web. Further, the fluid can be considered to have constant density and the flow in consequence can be considered as isochoric. The channel dimensions will be assumed constant over the whole length, which has been implicitly assumed, and the events at the inlet will be disregarded. Finally, isothermal conditions will be assumed. Under these conditions the velocity components u parallel to, and w normal to the web depend on y alone. Therefore there is a shear flow between a stationary wall and one moving tangentially to it (Fig. 3.11). We resolve the drag velocity $r\omega$ of the moving wall for greater convenience into a component $r\omega \cos\beta$ parallel to the web, and another $r\omega \sin\beta$ normal to it. Because the fluid adheres to the walls the following boundary conditions apply:

$$u(0) = 0, \qquad w(0) = 0 \tag{3.53}$$

$$u(h) = r\omega \cos\beta, \qquad w(h) = r\omega \sin\beta \tag{3.54}$$

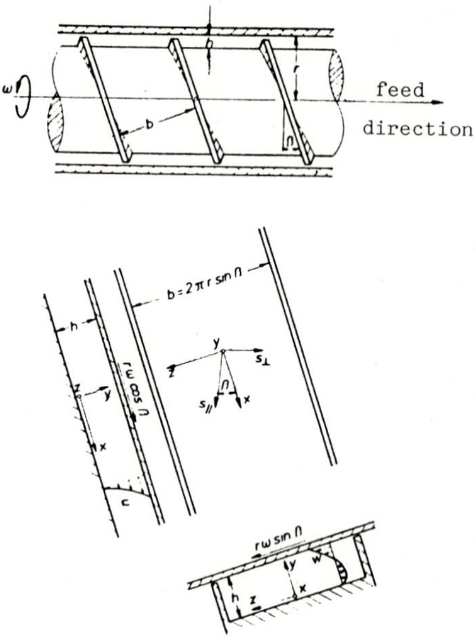

Fig. 3.11 Principle of a screw extruder and 'unwinding' of the flow channel

It is not possible with the simplifications given above to take into account the adhesion of the fluid to the sides of the web. Nevertheless the presence of the lateral boundary has an effect on the flow. Because the web is impervious the volume flux of course vanishes in the z-direction. Therefore

applies. $$\int_0^h w(y)dy = 0 \tag{3.55}$$

Accordingly the velocity profile w(y) must not only have a positive, but also a negative part. Considering the boundary conditions the shape qualitatively sketched in the bottom diagram of Fig. 3.11 results. The way in which the velocity profile is curved indicates that the pressure over the cross-section cannot be constant, but increases from the right- to the left-hand web ($\partial p/\partial z > 0$).

The evaluation of the equations of motion for the above shear flow follows from what has been said at the start of Section 3.2. Hence it is evident that the pressure gradient is constant and its components are equal to the derivative of the stress components τ_{xy} and τ_{zy} in the y-direction. Hence

$$\tau_{xy} = \tau_1 + \frac{\Delta p}{\ell} y \sin\beta, \qquad \tau_{zy} = \tau_2 + \frac{\partial p}{\partial z} y \tag{3.56}$$

Thus the pressure rise in the x-direction is represented by the pressure difference Δp and the channel length $\ell/\sin\beta$. The constants τ_1 and τ_2 have the significance of stress components at the stationary wall. Because this is a steady shear flow, the constitutive law in the form of (2.16) applies. Thus one has to insert for τ the total stress $\sqrt{\tau_{xy}^2 + \tau_{zy}^2}$ resulting from the components, and correspondingly $\sqrt{(du/dy)^2 + (dw/dy)^2}$ for $\dot{\gamma}$. If one multiplies the constitutive equation by

$$\mathbf{e} = \frac{1}{\dot{\gamma}}\left(\frac{du}{dy}\mathbf{e}_x + \frac{dw}{dy}\mathbf{e}_z\right) \tag{3.57}$$

the unit vector in that direction in which the resulting shear stress acts (different vectors **e** apply to the various layers), and notes the resolution of forces $\tau\mathbf{e} = \tau_{xy}\mathbf{e}_x + \tau_{zy}\mathbf{e}_z$, then one obtains the two equations

$$\frac{du}{dy} = \frac{\tau_*}{\eta_*} f\left(\frac{\sqrt{\tau_{xy}^2 + \tau_{zy}^2}}{\tau_*}\right) \frac{\tau_{xy}}{\sqrt{\tau_{xy}^2 + \tau_{zy}^2}} \tag{3.58}$$

$$\frac{dw}{dy} = \frac{\tau_*}{\eta_*} f\left(\frac{\sqrt{\tau_{xy}^2 + \tau_{zy}^2}}{\tau_*}\right) \frac{\tau_{zy}}{\sqrt{\tau_{xy}^2 + \tau_{zy}^2}} \tag{3.59}$$

Because for the given flow function and from equation (3.56) the right-hand sides are known functions of y, one can integrate and hence calculate the velocity field. In this two more constants of integration join the constants τ_1, τ_2 and $\partial p/\partial z$ to be determined. The five conditions (3.53) to (3.55) are involved in the determination of these five constants. The two homogeneous boundary conditions (3.53) can be taken into account explicitly. For the representation of the remaining conditions it is expedient to introduce an abbreviation for $f(\sigma)/\sigma$

$$\varphi(\sigma) := \frac{f(\sigma)}{\sigma} \tag{3.60}$$

Then equations (3.54) and (3.55), the latter after partially integrating once are transformed into the following expressions:

$$S\cos\beta = \int_0^1 (\sigma_1 + K\zeta\sin\beta)\cdot\varphi\left(\sqrt{(\sigma_1 + K\zeta\sin\beta)^2 + (\sigma_2 + \kappa\zeta)^2}\right)d\zeta \tag{3.61}$$

$$S\sin\beta = \int_0^1 (\sigma_2 + \kappa\zeta)\cdot\varphi\left(\sqrt{(\sigma_1 + K\zeta\sin\beta)^2 + (\sigma_2 + \kappa\zeta)^2}\right)d\zeta \tag{3.62}$$

$$0 = S\sin\beta - \int_0^1 (\sigma_2 + \kappa\zeta)\zeta\cdot\varphi\left(\sqrt{(\sigma_1 + K\zeta\sin\beta)^2 + (\sigma_2 + \kappa\zeta)^2}\right)d\zeta \tag{3.63}$$

The parameters S, K, σ_1, σ_2 and κ in these equations are all dimensionless: S is a suitably chosen drag velocity parameter, K is a dimensionless measure of the pressure rise

$$S := \frac{\eta_* r\omega}{\tau_* h}, \qquad K := \frac{h\,\Delta p}{\ell\,\tau_*} \tag{3.64}$$

σ_1, σ_2 and κ are dimensionless quantities ($\sigma_1 := \tau_1/\tau_*$, $\sigma_2 := \tau_2/\tau_*$, and $\kappa := (h/\tau_*)\partial p/\partial z$) related to the constants τ_1, τ_2 and $\partial p/\partial z$.

We now introduce the volume $\dot{V}$ per unit time flowing through the extruder, which is given by integrating the velocity field u(y) over the channel cross-section, as well as the dimensionless

$$Q := \frac{\eta_*\dot{V}}{2\pi r\tau_* h^2} \tag{3.65}$$

associated with it.

By one partial integration, using equation (3.58) with (3.56), and noting that for the width of the channel $b = 2r\sin\beta$ applies, one obtains the following formula for Q:

$$\frac{Q}{\sin\beta} = S\cos\beta - \int_0^1 (\sigma_1 + K\zeta\sin\beta)\zeta\cdot\varphi\left(\sqrt{(\sigma_1 + K\zeta\sin\beta)^2 + (\sigma_2 + \kappa\zeta)^2}\right)d\zeta \tag{3.66}$$

The relationship of the right-hand side to that of equation (3.63) is not accidental, but depends on the fact that here it is the volume flux in the x-direction which is involved, but in (3.63) it is the volume flux in the z-direction.

The properties of the extruder are found from equations (3.61), (3.62), (3.63) and (3.66). If one eliminates the parameters σ_1, σ_2 and κ, which in general must be done numerically, a relationship between Q, K and S results, in which the angle of pitch β appears. Thus one obtains for a Rabinowitsch fluid (for which $\varphi(\sigma) = 1 + \sigma^2$) the continuous curves for the flow characteristics shown in Fig. 3.12).

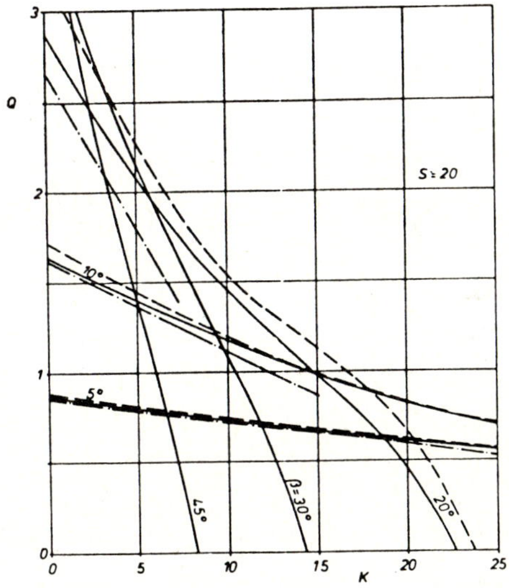

Fig. 3.12 Flow characteristics for constant drag velocity parameter for various pitch angles β ; Rabinowitsch model;

- - approximation by neglecting the transverse motion

—·— pseudo-Newtonian approximation in accordance with equation (3.74)

In order to understand the influence of the angle of pitch, the connection between the pressure parameter K and the dimensionless rate of flow Q is plotted for various angles β for constant velocity parameters

S. This shows that the flow characteristics fall off more rapidly as β increases. For a steep pitch of the helix therefore small changes in the pressure difference produce large changes in the rate of flow. Because this is mostly unfavourable for practical purposes, one generally uses screws with relatively low pitch angles ($\beta \lesssim 20^\circ$). The assumed numerical values ($S = 20$, $K \leq 25$) are based on the material values for a polymer melt at 190°C, a maximum pressure rise of 500 bars, practical length proportions (ℓ/h between 80 and 300, r/h between 5 and 15) and usual rotation speeds (between 20 and 500 r.p.m.).

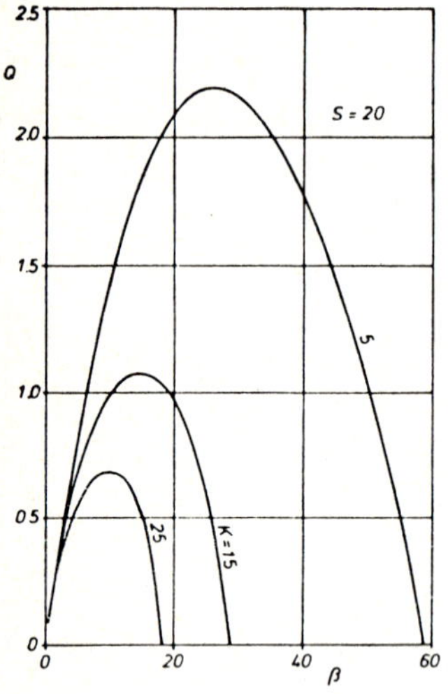

Fig. 3.13 Dimensionless flux as a function of helix angle; Rabinowitsch model

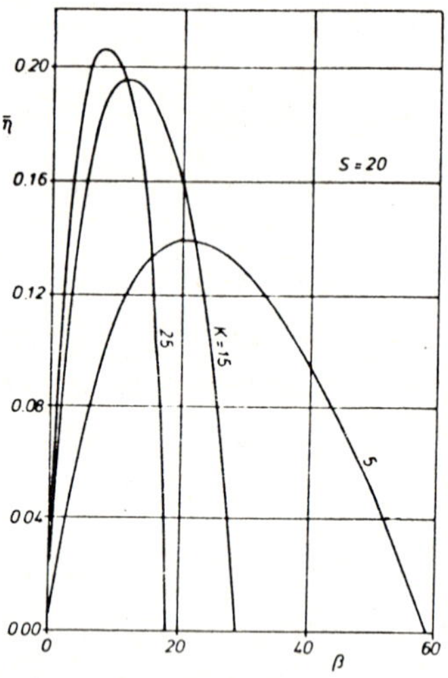

Fig. 3.14 Efficiency as a function of helix angle; Rabinowitsch model

Fig. 3.13 shows the rate of flow as a function of the helix pitch for various fixed values of K. Because for $\beta \to 0$ the length of the channel increases without limit and therefore the pressure rise parallel to the sides approaches zero, in this limiting case we have here a pure drag flow in the direction of rotation, for which the fluid circulates between screw and housing, but does not move forward in the axial direction. The rate of flow therefore vanishes for $\beta = 0$, so that in Fig. 3.13 all curves pass through the origin of the coordinates.

Without counterpressure (K = 0) the fluid is also in the limiting case $\beta = 90^\circ$ transported in the circumferential direction. The rate of flow therefore vanishes for a second time for finite pitch angle and takes up a maximum value dependent on K and S between the two zero points. It is obvious that with increasing counterpressure the second zero intersection is displaced towards smaller pitch angles, and hence the maximum output falls off (Fig. 3.13).

Fig. 3.14 gives the efficiency $\bar{\eta}$, which is once more defined as the ratio of useful work $\dot{V}\Delta p$ to the energy P_a supplied. The latter depends on the shear stress components at the moving wall, the associated velocity components and the wetted area: $P_a = 2\pi r \ell (\tau_{xy} \cdot r\omega \cos\beta + \tau_{zy} \cdot r\omega \sin\beta)_{y=h}$. By use of the previously given dimensionless expressions the following holds:

$$\bar{\eta} = \frac{Q \frac{K}{S}}{(\sigma_1 + K \sin\beta)\cos\beta + (\sigma_2 + \kappa)\sin\beta} \tag{3.67}$$

Fig. 3.14 shows that the efficiency of a polymer melt extruder is very low. We explained in Section 3.2.1 that the efficiency of a drag pump for pseudoplastic fluids cannot exceed the value 1/3. Additional dissipation losses arise in the extruder because of the cross-flow motion perpendicular to the direction of transport, which further reduces the efficiency. For given values of the pressure and velocity parameters the efficiency increases up to a maximum for a quite well defined, obviously small pitch angle. From this point of view it follows that for polymer melt extruders β should not be chosen too large.

Fig. 3.15 shows for some previously given situations the velocity profiles in directions parallel and perpendicular to the web. Normalising has the result that for $y/h = 1$ the values $S \cdot \cos\beta$ and $S \cdot \sin\beta$ occur at the moving wall. The velocity profile w(y) has at any given time the mean value zero (equation (3.55)), and therefore consists of two zones with opposite signs. Concerning the velocity profile u(y), for fixed values of K and S above a certain pitch angle, back flow occurs near the stationary wall. Under the chosen conditions the back flow for $\beta = 30^\circ$ has become so strong that the mean value and hence the rate of flow collectively turn out to be negative, so that this in fact leads to no pump action at all (curve K = 15 in Fig. 3.13).

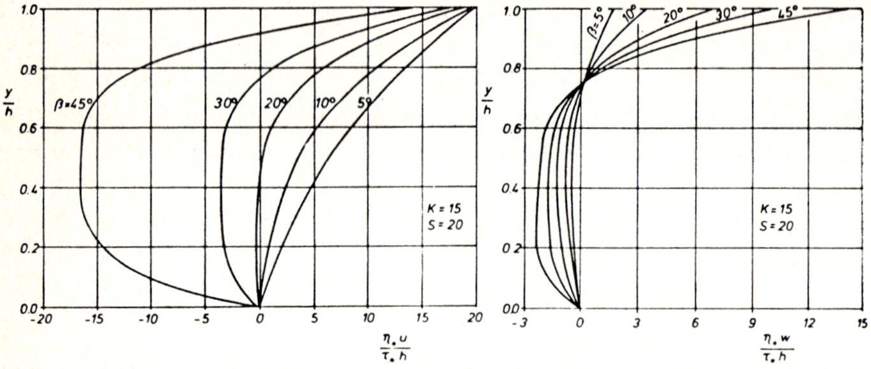

Fig. 3.15 Velocity profiles of an extruder flow; Rabinowitsch model

The fact mentioned above, that the parameter β in general is relatively small conveniently provides the opportunity to simplify the calculation of the extruder flow. Two possibilities present themselves here, which are based on different concepts. In both cases the starting point is that with a sufficiently small value of β the cross-flow in the z-direction is low compared with the main flow in the x-direction. It is therefore obvious to neglect wholly the effect of the cross-flow, namely τ_{zy}, in the conditional equation for u(y). This amounts to putting $\sigma_2 = \kappa = 0$ in equations (3.61) and (3.66), but nowhere else. Hence the calculation of the volume flux reduces to the shortened equations (3.61) and (3.66), the solution of which has been discussed in detail in Section 3.2.1.

Naturally the cross-flow can also be determined from the remaining equations, but this is omitted so long as one is only interested in the flow characteristics. In order to check the goodness of the approximation the result of the simplified theory has been plotted in Fig. 3.12. Thus it appears that for $\beta = 5^\circ$ there is good agreement with the general theory. For increasing angle the departures from this agreement occur increasingly. Hence it is feasible that for fixed rate of flow too great values are associated with the pressure parameter, because with the neglect of the cross-flow its deleterious effect disappears, so that the pressure loss is underestimated, and for given volume flux the attainable pressure rise is therefore overestimated.

A second possibility of simplification consists of considering the extruder flow as a weakly disturbed pure drag flow in the direction of rotation, because for small pitch angles only low pressure gradients arise. As was explained in Section 2.1, the critical factor is the differential viscosity $\hat{\eta}$ for the additional motion parallel to the shear direction of the basic flow, and for the motion normal thereto the usual viscosity η applies. Both quantities depend on the shear rate $(r\omega/h)$ of the undisturbed flow, but are constant across the cross-section of the channel. The effect of friction can therefore be considered simply in terms of two constant viscosity values. For sufficiently small pitch angles the motion of the real non-Newtonian fluid satisfies the same formulas as a Newtonian fluid, but note that for the description of the motion parallel to ($\parallel$) and normal to ($\perp$) the undisturbed shear direction (Fig. 3.11) different viscosity values are to be inserted. Thus the appropriate velocity profiles are approximated by parabolas and the following holds for the volume flow per unit width (cf. equation (3.38):

$$\dot{v}_{\parallel} = r\omega\frac{h}{2} - \frac{h^3}{12\hat{\eta}}\frac{\partial p}{\partial s_{\parallel}} \tag{3.68}$$

$$\dot{v}_{\perp} = -\frac{h^3}{12\eta}\frac{\partial p}{\partial s_{\perp}} \tag{3.69}$$

The quantities $\dot{v}_{\parallel}$ and $\dot{v}_{\perp}$ are consequently related to the volume flux $\dot{V}$ in the direction parallel to the web:

$$\frac{\dot{V}}{b} = \dot{v}_{\parallel}\cos\beta + \dot{v}_{\perp}\sin\beta \tag{3.70}$$

The condition that the web be impermeable leads to the relationship

$$0 = \dot{v}_{\parallel}\sin\beta - \dot{v}_{\perp}\cos\beta \tag{3.71}$$

The two pressure derivatives occurring in equations (3.68) and (3.69) can be expressed by the quantities used earlier, with the help of the transformation formulas

$$\frac{\partial p}{\partial s_{\parallel}} = \frac{\Delta p \sin\beta}{\ell}\cos\beta + \frac{\partial p}{\partial z}\sin\beta \tag{3.72}$$

$$\frac{\partial p}{\partial s_\perp} = \frac{\Delta p \sin\beta}{\ell} \sin\beta - \frac{\partial p}{\partial z} \cos\beta \tag{3.73}$$

If one eliminates from equations (3.68) to (3.73) the volume flows $\dot{v}_\parallel$ and $\dot{v}_\perp$, the two pressure changes concerned, and $\partial p/\partial z$, then one obtains

$$\frac{\dot{V}}{2\pi r \sin\beta} = \frac{1}{1 + \left(\frac{\eta}{\hat{\eta}} - 1\right)\sin^2\beta} \left[\frac{r\omega h}{2}\cos\beta - \frac{h^3 \Delta p}{12\hat{\eta}\,\ell}\sin\beta\right] \tag{3.74}$$

Therefore the volume flux, as is characteristic of a Newtonian fluid, and hence not expected to be otherwise, decreases linearly with increasing pressure difference. The flow characteristics concerned are therefore straight lines (Fig. 3.12). Their slope depends on the pitch of the screw, but also on the drag velocity parameter because of the velocity dependent viscosity values. The rotation speed therefore affects the rate of flow in a more complex way than the increase of pressure. Fig. 3.12 shows that the approximation described provides good results for quite small pitch angles. For greater pitch angles the departures become more noticeable, and the approximation to linearity can only be used for moderate pressure rise.

3.2.3 Fluid dynamics theory of the roller

In describing the processes which occur between rotating *rollers* we shall consider the gap between two similar rollers rotating in opposite directions (radii r, angular velocities ω). The rollers entrain the outer layer of the medium passing between them, and by internal friction all the material is drawn continuously through the operating gap (Fig. 3.16). Hence the medium emerges from the gap with a thickness less than that on the inlet side. We assume that the flow normal to the plane of the drawing is prevented, so that the width b of the medium remains constant and there is a planar movement alone. The aim of the following treatment is to determine the pressure distribution and resulting from that the force being applied by the rollers, and the required torque and the propulsion power. For this it is necessary to study the velocity distribution and the stress distribution in the gap between the rollers. Because generally the gap height h_0 at the narrowest part is considerably smaller than the radius of the rollers, and the rollers are in contact with the medium only in the vicinity of the throat, we can assume that the part of the gap filled with the medium is narrow. Hence the circular section rollers can be replaced by parabolas of curvature 1/r, and accordingly the local gap height h by a quadratic expression of the longitudinal coordinate x:

$$h = h_0 + \frac{x^2}{r} \tag{3.75}$$

The fact that the gap is narrow allows us to regard the motion through it as a plane shear flow. We therefore use for a quantitative description of the roller process the formulas already derived earlier for plane shear flows, while noting that the above mentioned constant quantities, particularly the gap height h and the pressure gradient dp/dx now depend on x.

One must of course clearly understand that this simplification also implies assumptions about the properties of the material, which for steady shear flows do not arise and therefore have not so far been discussed. Because the gap height varies with x a particle passing through the gap experiences almost no motion of constant stretch history, so it is affected in addition to flow properties for example, also by the memory of the material. When we consider the flow only, we therefore presume that the material has a short memory. One can in addition show that in the case under consideration the first normal stress difference also plays a part, but its effect is less the narrower the gap, so that one can justify neglecting it.

We start from the assumption of the medium adhering to the rollers, and therefore put $u_0 = u_h = r\omega$ for the drag velocity of the two walls. Hence the right-hand side of equation (3.35) vanishes, from which one would conclude that the shear stresses at the walls are equal, apart from their algebraic sign. This is also clearly evident, because the centre line between the rollers obviously represents a line of symmetry. Putting $\tau_0 = -\tau_h$, it follows from equation (3.32) that the value for the wall shear stress is $\tau_h = (h/2).dp/dx$. Thus one obtains for the rate of flow relative to the width (equation (3.37)):

$$\frac{\dot{V}}{b} = r\omega h - \frac{2\tau_*^3}{\eta_* \left(\frac{dp}{dx}\right)^2} \int_0^{\frac{h}{2\tau_*}\frac{dp}{dx}} \sigma f(\sigma)\, d\sigma \tag{3.76}$$

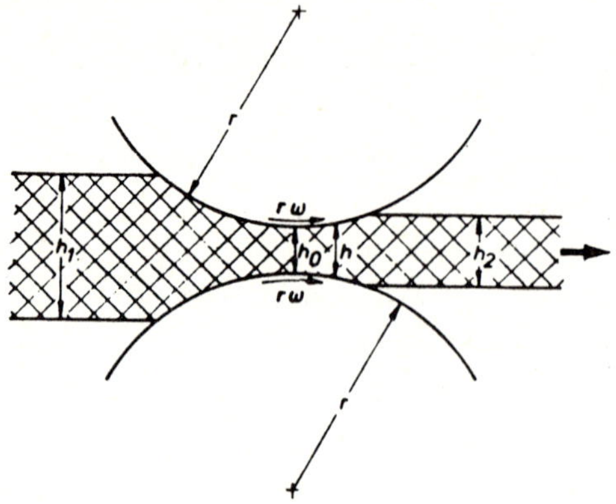

Fig. 3.16 Roller process principle

We now consider the position where the material leaves the rollers ($x = x_2$, Fig. 3.17). On the one hand the shear stress vanishes at the walls there; on the other hand the ambient pressure prevails, which we can put at zero, without loss of generality.

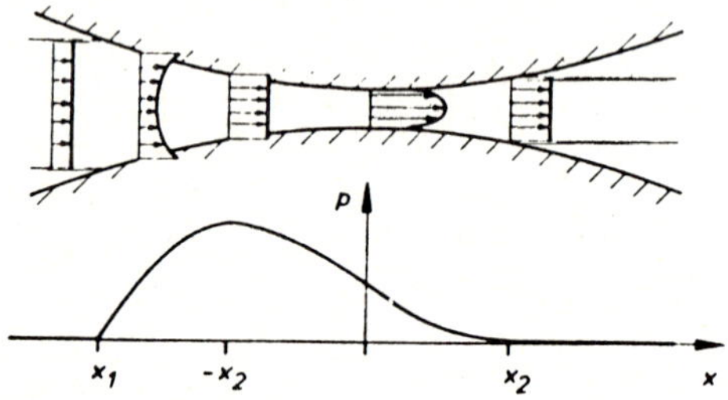

Fig. 3.17 Velocity distribution and pressures during rolling

Because $\tau_h = (h/2) \cdot dp/dx$ holds, the pressure gradient simultaneously vanishes. We therefore obtain as criteria the conditions $p = dp/dx = 0$ for $x = x_2$ for the exit. A pure drag flow occurs at the place where the pressure gradient vanishes. Because of the symmetry of the situation the velocity is constant over the cross-section. Hence it follows that the material passed through the rollers leaves the operating gap at constant velocity $r\omega$ over the cross-section, and its thickness h_2 is therefore related to the rate of flow in a simple way as:

$$\frac{\dot{V}}{b} = r\omega h_2 \qquad (3.77)$$

One can therefore consider the coordinate x_2 as a measure for the rate of flow for given geometrical proportions and given roller rotation speeds. The greater x_2 is the greater is the amount of material passing through the operating gap per unit time. In equation (3.76) the left-hand side is constant for reasons of continuity; h and dp/dx are quantities which vary with x. Because the coordinate x in equation (3.76) is not stated explicitly, the local pressure gradient depends only on the local gap height. Because h(x) is a symmetrical function relative to the narrowest position, the same also holds for dp/dx. Therefore the pressure gradient also vanishes for $x = -x_2$. Hence there is a maximum, because for $h(x) < h_2$, i.e. in the range $-x_2 < x < x_2$ the pressure flow superposed on the pure drag flow makes a positive contribution to the volume flux for reasons of continuity. Hence the pressure decreases in the x-direction. For $x < -x_2$ the opposite happens. From these considerations there follows the velocity profile shown qualitatively in Fig. 3.17, and the pressure distribution noted therein. Thus a difficulty arises at the inlet of the medium into the operating gap (for $x = x_1$). It is clear that at this position the pressure is equal to the ambient pressure ($p = 0$ for $x = x_1$). This condition determines the entry point x_1 and therefore the required thickness h_1 of the material passed through the rollers for a given rate of flow and for a given roller velocity. Hence there results a break in the pressure distribution. In front of the entry point into the operating gap the material is of course under constant external pressure; behind it is dragged along against rising pressure. This acts on the velocity field in such a way that on the left of the entry point the material runs in at constant velocity.

Conforming to the conservation of mass because $h_1 > h_2$ it is less than the roller velocity, but on the right-hand side there occurs a combined pressure-drag flow with a non-uniform velocity profile. The transition does not in fact occur abruptly, but comes about over an inlet path of finite length. The theory as given here neglects this inlet effect, which for very viscous substances is limited in other respects to a region which has the characteristic dimensions of the gap width.

For the quantitative evaluation of the formulas given above it is expedient to express the longitudinal coordinate x, the drag velocity $r\omega$ and the pressure in suitable dimensionless terms:

$$\xi := \frac{x}{\sqrt{h_0 r}}, \qquad S := \frac{\eta_* r\omega}{\tau_* h_0}, \qquad \bar{p} := \frac{h_0^2 p}{\eta_* r\omega \sqrt{h_0 r}} \tag{3.78}$$

Thus equation (3.76) is transformed by means of (3.77) and (3.75) into

$$\int_0^{\frac{S}{2}(1+\xi^2)\frac{d\bar{p}}{d\xi}} \sigma f(\sigma)\, d\sigma = \frac{S^3}{2}(\xi^2 - \xi_2^2)\left(\frac{d\bar{p}}{d\xi}\right)^2 \tag{3.79}$$

If the flow function for the substance is known, this relationship gives the pressure gradient $d\bar{p}/d\xi$ as a function of position ξ, whereby two independent numerical parameters appear, namely the velocity parameter S and the dimensionless coordinate ξ_2 of the release point, which is related in a simple way to the rate of flow (equation (3.77)). By integrating and taking into account the boundary conditions $\bar{p} = 0$ for $\xi = \xi_2$, one obtains the pressure distribution for the predetermined values of S and ξ_2. The point of interest here is only the range $\xi_1 < \xi < \xi_2$, for which ξ_1 is obtained from the zero intercept on the negative ξ-axis. In this way one obtains a relationship between the inlet coordinate ξ_1 and the exit coordinate ξ_2 for constant velocity parameter S.

Fig. 3.18 shows this relationship for a Rabinowitsch fluid. The result for a Newtonian fluid is included here as a special case (S = 0). Equation (3.79) was the starting point of the numerical calculation, which in this case reduces to the algebraic relationship

$$\frac{d\bar{p}}{d\xi} + \frac{3}{20} S^2(1+\xi^2)^2 \left(\frac{d\bar{p}}{d\xi}\right)^3 = 12 \frac{\xi^2 - \xi_2^2}{(1+\xi^2)^3} \tag{3.80}$$

From here by means of a suitable iterative method for constant values of S and ξ_2 the quantity $d\bar{p}/d\xi$ is calculated as a function of ξ. The condition that the mean value of $d\bar{p}/d\xi$ vanishes in the range which is of interest fixes ξ_1. The corresponding pressure distributions have the expected properties, which are shown in Fig. 3.17.

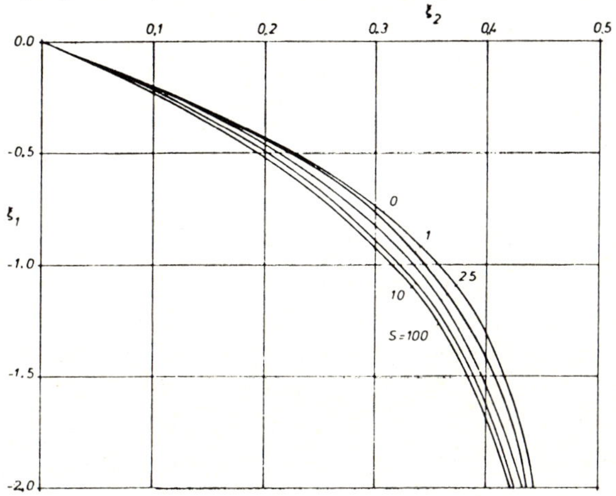

Fig. 3.18 Relationship between inlet and exit coordinates; Rabinowitsch model

Of particular interest in practice are the force F and the required power P, which are obtained by integrating over the pressure and over the shear stress distributions at the two rollers. They are calculated from the pressure field thus:

$$\frac{F}{\eta_* b r^2 \omega / h_0} = - \int_{\xi_1}^{\xi_2} \xi \frac{d\bar{p}}{d\xi} d\xi \tag{3.81}$$

$$\frac{P}{\eta_* b r^2 \omega^2 \sqrt{r/h_0}} = \int_{\xi_1}^{\xi_2} \xi^2 \frac{d\bar{p}}{d\xi} d\xi \tag{3.82}$$

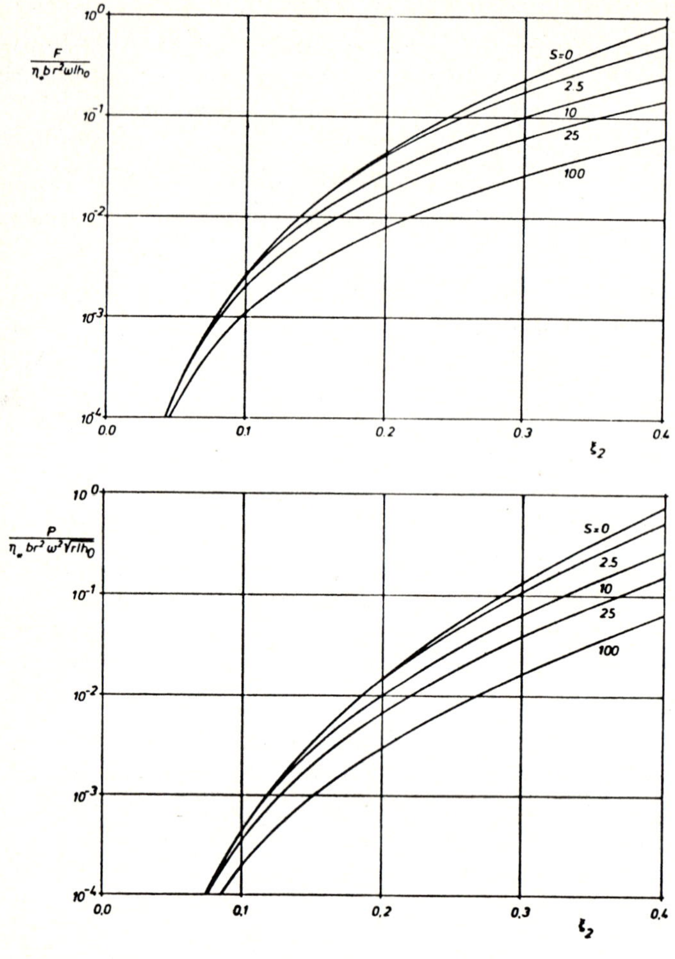

Fig. 3.19 Pressure force and required power of a roller for a pseudoplastic material; Rabinowitsch model

The right-hand side of equation (3.81) was obtained by transforming the pressure integral. Thus on account of the pressure gradient symmetry the upper limit can also be replaced by $-\xi_2$. To understand the meaning of the right-hand side of equation (3.82) remember that for the shear stress at the rollers $\tau_h = 0.5\, h\, dp/dx \sim 0.5(1 + \xi^2) d\bar{p}/d\xi$ holds, and the integral over $d\bar{p}/d\xi$ vanishes. The factor 0.5 drops out because two rollers are in operation. The numerical evaluation of equations (3.81) and (3.82) in the case of a Rabinowitsch fluid leads to the result shown in Fig. 3.19. The parameter S = 0 corresponds to the limiting case of Newtonian fluid behaviour.

It is evident that the force and the required power increase when the rate of flow under otherwise constant conditions is increased.

The exit coordinate ξ_2 is, as is well known, a measure of the rate of flow. It is also evident that for a pseudoplastic substance the force and the required power decrease if the non-Newtonian parameter S increases, because the fluid becomes less and less viscous with increasing distance from the Newtonian limiting case.

3.2.4 Flow in journal bearings

As a further application we consider the situation shown in Fig. 3.20. The viscous fluid is located in the annular gap between a static circular section housing and a rotating circular section shaft. If the movement is concentric the pressure in the fluid is constant, and the resultant force transferred from the fluid on to the shaft vanishes in consequence. The shaft can therefore be located in the centre position only in the unloaded state. When an external load acts on it the shaft moves away from the centre position, so that the lubricating substance is entrained through a gap of varying width. Thus there arises a non-uniform pressure distribution with a resultant force on the shaft which is in equilibrium with the external load in the steady case.

The following relationships are derived from Fig. 3.20: r is the radius of the shaft, U is its circumferential velocity, h_0 is the width of gap for the central position and εh_0 is the displacement of the loaded shaft from the central position, so that the dimensionless quantity ε can be denoted as an eccentricity $(0 \leqq \varepsilon < 1)$. The gap width is called $h(\varphi)$, and the angular coordinate φ is reckoned from the position of the greatest gap in the direction of rotation of the shaft $(0 \leqq \varphi \leqq 2\pi)$. Assuming that the gap is narrow $(h_0 << r)$ the following holds:

$$h(\varphi) = h_0(1 + \epsilon \cos \varphi) \qquad (3.83)$$

In this case one can also express the radius of the housing $r + h_0$ in the form $r + h - \varepsilon h_0 \cos\varphi$ (Fig. 3.20). By equating the two expressions one obtains (3.83).

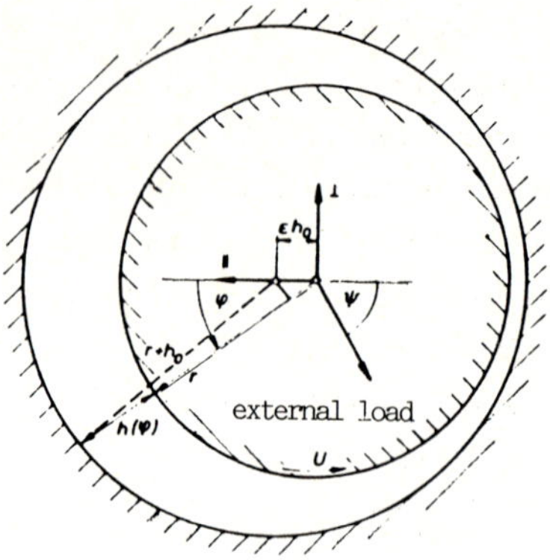

Fig. 3.20 Journal bearing

With the same restrictions as in Section 3.2.3 the motion is regarded as a plane shear flow locally, and only the viscometric flow properties of the lubricant are considered. We therefore use the formulas derived above, particularly equations (3.32), (3.35) and (3.37). Hence, corresponding to the actual situation, the quantities u_0, u_h and dp/dx are replaced by 0, U and $(1/r)dp/d\varphi$ respectively. Besides the gap width h there are also the pressure p and the wall shear stresses τ_0 at the housing and τ_h at the shaft dependent on position, i.e. the quantities varying with φ. The flux $\dot{V}/b$ on the other hand must be the same at each position φ and therefore it plays the part of a constant, even if initially unknown.

If one eliminates by use of equation (3.32) the quantity τ_0 from equations (3.35) and (3.37), then two relationships arise, in which five dimensionless quantities are involved besides the flow function $f(\sigma)$ of the lubricant, namely

$$\frac{\eta_* U}{\tau_* h_0}, \quad \frac{\eta_* \dot{V}}{\tau_* b h_0^2}, \quad \frac{h_0}{\tau_* r}\frac{dp}{d\varphi}, \quad \frac{\tau_h}{\tau_*}, \quad \frac{h}{h_0}$$

The first two parameters are constants. The other three depend on position. The reader can determine by himself the two above-mentioned relationships explicitly. We content ourselves here to mention some fundamental considerations, for which the exact form of the two equations is not needed. If one considers the pressure parameter as eliminated, then a relationship arises between the remaining four dimensionless ones. It is hence clear that the wall shear stress τ_h, apart from positionally constant quantities only depends on the variable gap $h(\varphi)$ and therefore has the same symmetry as $h(\varphi)$. Because the gap width is a symmetrical function with respect to the narrowest part ($\varphi = \pi$), this accordingly occurs also in the shear stress distribution (Fig. 3.21). Similar reasoning leads to the same statement for the pressure change $dp/d\varphi$. Hence the pressure itself, or more accurately the difference $p(\varphi) - p(\pi)$, is an odd function with respect to the narrowest part. Now the original value for the pressure must follow for one revolution in the gap between shaft and housing. The pressure distribution therefore satisfies the condition

$$p(2\pi) - p(0) = \int_0^{2\pi} \frac{dp}{d\varphi} d\varphi = 0 \tag{3.84}$$

It follows from the previously mentioned symmetry property that the pressures at the places of the greatest and least gap are equal, i.e. $p(0) = p(\pi)$. Moreover one can state that the pressure in the range $0 \leqq \varphi \leqq \pi$ must first rise and then decrease again.

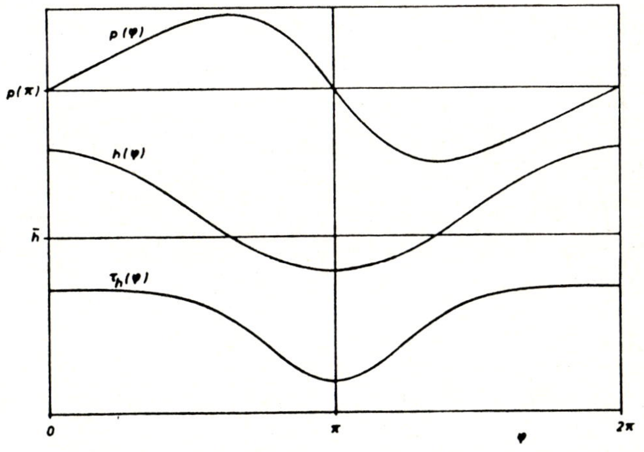

Fig. 3.21 Gap, wall shear stress and pressure as functions of the angular coordinate (qualitative)

Where the pressure distribution has a maximum (or a minimum) there is a pure drag flow with a linear velocity profile. Let $\bar{h}$ be the gap width concerned. To satisfy the continuity equation it is clear that the pressure effect reduces the pure drag effect at the place where the local gap width exceeds the value $\bar{h}$ and increases it at the place where $h(\varphi)$ is less than $\bar{h}$. Therefore it holds that $dp/d\varphi > 0$ for $h < \bar{h}$ and $dp/d\varphi < 0$ for $h > \bar{h}$. This gives the pressure distribution in the lubricated gap, qualitatively shown in Fig. 3.21. On the lower side of the shaft overpressure is dominant, on the upper side underpressure.

We shall resolve the force resulting from the stress distribution at the wall into a component parallel to ($\parallel$) and normal to ($\perp$) that direction in which the centre part of the shaft is shifted (Fig. 3.20). The symmetry properties of pressure and wall shear stress described above imply that the component $F_\parallel$ vanishes. The resultant force is therefore normal to the line joining the mid-points. In other words: for an external loading the shaft is not displaced parallel to the load, but normal to it. For the angle ψ shown in Fig. 3.20 therefore $\psi = 90°$. The force transferred by the fluid to the shaft is obtained by integrating the stresses operating, so that initially contributions not only from the pressure but also from the shear stress must be expected:

$$F_\perp = b \int_0^{2\pi} p(\varphi) \cdot \sin\varphi \, r d\varphi + b \int_0^{2\pi} \tau_h(\varphi) \cdot \cos\varphi \, r d\varphi \tag{3.85}$$

By partial integration of the first term and considering the periodicity conditions (3.84) for the pressure, the right-hand side can be put in the form of a simple integral of the type of the second term, in which instead of τ_h the sum $(dp/d\varphi + \tau_h)$ appears. Now the values of τ_h are smaller by the factor h_0/r than the values of $dp/d\varphi$. In this context remember equation (3.32). For a narrow gap ($h_0/r \ll 1$) we can therefore limit ourselves to that part of the load capacity which arises from the pressure:

$$F_\perp = b\, r \int_0^{2\pi} \frac{dp}{d\varphi} \cos\varphi \, d\varphi \tag{3.86}$$

It is usual to bring in the *Sommerfeld number* as the dimensionless number for the load capacity F (here $F = F_\perp$):

$$So := \frac{F}{\eta_* U b \frac{r^2}{h_0^2}} \tag{3.87}$$

For a Newtonian lubricating oil of viscosity η_* this quantity depends only on the eccentricity ε, and according to Reynolds and Sommerfeld

$$So_N = \frac{12\pi\epsilon}{(2+\epsilon^2)\sqrt{1-\epsilon^2}} \tag{3.88}$$

For lubricants with non-Newtonian fluid properties the dependence of the Sommerfeld number on ε cannot be determined analytically. In addition at least one more numerical parameter enters into the relationship. Omitting the details of the numerical calculations, consider the results given in Fig. 3.22 for a pseudoplastic Prandtl-Eyring model. They are based, as already noted, on the equations (3.32), (3.35) and (3.37), from which the two wall shear stresses τ_0 and τ_h in the case of a Prandtl-Eyring fluid can be analytically eliminated (equation (3.48)). The volume flux is determined by the periodicity condition (3.84). Thus the dimensionless pressure rise $(h_0/(\tau_* r))dp/d\varphi$ besides depending on $\varepsilon \cos\varphi$ depends only on the numerical values of the drag velocity parameter

$$S := \frac{\eta_* U}{\tau_* h_0} \tag{3.89}$$

The Sommerfeld number accordingly depends on ε and S. The result (3.88) for a Newtonian fluid is considered as a special case with S = 0. Fig. 3.22 shows that the load capacity of the journal bearing decreases for constant eccentricity if the non-Newtonian parameter S increases. This is explained by the fact that a pseudoplastic fluid becomes less and less viscous the more it is sheared. Conversely for a dilatant fluid the load capacity would increase with greater separation from the lower limiting Newtonian case.

Besides the load capacity, in practice the driving torque M is of particular interest, which is obtained by integrating the shear stress distribution:

$$M = b\, r^2 \int_0^{2\pi} \tau_h(\varphi)\, d\varphi \tag{3.90}$$

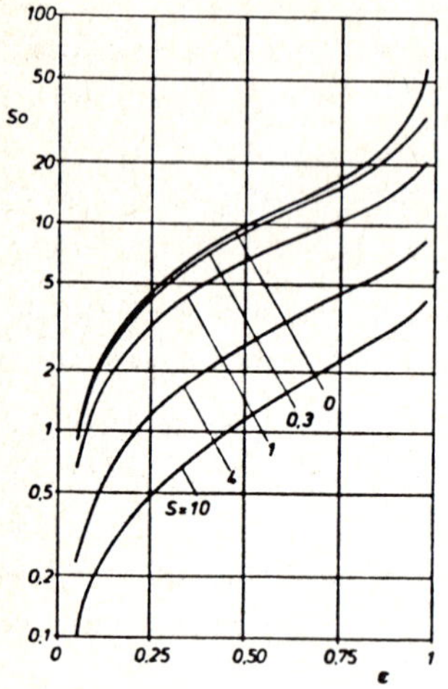

Fig. 3.22 Load capacity of the journal bearing; Prandtl-Eyring fluid

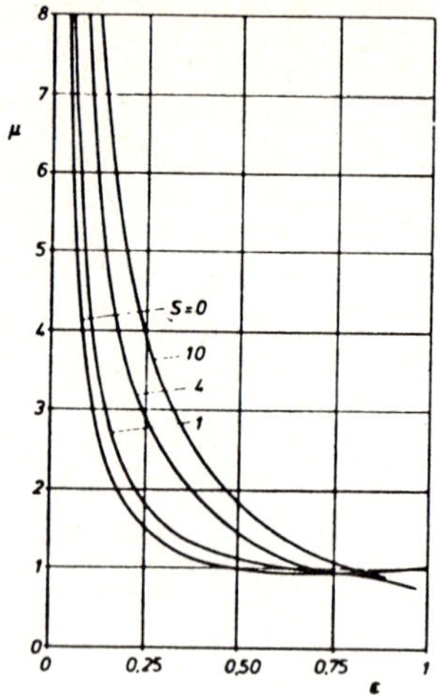

Fig. 3.23 Coefficient of friction of the journal bearing; Prandtl-Eyring fluid

It is convenient to divide the torque by the quantity Fh_0, hence writing the coefficient of friction as

$$\mu := \frac{M}{F\,h_0} \tag{3.91}$$

The expression $3\mu = 2\varepsilon + 1/\varepsilon$ holds for a Newtonian lubricating oil. This result corresponds to the graph in Fig. 3.23 for the parametral value S = 0. In the case of a pseudoplastic lubricant the coefficient of friction for constant and not too great eccentricity increases as the drag velocity parameter S increases (Fig. 3.23). For dilatant fluids the opposite applies. It would of course be rash to conclude from this that dilatant fluids are better lubricants. For fixed dimensions of the bearing and constant angular velocity different eccentricities arise for a given external load F for oils of different fluid properties. If one takes as a basis substances that have the same zero-shear viscosity η_*, then for the above described conditions the Sommerfeld number remains constant. One can learn from Figs. 3.22 and 3.23 that the coefficient of friction for constant Sommerfeld number decreases with the eccentricity. In consequence the journal displacement is greater with the use of a pseudoplastic lubricant than with a Newtonian oil, but the coefficient of friction and therefore the driving torque are smaller than in the Newtonian case, and for the latter again smaller than for a dilatant fluid of the same zero-shear viscosity.

We shall now state that the relationships shown in Figs. 3.22 and 3.23 for small eccentricities can be written as mathematical expressions. Note however that for $\varepsilon \ll 1$ a weakly disturbed drag flow occurs with the spatially constant shear rate U/h_0, and remember that for the disturbance, i.e. for the pressure effect, the differential viscosity concerned is critical, so the Newtonian expression (3.88) must hold for the load capacity, if one replaces η_* by $\hat{\eta}$. Because the Sommerfeld number has been derived from equation (3.87) using η_*, a factor $\hat{\eta}/\eta_*$ enters into the right-hand side of equation (3.88) and hence for $\varepsilon \ll 1$

$$So = 6\pi\,\epsilon\,\frac{\hat{\eta}}{\eta_*} \tag{3.92}$$

Hence for the load capacity the result is: $F = 6\pi\varepsilon\hat{\eta}Ubr^2/h_0^2$ (cf. equation (3.87)). For almost coaxial operation $\tau_h = \eta\,U/h_0$ can be used for the shear stress. The torque concerned is then $M = 2\pi r^2 b\tau_h = 2\pi r^2 b\eta U/h_0$. For small eccentricity therefore the coefficient of friction in equation (3.91) is found from

$$\mu = \frac{1}{3\epsilon}\frac{\eta}{\hat{\eta}} \tag{3.93}$$

Note that the factors $\hat{\eta}/\eta_*$ in equation (3.92) and $\eta/\hat{\eta}$ in equation (3.93) are functions of the velocity parameter S. For a Prandtl-Eyring fluid $\hat{\eta}/\eta_*$ becomes smaller with increasing S and $\eta/\hat{\eta}$ increases, whereby the ordered arrangement of the graphs S = constant in Figs. 3.22 and 3.23 resulted.

We shall now return briefly to the question of what conditions allowed the inertia of the fluid to be neglected, as has been done. In reality the velocity of the points varies while traversing the gap, so that accelerations and hence associated inertia forces occur. The acceleration of the particles in the direction of motion is evaluated from the expression $u\partial u/\partial x \sim U^2/r$. By multiplying by the density ρ one obtains a characteristic inertia force per unit volume of the quantity $\rho U^2/r$. Opposing that is a characteristic friction force per unit volume of an amount $\partial\tau/\partial y \sim \eta_* U/h_0^2$. If one forms the ratio of the two expressions, then one obtains the product of the Reynolds number $\rho Ur/\eta_*$ and the geometric term $(h_0/r)^2$. In order to be able to disregard the effect of the inertia, this *reduced Reynolds number* must be small compared with unity. This condition is generally fulfilled because $h_0/r \ll 1$, also if the Reynolds number itself has values of the order of unity or more.

Until now we have implied that the lubrication gap is completely filled with oil. If this is so an underpressure occurs on the top side of the shaft, which possibly lets the film of oil break up under heavier loads and thus impedes even running. One therefore incidentally uses an only partly filled bearing so that at one part of the shaft (in the angle sector $\varphi_1 \leqq \varphi \leqq \varphi_2$) a pressure cushion forms, but otherwise constant ambient pressure prevails. Therefore in the calculation of the force component $F_\perp$ from equation (3.86) the integrating limits can also be replaced by φ_1 and φ_2, both angles being unknown a-priori. The pressure distribution in the lubricated gap has qualitatively the same properties as for rolling (Fig. 3.17). Therefore $p(\varphi_2) = p(\varphi_1)$ and $dp/d\varphi = 0$ for $\varphi = \varphi_2$. Because in contrast to the completely filled gap the symmetry of the pressure distribution is distorted, a component of the load capacity $F_\parallel$ also arises, which can be calculated from the pressure distribution as follows:

$$F_\parallel = -br \int_{\varphi_1}^{\varphi_2} [p(\varphi) - p(\varphi_1)] \cos\varphi \, d\varphi = br \int_{\varphi_1}^{\varphi_2} \frac{dp}{d\varphi} \sin\varphi \, d\varphi \tag{3.94}$$

Therefore the angle ψ shown in Fig. 3.20 between the direction of the external load and the line joining the mid-points is generally less than 90°. The angle is related in a simple way to the force components: $\tan\psi = F_\perp/F_\parallel$. By using equations (3.86) and (3.94) it follows that

$$\int_{\varphi_1}^{\varphi_2} \frac{dp}{d\varphi} \cos(\varphi + \psi) \, d\varphi = 0 \tag{3.95}$$

In the numerical evaluation of the relationships stated above it was assumed that the required amount of oil is conveyed horizontally and that the external load always acts perpendicularly, so that $\varphi_1 + \psi = 90^\circ$. For a Newtonian fluid the size of the angle ψ is determined only by the value of the eccentricity ε; consider the 'Gümbel curve' in Fig. 3.24 denoted by S = 0. In the case of a non-Newtonian lubricant the angle besides depending on the eccentricity ε also depends on the velocity parameter S. Fig. 3.24 shows the connection for a pseudo-plastic Prandtl-Eyring fluid.

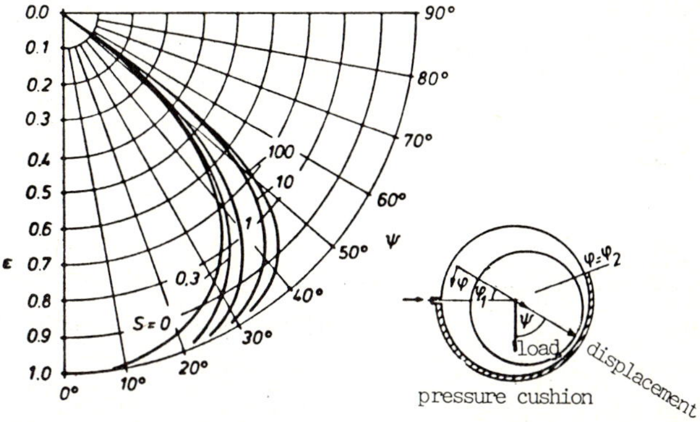

Fig. 3.24 Load-displacement angle in a half-filled bearing; Prandtl-Eyring model

3.3 Radial flow between two parallel planes

It sometimes happens that a viscous fluid flows radially outwards in a gap of constant height. We first consider the situation shown in Fig. 3.25. The fluid is injected axially below a body with a circular lower surface, and flows radially outwards. Thus there arises a *pressure cushion* between the body and the wall, so that the pressures occurring depend on the rheological properties of the lubricant, the rate of flow $\dot{V}$, the radius r_0 of the body and on the gap height h.

If the force F with which the body is pressed against the wall is given, the gap distance adjusts itself in such a way that the resultant pressure force maintains the equilibrium. Therefore this is a case of hydrodynamic lubrication.

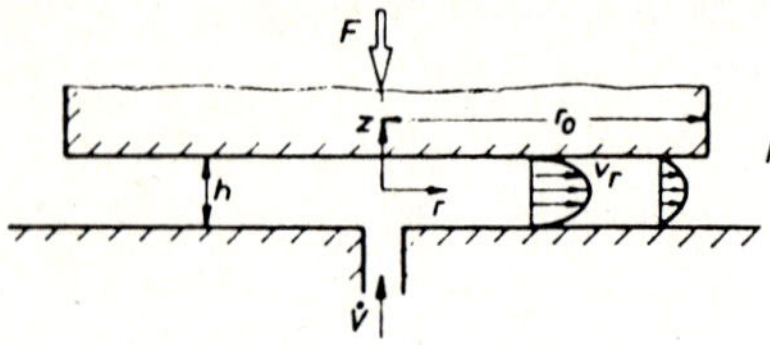

Fig. 3.25 Flow through a radial gap

One can immediately see that the flow velocity v_r decreases towards the outside. For reasons of continuity the same volume flux passes through each cylindrical control surface of area $2\pi rh$, i.e. the expression

$$\dot{V} = 2\pi r \int_{-h/2}^{h/2} v_r dz \tag{3.96}$$

does not depend on r. The velocity averaged over the channel height accordingly decreases as 1/r outwards. Hence the shear rate varies with time for an individual fluid particle, so that the flow is certainly not viscometric. The fluid particle also experiences in addition to the shear another kind of deformation. It is clear that a particle in the middle of the channel is not at all sheared for reasons of symmetry, but because of the decrease in velocity outwards is compressed in a radial direction, and because of the strictly radial flow direction is elongated in the circumferential direction. In the middle of the channel therefore there is a pure elongational flow; at the channel walls on the contrary there is a pure shear flow. For this reason in a radially permeated gap both the shearing behaviour of the fluid and its rheological elongation behaviour will also have an effect. The memory of the fluid also plays a part because the flow is unsteady for any material point. Nevertheless it may be correct to a first approximation to consider only the viscometric shear behaviour. If the gap distance is small compared with the radius of the body so that ($h \ll r_0$), then the fluid is located, apart from the immediate neighbourhood of the axis, in a weakly divergent channel compared with the gap distance. Considerable velocity variations occur in the z-direction, but on the other hand in the r-direction there are only comparatively small changes, i.e. the elongation rates are small compared with the shear rate. The shearing deformation of the fluid particle is dominant. Moreover if the memory of the fluid particle is so short that a recollection of the past results only for such states as differ only slightly from the present state, the flow is 'quasi-viscometric', i.e. the extra-stresses are predominantly determined by the local shear rate $\partial v_r/\partial z$, so that the constitutive equations for plane steady shear flows apply, particularly the flow law in the form (2.16). Thus $\dot{\gamma}$ corresponds to the 'essential' velocity derivative $\partial v_r/\partial z$, and ignoring inertia forces which occur in an accurately analysed divergent channel, the shear stress $\tau_{(rz)}$ is connected with the local pressure gradient according to $\partial\tau/\partial z = dp/dr$ (equation (3.30)). Because z is measured from the centre of the channel, where the shear stress vanishes, it follows that

$$\tau = z\frac{dp}{dr} \tag{3.97}$$

and therefore the flow law (2.16) takes on the following form:

$$\frac{\partial v_r}{\partial z} = \frac{\tau_*}{\eta_*} f\left(\frac{z}{\tau_*}\frac{dp}{dr}\right) \tag{3.98}$$

If one partially integrates the right-hand side of equation (3.96), notes the condition at the walls ($v_r = 0$ for $z = \pm h/2$) to take into account the adhesion, and eliminates the velocity derivative $\partial v_r/\partial z$ which thereby occurs by using equation (3.98), then one obtains

$$\dot{V} = -4\pi r \frac{\tau_*}{\eta_*} \int_0^{h/2} z f\left(\frac{z}{\tau_*}\frac{dp}{dr}\right) dz \tag{3.99}$$

For a given flow function $f(\sigma)$ and a given flow rate V this is in general a non-linear algebraic equation for dp/dr as a function of r. By integrating over the pressure distribution one obtains the load capacity of the pressure cushion

$$F = 2\pi \int_0^{r_0} r[p(r) - p_0]dr = -\pi \int_0^{r_0} r^2 \frac{dp}{dr} dr \tag{3.100}$$

For a Newtonian fluid of constant viscosity η_* $f(\sigma)$ would be put equal to σ. In this case equations (3.99) and (3.100) reduce to

$$\frac{dp_N}{dr} = -\frac{6\eta_* \dot{V}}{\pi h^3 r} \tag{3.101}$$

$$F_N = \frac{3\eta_* \dot{V} r_0^2}{h^3} \tag{3.102}$$

To illustrate the results for non-Newtonian fluids it is expedient to introduce the dimensionless terms

$$R := \frac{r}{r_0}; \qquad K := -\frac{h}{\tau_*}\frac{dp}{dr}; \qquad Q := \frac{\eta_* \dot{V}}{2\pi r_0 \tau_* h^2} \tag{3.103}$$

Equation (3.99) then becomes

$$Q = \frac{2R}{K^2}\int_0^{K/2} \sigma\, f(\sigma)\, d\sigma \tag{3.104}$$

If one has thus determined the pressure parameter K as a function of the position R, and the dimensionless volume flux Q (in fact K obviously depends only on R/Q), then the load capacity can be obtained from the expression

$$\frac{F}{F_N} = \frac{1}{6Q}\int_0^1 R^2 K(R;Q)\, dR \tag{3.105}$$

This is the same as the dimensionless version of equation (3.100). The pressure distribution now follows from the relationship

$$\frac{p(R) - p_0}{\tau_*} = \frac{r_0}{h}\int_R^1 K(R;Q)\, dR \tag{3.106}$$

The numerical evaluation of the expressions (3.104) to (3.106) for a fluid model which has the flow function $f(\sigma) = \sigma + \sigma^3$ (pseudoplastic Rabinowitsch fluid) leads to the pressure distribution in the gap shown in Fig. 3.26 and the diagram connected with it for the resulting force as a function of the rate of flow parameter Q. In the limiting case of $Q \to 0$ the Newtonian results apply (equations (3.101) and (3.102)), because with a sufficiently slow flow the zero-shear viscosity η_* of the fluid makes itself noticeable.

With increasing rate of flow the pressure values related to the rate of flow and therefore the normalised load capacity of the pressure cushion decrease more and more. The reason is the pseudoplastic behaviour of the fluid, according to which the viscosity of the fluid becomes less and less with increasing shear stress.

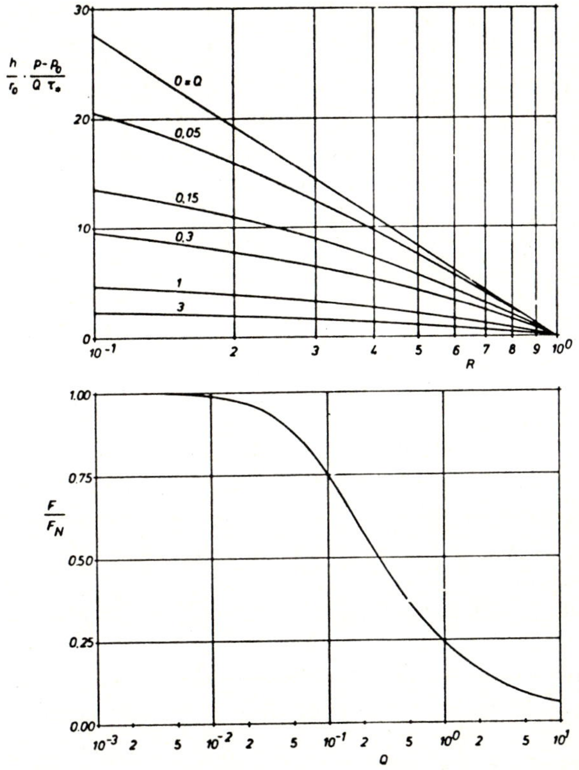

Fig. 3.26 Pressure distribution and load capacity; Rabinowitsch model

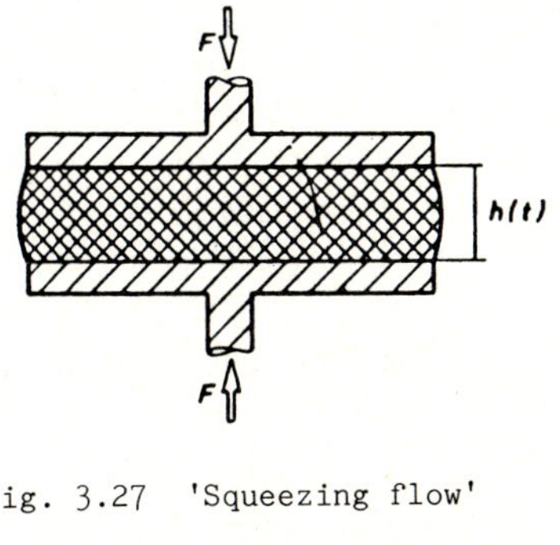

Fig. 3.27 'Squeezing flow'

Fig. 3.27 shows a similar situation in a gap with a radial flow in it. The gap is formed by two equal size parallel circular discs pressed together by a certain force, so that the gap volume is continuously reduced, and the fluid originally present in the gap is pressed outwards. This '*squeezing flow*' is interesting for various reasons. It can for example serve to determine the flow properties of highly viscous substances. The variation with time of the distance between plates, particularly the 'half-value time' during which the distance is reduced to half the original value represents individual friction properties of the compressed fluid. The form of the flow is important in hydrodynamic lubrication. Finally it has a certain attractiveness from the continuum mechanics point of view, because it consists of a combination of shear flow and elongational flow.

This flow is more complex than the other one already described, because in addition to the dependence of the field quantities on position r, z, there is a dependence on time. The flow field is unsteady, particularly the inter-plate distance varies with time t, $h = h(t)$, when $dh/dt < 0$ can be assumed. In the element of time Δt the discs approach each other along the path $-\Delta t\, dh/dt$ and hence displace within the cylinder of radius r the volume $-\pi r^2 \Delta t\, dh/dt$. From this consideration it is clear that through the cylindrical control surface at a distance r from the axis of symmetry the volume

$$\dot{V} = -\pi r^2 \frac{dh}{dt} \tag{3.107}$$

flows through per unit time. Using the expression (3.96) valid again it follows that

$$-r \frac{dh}{dt} = 2 \int_{-h/2}^{h/2} v_r dz \tag{3.108}$$

One might think that v_r increases linearly with r. For reasons of continuity v_z would be independent of r, and the following would apply:

$$v_r = \frac{r}{2} \frac{\partial V(z,t)}{\partial z}, \qquad v_z = -V(z,t) \tag{3.109}$$

Thus material surfaces which are oriented parallel to the discs (planes z = constant) would remain plane surfaces at all times. The kinematic statement (3.109) leads directly to a contradiction. It implies that the shear rate $\dot{\gamma} := \partial v_r/\partial z$ at each moment is defined by a product of a function dependent only on r, and one dependent only on z. The same follows for the shear stress τ from equation (3.97). This means that for an inelastic fluid with an arbitrary flow law $\tau = F(\dot{\gamma})$ both statements are generally incompatible with each other. Some substances are exceptions to this, whose flow properties can be defined after Ostwald-de Waele in terms of a power law $(F(\dot{\gamma}) \sim \dot{\gamma}^n)$ including Newtonian fluids. For these substances alone does the kinematic description given by equation (3.109) apply. For more general fluid properties the expression (3.109) is not correct. The flow field has more complex properties, with the result that originally plane material surfaces parallel to the plates not only elongate during the consequent deformation, but also undergo bending.

A further difficulty arises from the fact that for a given normal thrust primarily only a statement about the pressure integral is possible. Equation (3.100) does not in fact permit the calculation of the pressure field itself. If the contact force F is held constant with time, one could presume from the relationship given by (3.100) that the pressure field shall also be independent of time. Now equation (3.108) after a partial integration and using the flow law (3.98) can be expressed as

$$r\dot{h} = \frac{4\tau_*}{\eta_*}\int_0^{h/2} z\, f\left(\frac{z}{\tau_*}\frac{dp}{dr}\right)dz \tag{3.110}$$

$\dot{h}$ is the derivative of h with respect to time, and z is measured from the centre of the channel as before.

If the pressure field remains constant with respect to time, the right-hand side depends on t only through the limit of integration h(t)/2. By differentiation with respect to time it follows that

$$r\frac{\ddot{h}}{h\dot{h}} = \frac{2\tau_*}{\eta_*} f\left(\frac{h}{2\tau_*}\frac{dp}{dr}\right) \tag{3.111}$$

The left-hand side has obviously the form of a product with respect to r and t, likewise the argument of the flow function on the right-hand side. But from this it follows that equation (3.111) cannot be correct for a general flow law, again with the exception of Newtonian and Ostwald-de Waele fluids. One can best explain the contradiction by means of an example, say for $f(\sigma) = \sigma + \sigma^3$. One can only conclude from this that the assumption that the pressure field is independent of time, was false. For actual non-Newtonian fluids the pressure distribution in the gap therefore varies with time, although the resulting pressure integral remains constant with respect to time.

In the limiting case of low force and with such low flow rates that only the lower Newtonian viscosity limit η_* is operating ($f(\sigma) = \sigma$), it follows from equation (3.110) that $dp/dr = c \cdot r$, and after one integration $h = h_0(1 + ct/3\eta_*)^{-1/2}$, in which h_0 is the disc separation at time $t = 0$. The constant c is obtained from equation (3.100) thus: $c = 4F/\pi r_0^4$. As a whole therefore the following time law results for the disc separation:

$$\frac{h}{h_0} = \left(1 + \frac{4Fh_0^2}{3\pi\eta_* r_0^4}t\right)^{-\frac{1}{2}} \tag{3.112}$$

Accordingly the 'half-value time' for a Newtonian fluid, in which the disc separation decreases to $h_0/2$ under the effect of a normal thrust constant with time, is directly proportional to the viscosity and inversely proportional to the normal thrust, which one could have found by simple dimensional analysis.

3.4 Pipe flow

The fully developed steady pressure flow through a circular section pipe is distinguished by the fact that the velocity field has only one axial component $u(r)$ (cf. Section 1.6.2, in which the kinematics of this form of viscometric flow were discussed). Accordingly the shear rate $\dot{\gamma} = du/dr$ and the shear stress τ only depend on the axis distance r. Using the theorem of momentum on the cylindrical control volume of height dz shown in Fig. 3.28 it should be noted that the inlet and outlet momentum fluxes are of equal magnitude, and therefore the sum of the forces acting on the element vanishes. Neglecting volume forces (horizontal pipe) therefore $2\pi r dz \tau(r) = \pi r^2[p(z+dz) - p(z)]$, or

$$\frac{2}{r}\tau(r) = \frac{dp(z)}{dz} = -\frac{\Delta p}{\ell} \tag{3.113}$$

Obviously this relationship can apply only at each position r, z if both sides are constant. The pressure then varies linearly in the axial direction, so that dp/dz can be represented by the pipe length ℓ and the (positive) pressure difference Δp associated with it. If the pipe is not horizontal then the component of the gravitational force parallel to the axis is added to the driving pressure gradient $\Delta p/\ell$.

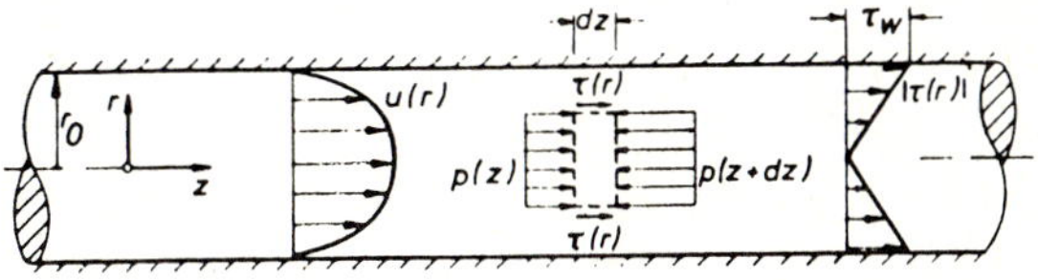

Fig. 3.28 Fully developed pipe flow

Equation (3.113) shows that the absolute value of the shear stress between the cylindrical layers increases linearly with the distance r from the axis of the cylinder. From this one learns that the amount of shear stress τ_w at the wall depends simply on the pressure drop $\Delta p/\ell$ and on the pipe radius r_0:

$$\tau_w = \frac{\Delta p\, r_0}{2\ell} \tag{3.114}$$

By using this quantity one can write equation (3.113) more briefly in the form

$$\tau = -\tau_w \frac{r}{r_0} \tag{3.115}$$

If one puts this result into the flow law (2.16) one obtains the expression below as the conditional equation for the velocity field:

$$\frac{du}{dr} = -\frac{\tau_*}{\eta_*} f\left(\frac{\tau_w}{\tau_*}\frac{r}{r_0}\right) \tag{3.116}$$

By considering the adhesion condition at the pipe wall, $u(r_0) = 0$, one obtains the following expression for the velocity profile:

$$u(r) = \frac{\tau_*}{\eta_*} \int_r^{r_0} f\left(\frac{\tau_w}{\tau_*}\frac{x}{r_0}\right) dx \tag{3.117}$$

The volume flux $\dot{V}$, which collectively moves along the pipe, is found by integrating the velocity field over the cross-section:

$$\dot{V} = 2\pi \int_0^{r_0} r\, u(r)\, dr = -\pi \int_0^{r_0} r^2 \frac{du}{dr}\, dr \tag{3.118}$$

The second equality follows after one partial integration. For a Newtonian fluid of viscosity η_* (subscript N) $f(\sigma) = \sigma$ holds and equation (3.117) reduces to a quadratic expression in r. This is $u_N(r) = \tau_w(r_0^2 - r^2)/2\eta_* r_0$, which shows that there is a parabolic velocity profile. The maximum velocity (at the axis) and the associated volume flux are therefore connected with the quantities involved by

$$u_N(0) = \frac{\tau_w r_0}{2\eta_*}, \qquad \dot{V}_N = \frac{\pi \tau_w r_0^3}{4\eta_*} \tag{3.119}$$

These expressions can serve as reference quantities for velocity and flow rate when considering non-Newtonian fluids. We therefore compare the flow of a fluid that has non-linear properties with that of a Newtonian fluid, for which the pipe radius and the pressure drop are the same in both cases, and the reference quantity η_* in the flow law (2.16), which in general describes the zero-shear viscosity of the actual fluid is chosen as the Newtonian reference viscosity. By using the quantity $u_N(0)$ equation (3.117) can be written in the following dimensionless form:

$$\frac{u(r)}{u_N(0)} = \frac{2}{\sigma_w^2} \int_{\sigma_w \frac{r}{r_0}}^{\sigma_w} f(\sigma)\, d\sigma \tag{3.120}$$

Hence the suitably normalised velocity in the pipe besides depending on the dimensionless coordinate r/r_0 also depends on the form of the flow function $f(\sigma)$ and the dimensionless value of the wall shear stress

$$\sigma_w := \frac{\tau_w}{\tau_*} = \frac{\Delta p\, r_0}{2\ell\, \tau_*} \tag{3.121}$$

which we can also denote as the 'pressure parameter'. Finally the relationship (3.118) for the flow rate obtained from equation (3.116) transforms into

$$\Phi := \frac{\dot{V}}{\dot{V}_N} = \frac{4}{\sigma_w^4} \int_0^{\sigma_w} \sigma^2 f(\sigma)\, d\sigma \tag{3.122}$$

The flux formula in this shortened form is valid for fluids having any flow properties. For a particular substance of known flow function $f(\sigma)$ equation (3.122) represents a generally non-linear relationship between the reduced flow rate $\Phi := \dot{V}/\dot{V}_N$ and the pressure parameter σ_w. If one uses one of the fluid models quoted in Table 2.1 to describe real fluid properties, then the integrals in equations (3.122) and (3.120) can be evaluated analytically, so that simple formulas arise.

For a Prandtl-Eyring model (for which $f(\sigma) = \sinh\sigma$) for example it follows that

$$\frac{u(r)}{u_N(0)} = \frac{2}{\sigma_w^2}\left[\cosh\sigma_w - \cosh\left(\sigma_w \frac{r}{r_0}\right)\right] \tag{3.123}$$

$$\Phi = \frac{4}{\sigma_w^4}\left[\sigma_w^2 \cosh\sigma_w - 2\sigma_w \sinh\sigma_w + 2\cosh\sigma_w - 2\right] \tag{3.124}$$

Fig. 3.29 is a diagram of these relationships. The increase in the volume flux related to the Newtonian reference value with increase in the pressure parameter (therefore with increasing pressure drop) is explained by the fact that a Prandtl-Eyring fluid becomes less and less viscous as the stress increases. This pseudoplastic behaviour

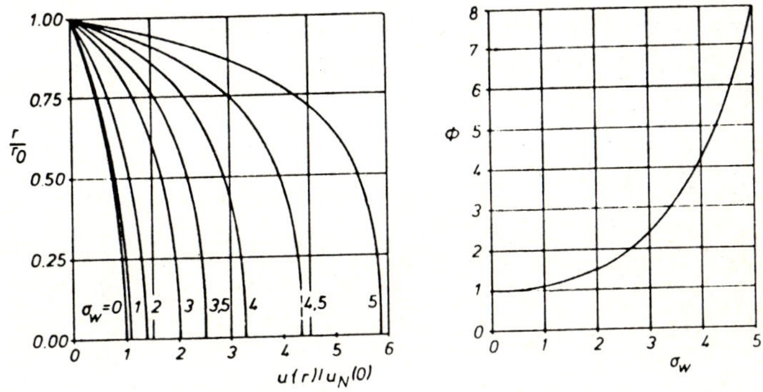

Fig. 3.29 Velocity profiles and pressure-flow rate relation; Prandtl-Eyring fluid

moreover makes itself felt in that with increasing pressure drop blunter velocity profiles appear than with the Newtonian reference fluid ($\sigma_w \to 0$); cf. the exhaustive discussion about this property in connection with Fig. 3.5.

Equation (3.122) gives the general relation between the dimensionless flow rate, the dimensionless wall shear stress, and the flow function of the flowing substance. We shall in passing take the view that the flow through the circular cylindrical capillary will serve to determine the fluid property $f(\sigma)$ (*capillary viscometer*). Hence the measured quantity is the flow rate for a known pressure drop. In order to determine the flow function from this pressure-flow rate characteristic, the expression (3.122) must be solved for $f(\sigma)$. For this one imagines the wall shear stress to be variable, by altering the pressure drop slightly. Hence the associated change of the reduced flow rate corresponds to the increment of the expression (3.122), and can be stated in the form

$$d\Phi = \left[-\frac{4}{\sigma_w}\Phi + \frac{4}{\sigma_w^2} f(\sigma_w) \right] d\sigma_w \tag{3.125}$$

Hence it follows that

$$f(\sigma_w) = \left[1 + \frac{1}{4}\frac{d \log \Phi}{d \log \sigma_w} \right] \sigma_w \Phi \tag{3.126}$$

Evaluation of the measurements in accordance with this equation leads to the viscometric flow function of the substance. As long as the flow law of a substance is unknown, one chooses for τ_* and η_* (which enters into $\dot{V}_N$) suitable and as simple as possible values, e.g. $\tau_* = 1\ N/m^2$, $\eta_* = 1\ Ns/m^2$. If one has measured the volume flux $\dot{V}$ for various values of wall shear stress τ_w (various values of $r_0 \Delta p/\ell$, cf. equation (3.114)), then one can assign to each measurement point the reduced quantities $\sigma_w := \tau_w/\tau_*$, and $\Phi := \dot{V}/\dot{V}_N$. The reference quantity $\dot{V}_N$ is determined by τ_w, r_0 and η_* (cf. equation (3.119)). One then applies the values of $\log \Phi$ to the relevant values of $\log \sigma_w$ and obtains from the resulting graph the derivative required in equation (3.126) at the positions σ_w obtained in the test. Thus the expression on the right-hand side of equation (3.126) can be formed for discrete positions σ_w. If one applies the result to σ_w, then one obviously obtains the flow function $f(\sigma_w)$, and after multiplication of the axes by the reference values τ_* and η_*/τ_* the flow graph in the form $\tau = F(\dot{\gamma})$. Using equation (2.5) one now obtains the viscosity η as a function of the shear rate $\dot{\gamma}$. If one identifies the reference quantity η_* henceforth with the zero-shear viscosity of the substance, then one arrives at the normalisation mentioned in Section 2.1, $df/d\sigma = 1$ for $\sigma = 0$. With a suitable value of τ_* another coefficient of the dimensionless flow function $f(\sigma)$ can be made equal to unity, as for the fluid models proposed by Rabinowitsch, Prandtl-Eyring, Ellis, Reiner-Philippoff, and Reiner in Table 2.1.

One conveniently defines by application of the mean velocity $\bar{u}:=\dot{V}/\pi r_0^2$ and the pipe diameter $d := 2r_0$ in the usual way as for Newtonian fluids a Reynolds number Re and a *resistance number* λ according to

$$\mathrm{Re} := \frac{\rho \bar{u} d}{\eta_*}, \qquad \lambda := \frac{\Delta p}{\frac{\rho}{2}\bar{u}^2} \cdot \frac{d}{\ell} \tag{3.127}$$

For a laminar pipe flow of a Newtonian fluid of viscosity η_* as is well known $\lambda = 64/\mathrm{Re}$ applies. In the case of a fluid which has non-Newtonian flow properties, λ depends, as one can easily prove, besides on Re in a simple way on the previously introduced dimensionless quantity Φ:

$$\lambda = \frac{64}{\Phi\,\mathrm{Re}} \tag{3.128}$$

Remember that Φ was introduced as a constant of proportionality in the relationship between $\dot{V}$ and $\dot{V}_N$ ($\dot{V} = \Phi\,\dot{V}_N$, equation (3.122)). From equation (3.128) the reciprocal value Φ^{-1} also plays the part of a correction factor in the resistance law, with which the relationship applicable to Newtonian fluids is to be modified, in order that it is valid for non-Newtonian fluids. For a given flow function this correction factor obtained from equation (3.122) can be expressed explicitly as a function of the pressure parameter σ_w.

In order to express the non-linear relationship between the axial pressure drop and the volume flux through the circular pipe in the dimensionless form, we had decided to relate the actual volume flux to that of the Newtonian reference flow with the same pressure drop, same pipe diameter, constant viscosity η_*; cf. equation (3.122). For many purposes it can be more expedient to use a reference value which is independent of pressure. Thus for example the quantity $\eta_*\bar{u}/\tau_*$ can also be regarded as a suitable dimensionless flow parameter. The pressure-flow rate relationship is then expressed in the form

$$\frac{\eta_* \bar{u}}{\tau_* d} = \frac{1}{2\sigma_w^3} \int_0^{\sigma_w} \sigma^2 f(\sigma)\, d\sigma \tag{3.129}$$

Hitherto an isothermal flow has been assumed; otherwise the temperature dependence of the viscosity would have been taken into account. Now however the flow field cannot be isothermal because heat is generated by dissipation within the viscous fluid, which for a finite thermal conductivity cannot be spontaneously conducted away. Thus there arises in the fluid a non-homogeneous temperature distribution, and in consequence the pure mechanical theory previously applied is only a more or less useful approximation to the actual process. In respect of an opinion about the goodness of the approximation and about the question of which reference temperature is to be logically brought in for the viscosity values within the framework of the isothermal treatment, the temperature field is also of interest, particularly the greatest temperature difference occurring in the fluid.

In the treatment of this dissipation effect we limit ourselves to the simplest case and assume that the same thermal conditions prevail overall in the pipe material independent of position. The fluid particles adhering to the wall then have a constant temperature Θ_w. Under such conditions it is to be expected that for the fully developed flow considered here the temperature field only depends on r, but does not vary in the axial direction. Thermal effects at the inlet are therefore not considered. Because $\Theta = \Theta(r)$ and $v_r = 0$, $D\Theta/Dt = 0$ holds, i.e. there is no convectional transfer of energy. This goes without saying because the temperature is constant along the path lines (r = constant). The energy balance (1.126) thus reduces to $0=\Phi+\mathrm{div}(\lambda \,\mathrm{grad}\,\Theta)$, or

$$\tau \frac{du}{dr} + \frac{1}{r}\frac{d}{dr}\left(r\lambda\frac{d\Theta}{dr}\right) = 0 \tag{3.130}$$

which means that the energy dissipated per unit time in an element of volume flows as heat via the surface of the element. If one inserts the expressions (3.115) and (3.116) applying to a pipe flow into the dissipation term, then it follows that

$$\frac{d}{dr}\left(r\lambda\frac{d\Theta}{dr}\right) = -\frac{\tau_w \tau_*}{r_0 \eta_*} r^2 f\left(\frac{\tau_w}{\tau_*}\frac{r}{r_0}\right) \tag{3.131}$$

This is in general a non-linear differential equation for $\Theta(r)$, because the thermal conductivity λ and the zero-shear viscosity η_* (but not the reference shear stress τ_*) are temperature dependent quantities, as is well known. Assuming that the temperature differences in the fluid caused by dissipation are small compared with the reference temperature Θ_w, we can, as previously for the calculation of the velocity field, approximate the zero-shear viscosity η_* and accordingly the thermal conductivity λ by constant mean values. This moreover corresponds to a limitation to the first term of the series expansion of the actual temperature field with respect to powers of the dimensionless coefficient $\tau_w^2 r_0^2/(\lambda\eta_*\Theta_w)$. If one considers λ and η_* to be constants, then one can integrate equation (3.131) once:

$$r\frac{d\Theta}{dr} = -\frac{\tau_w^2 r_0^2}{\lambda\eta_*}\cdot\frac{1}{\sigma_w^4}\int_0^{\sigma_w \frac{r}{r_0}} \sigma^2 f(\sigma)\, d\sigma \tag{3.132}$$

The lower limit of the integral results from the condition that the left-hand as well as the right-hand side must vanish at the symmetry axis ($r = 0$). One learns from this expression that because $f(\sigma) > 0$ for $\sigma > 0$, $d\Theta/dr < 0$ for $r > 0$ always applies. Thus the maximum temperature occurs on the axis; from there to the wall of the pipe the fluid becomes cooler. The inner layers are therefore warmer than the outer ones.

Equation (3.132) also shows that the local temperature gradient, particularly that at the pipe wall is not at all affected by the thermal boundary conditions. This is quite understandable because the total energy dissipated inside a cylinder of radius r must of course flow away across the cylindrical surface outwards, and thereby creates a local value of the heat flux independent of the thermal boundary conditions.

By a second integration one obtains for the temperature profile the explicit expression

$$\frac{\Theta(r) - \Theta_w}{\dfrac{\tau_w^2 r_0^2}{\lambda\eta_*}} = \frac{1}{\sigma_w^4}\int_{\sigma_w \frac{r}{r_0}}^{\sigma_w} \frac{1}{\sigma'}\int_0^{\sigma'} \sigma^2 f(\sigma)\, d\sigma\, d\sigma' \tag{3.133}$$

In a suitably standardised form therefore the temperature field of a pipe flow induced by dissipation depends critically on the flow function $f(\sigma)$ of the substance, and on the dimensionless pressure parameter σ_W.

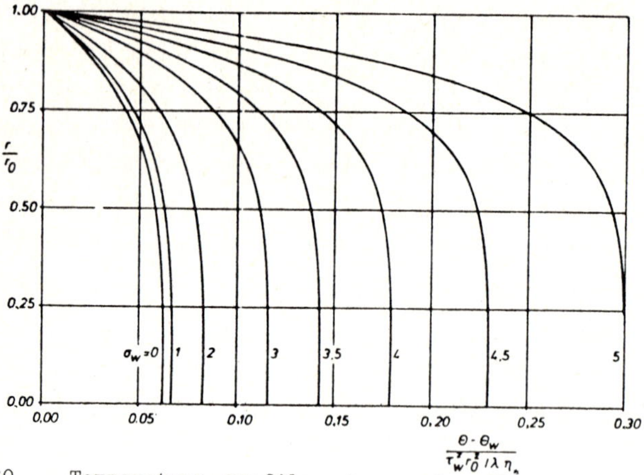

Fig. 3.30 Temperature profiles in a pipe flow in accordance with equation (3.133); Prandtl-Eyring model

Fig. 3.30 shows the result of an evaluation of the formula for the Prandtl-Eyring model. Hence the limiting case of Newtonian flow behaviour ($f(\sigma) = \sigma$) is contained with $\sigma_W \to 0$. In this case the right-hand side reduces to the expression $[1 - (r/r_0)^4]/16$, so that here the following maximum temperature difference occurs:

$$\Theta_N(0) - \Theta_w = \frac{\tau_w^2 r_0^2}{16\lambda\eta_*} = \frac{\eta_*}{\lambda}\,\bar{u}_N^2 \tag{3.134}$$

The second equality results from the use of equation (3.119). Hence for a polymer melt with $\eta_* = 10^4$ Ns/m^2 and $\lambda = 0.24$ N/sK at an average velocity $\bar{u} = 5$ cm/s a maximum temperature difference of 104 K is calculated, provided that only the zero-shear viscosity is affecting the melt. Under these conditions one can naturally question the accuracy of describing the temperature dependent quantities η_* and λ in terms of constant mean values. For a cold automobile oil of $\eta_* = 0.8$ Ns/m^2 and $\lambda = 0.15$ N/sK at the same mean velocity as before, a maximum temperature difference of only about 0.01 K results. Here therefore the dissipation has almost no effect in practice, and the treatment as an isothermal flow is completely justified.

3.5 Helical flow

In the treatment of the extruder flow in Section 3.2.2 the curvature of the walls was neglected and the motion was regarded as a plane shear flow (Fig. 3.11). This simplification was based on the assumption $h \ll r$, which in practice is not entirely fulfilled. If one does not make this limitation, but retains all the other assumptions, then the observer who is rotating with the screw sees a steady helical flow with a velocity field of the type expressed by the equation (1.86). Because such helical flows also play a definite part in other situations the necessary fundamental relationships used in their calculation will now be derived and briefly explained.

For a flow field as expressed by equation (1.86), besides the velocity components the extra-stresses and acceleration components depend only on the axial distance r. If volume forces can be neglected, then the pressure gradient is a function of r alone (equation 1.116)). This statement implies that the pressure field contains a term linear in z and one linear in φ, so that the coefficients are in fact constant. Thus for a steady helical flow the pressure drop in the axial direction and the pressure change along the circumference are accordingly constant:

$$\frac{\partial p}{\partial z} = -k, \qquad \frac{\partial p}{\partial \varphi} = -a \tag{3.135}$$

Now the fluid particles are neither accelerated in the φ-, nor in the z-direction. The equations of motion associated with the two directions therefore reduce to

$$\frac{a}{r} + \frac{d\tau_{r\varphi}}{dr} + \frac{2}{r}\tau_{r\varphi} = 0 \tag{3.136}$$

$$k + \frac{d\tau_{zr}}{dr} + \frac{1}{r}\tau_{zr} = 0 \tag{3.137}$$

(cf. Formula Appendix for the cylindrical coordinates).

Two new constants b, c enter with the integration. The result is:

$$\tau_{r\varphi} = \frac{c}{r^2} - \frac{a}{2}, \qquad \tau_{zr} = \frac{b}{r} - \frac{k}{2}r \tag{3.138}$$

The third equation of motion can be omitted here. It defines how the pressure, or more accurately, how the normal stress varies in the radial direction. The two shear stress components $\tau_{r\varphi}$ and τ_{zr} add together thus: $\tau_{r\varphi}^2 + \tau_{zr}^2 = \tau^2$, to give the resultant shear stress τ, which affects the cylindrical sliding surfaces. Using equation (3.138) one obtains

$$\tau = \sqrt{\left(\frac{c}{r^2} - \frac{a}{2}\right)^2 + \left(\frac{b}{r} - \frac{k}{2} r\right)^2} \tag{3.139}$$

Therefore the position coordinate r explicitly determines the shear stress. Because this is a viscometric flow the material law (2.36) holds. For the two stress components of interest here one can derive from this the relationships $\tau_{r\varphi} = \eta(\dot{\gamma}) r d\omega/dr$ and $\tau_{zr} = \eta(\dot{\gamma}) du/dr$ (cf. Problem 2.4). Hence $\dot{\gamma}$ is the expression given in equation (1.89). If one replaces η by $\tau/\dot{\gamma}$, uses the flow law of the material in the form (2.16) and inserts the expressions (3.138) for the stresses, then one obtains the following equations for the two velocity fields:

$$\eta_* \frac{d\omega}{dr} = \frac{1}{r}\left(\frac{c}{r^2} - \frac{a}{2}\right) \frac{f(\tau/\tau_*)}{\tau/\tau_*} \tag{3.140}$$

$$\eta_* \frac{du}{dr} = \left(\frac{b}{r} - \frac{k}{2} r\right) \frac{f(\tau/\tau_*)}{\tau/\tau_*} \tag{3.141}$$

For given flow properties the right-hand sides are known functions of r from equation (3.139). Therefore the determination of the two velocity fields is reduced to the calculation of the two integrals. Thus the two newly appearing constants of integration and the four constants a, b, c, k are in the actual case determined by the boundary conditions for the pressure and for the velocity components.

Problems

3.1 Fig. 3.31 shows the layout of a wire coating tool. It consists essentially of a melt chamber in which the overpressure is Δp, and a circular section cylindrical orifice (length L, diameter $2(r_0 + h)$ through which the wire (diameter $2r_0$) which has to be insulated passes at velocity U. Because of pressure and drag effects the polymer melt moves continuously through the annular gap between the orifice and the wire outwards, and finally forms an insulating layer of thickness s after setting. We shall assume that $h \ll r_0$, so that the gap can be regarded as a plane channel. The dimensionless flow function $f(\sigma)$ and the material constants η_* and τ_* of the melt area are known.

For given fixed quantities L, r_0, h and U find Δp so that an insulating layer of given thickness s is formed. Describe how this problem can be solved, and state by using the concept of differential viscosity an analytical approximation formula for the relationship between Δp and s.

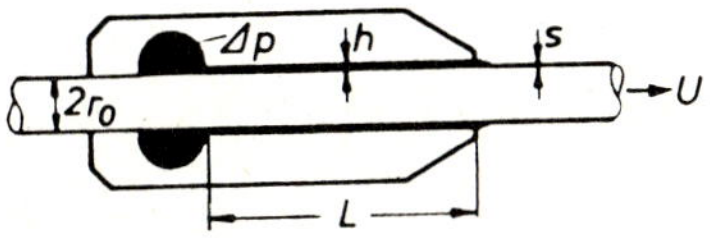

Fig. 3.31 Wire coating tool

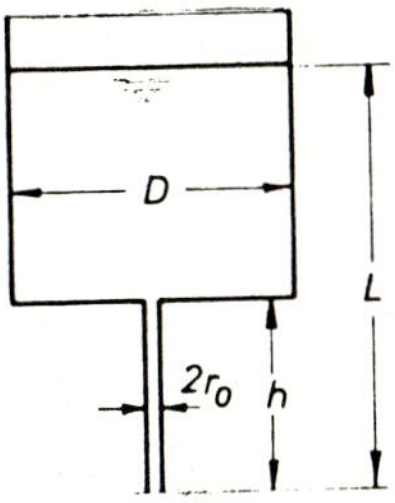

Fig. 3.32 Illustration of Problem 3.6

3.2 Calculate for an extruder flow in the frame of a pseudo-Newtonian approximation, which is based on two constant viscosity values η and $\hat{\eta}$, the pressure rise transverse to the web, i.e. establish a formula for $\partial p/\partial z$ analogous to equation (3.74), and compare it with the result for a Newtonian fluid.

3.3 The thickness of the extruder in a roller process is to be reduced in a definite way. Find the angular velocity ω of the rollers and their shortest distance apart so that the sheet thickness can be changed at will. The data are h_1 and h_2 (see Fig. 3.16), the roller radius r and the flow rate per unit width $\dot{V}/b$. Assume that the flow properties of the extrudate are known, so that numerical values for η_* and τ_*, and diagrams as shown in Figs. 3.18 and 3.19 are known. How do you determine ω and h_0? Finally how does one determine the thrust force and the required power?

3.4 Find the pressure distribution in the working gap during the rolling of a pseudoplastic substance which has yield stress τ_* and differential viscosity η_* (Bingham model) and the relationship between the dimensionless inlet coordinate ξ_1, and the exit coordinate ξ_2. Thus the results in the special case of a rigid plastic substance ($\eta_* \to 0$) can be found analytically.

Therefore starting from equation (3.79) first obtain a differential equation for $\bar{p}(\xi; S, \xi_2)$ and solve this for the limiting case $S = 0$. Note that the product $S\bar{p}$ remains finite for the limit $\eta_* \to 0$ (equation (3.78)). The calculation of the pressure distribution also gives the relationship between ξ_1 and ξ_2.

3.5 There are two immiscible viscous fluids in a plane channel of depth h between one stationary and one moving wall with a constant velocity U ('coextrusion'). The two layers of fluid have constant thickness h_1 and $h - h_1$ respectively. A pressure rise in the direction of motion is not initially present; inlet effects have died away. Show that under these conditions the shear stresses in both fluids have the same constant values, say τ_0. Let the viscosities of the two fluids be η_1 and η_2.

Determine the shear stress, the velocity of the separating surface, and the volume fluxes of the two fluid layers as a function of the drag velocity, the two viscosities and the two layer thicknesses. Is it more advantageous in respect of the greatest possible flow rate to have the more viscous of the two fluids at the moving or the stationary wall?

How does the velocity profile vary when the flow is subject to a pressure rise in the direction of the motion? Determine within a linear theory for sufficiently small pressure rise dp/dx the velocity change of the surface of separation and the changes in the two volume fluxes as functions of dp/dx, h, h_1 and the differential viscosity values $\hat{\eta}_1$ and $\hat{\eta}_2$ of the two fluids associated with the reference flow.

3.6 A pseudoplastic Rabinowitsch fluid of known properties ρ, η_*, and τ_* flows out of an open circular cylindrical tank of diameter D through a circular section capillary (length h, radius r_0 with $r_0 << D/2$) by force of gravity (Fig. 3.32). Calculate the volume flux as a function of the actual fluid level above the nozzle outlet. Neglect the inertia forces as well as the inlet and outlet phenomena in the capillary. Regard the flow therefore as 'momentarily viscometric'. Finally determine the height L of the surface of the fluid as a function of time t, if the process at time t = 0 starts at $L = L_0 > h$. Record it in the form $\log(L/L_0)$ against t, and sketch the result also for a Newtonian fluid for which the viscosity η_N is given by

$$1 < \frac{\eta_*}{\eta_N} < 1 + \frac{1}{6}\left(\frac{\rho g L_0 r_0}{\tau_* h}\right)^2$$

Discuss on the basis of this figure the events during the 'race' between the two fluids in two similar containers.

The assumption $r_0 \ll D/2$ means that the fluid in the upper part of the arrangement can be regarded as at rest. The appropriate volume force acting in the perpendicular capillary is composed of the gravity force ρg and the pressure drop $\rho g(L - h)/h$, therefore it has the value $\rho gL/h$.

3.7 Show that for the pure pressure flow of a chosen non-Newtonian fluid along a straight channel the velocity profile has the same form as the meridian section of the rotational symmetric profile for the pressure flow through a circular pipe, supposing that the wall shear stress in both cases has the same value.

3.8 Determine for the flow of a Bingham plastic through a circular pipe under the force of an axial pressure drop (a) the volume flux $\dot{V}$ relative to the characteristic value $\dot{V}_N$ for a Newtonian fluid of viscosity η_*, as a function of τ_w/τ_* (qualitative sketch),
(b) the reduced velocity profile $u(r)/u_N(0)$,
(c) the reduced temperature profile $(\Theta(r) - \Theta_w)/(\Theta_N(0) - \Theta_w)$ which arises as a result of viscous heating, if the wall temperature Θ_w is constant and the viscosity η_* and the thermal conductivity λ of the substance are not dependent on the temperature.

3.9 In a circular section pipe of radius r_0 through which there is a laminar flow of a constant density Newtonian fluid, the pressure falls by Δp over a length L. The flow is not isothermal, but the temperature changes and with it the viscosity of the fluid in the axial direction. The function $\eta(z)$ can therefore be regarded as known.

Show that for a Newtonian fluid under these thermally non-homogeneous conditions the velocity profile in each cross-section is parabolic and is uniform everywhere for reasons of continuity. The pressure drop is not however constant, but depends on z. Calculate the pressure distribution p(z) and the volume flux $\dot{V}$ as a function of r_0, L, Δp and $\eta(z)$.

3.10 For the situation shown in Fig. 3.25 let a load F = 7.5 kN be supported by an oil pressure cushion which is placed above a circular surface of radius r_0 = 6 cm. The flow properties of the lubricant can be defined by a Rabinowitsch model which has the constants $\eta_* = 5.6 \cdot 10^4$ Ns/m^2 and $\tau_* = 1.5 \cdot 10^4$ N/m^2.

Determine by using diagram 3.26b the necessary flow rate $\dot{V}$ [mm^3/s], if the gap width shall be h = 1 mm.

3.11 We consider the steady shear flow of a non-Newtonian fluid of given flow properties (η_*, τ_*, $f(\sigma)$) between two coaxial circular cylinders (radii r_1 and r_2, $r_1 < r_2$), which occurs when the inner cylinder is drawn along at a given velocity u_1 in the axial direction, while the outer cylinder is at rest. There is no pressure gradient. Hence there is a special helical flow. Determine the four constants a, b, c, k in equations (3.140) and (3.141) or state conditional equations for them.

Then by limitation to a Newtonian fluid ($f(\sigma) = \sigma$) calculate the axial velocity field and the associated volume flux analytically. In particular discuss the two limiting cases $r_2 - r_1 \ll r_1$ and $r_1 \ll r_2$.

4 EFFECT OF NORMAL STRESS DIFFERENCES

It was shown in Chapter 2 that for a steady shear flow of simple fluids two kinds of extra-stresses occur. Firstly shear stresses arise, secondly there are the normal stresses in various directions, generally of varying magnitude. Thus it may be surprising to learn that it is possible to analyse the motion of steady shear flows by considering the shear properties alone. In fact however the velocity profile say of a fully developed channel flow or a pipe flow depends only on the flow function but not on the normal stress differences. Fluids that have the same flow function but which have different normal stress properties therefore flow in the same way through the channel or pipe. Correspondingly this applies to all other steady shear flows. Because they are controlled by the flow properties alone of the fluid concerned, one also conveniently speaks of 'partly controllable' flows.

If the normal stress differences also have no effect on the flow form, they operate as forces on the bounding walls and lead to some interesting effects characteristic of non-Newtonian fluids.

4.1 Cone-and-plate flow

We first of all turn to the matter of how one can determine the viscometric material functions $N_1(\dot{\gamma})$ and $N_2(\dot{\gamma})$ experimentally. In this context we consider a cone-and-plate flow as an example. It will be shown that one can determine with this arrangement in particular the sum $N_1(\dot{\gamma}) + 2N_2(\dot{\gamma})$. Other flow forms, particularly a torsional flow, permit a further independent approach to the two normal stress functions.

For a cone-and-plate flow which has the velocity field given by (1.93) and the shear rate given by (1.94) the extra-stresses depend only on the coordinate ϑ (cf. Fig. 1.12). The equation of motion in the flow direction provides the velocity profile which is dependent on the flow function. If one analyses the two other equations of motion one finds that they contradict each other in general. This shows that the motion of a fluid between a rotating cone and a flat plate cannot cause a pure shear flow as assumed. The contradiction vanishes however if the angle between the cone and the plate is very small and in addition the inertia forces stay negligibly small compared with the friction forces. The former can be obtained by use of a suitably blunt cone ($\beta \ll 1$). The latter is then in general obtained in the same way. $\rho R_0 \omega^2$ expresses the radial inertia force per unit volume; $N_1/R_0 \sim \nu_{10}\omega^2/\beta^2 R_0$ the characteristic extra-force in the radial direction. The ratio of the two expressions, the dimensionless quantity $\rho\beta^2 R_0^2/\nu_{10}$ depends only on the outer gap width βR_0 apart from material constants ρ and ν_{10}, but not on the angular velocity. For fluids with marked normal stress properties and for a narrow gap $\rho\beta^2 R_0^2/\nu_{10} \ll 1$ holds, i.e. the inertia of the fluid plays independently of the rotation speed only a subordinate role, and can therefore be disregarded.

We can therefore state that the cone-and-plate arrangement is a useful device for rheometry only for small included angles. Under these conditions however the shear rate $\dot{\gamma}$ and with it all extra-stresses can be regarded as spatially constant ($\dot{\gamma} = \omega/\beta$, in which ω describes the angular velocity of the cone when the plate is stationary, cf. Section 1.6.6). This simplifies the following considerations.

We now turn therefore to the two hitherto not evaluated equations of motion in the ϑ- and R-directions (see Formula Appendix). We start by assuming that the plate is horizontal. The force of gravity can then be neglected because of the small differences in height in the gap. Because the shear stress $\tau_{R\vartheta}$ vanishes, the flow is rotational symmetric; ($\partial/\partial\varphi = 0$) and the polar angle ϑ departs only slightly from $\pi/2$, ($\cot\vartheta \approx 0$); the equilibrium of forces in the ϑ-direction reduces to $\partial(-p + \tau_{\vartheta\vartheta})/\partial\vartheta = 0$, i.e. the normal stress

$$\sigma_{\vartheta\vartheta} := -p + \tau_{\vartheta\vartheta} \tag{4.1}$$

depends only on R. By neglecting the inertia (*creeping flow*) the balance of forces in the radial direction is given by

$$R\frac{\partial(-p + \tau_{RR})}{\partial R} + 2\tau_{RR} - \tau_{\vartheta\vartheta} - \tau_{\varphi\varphi} = 0 \tag{4.2}$$

If one inserts in this the two constitutive equations $\tau_{\varphi\varphi} - \tau_{\vartheta\vartheta} = N_1(\dot{\gamma})$ and $\tau_{\vartheta\vartheta} - \tau_{RR} = N_2(\dot{\gamma})$ (equations (2.24) and (2.25)) and notes that the shear rate is spatially constant one first obtains

$$R\frac{d\sigma_{\vartheta\vartheta}}{dR} = N_1(\dot{\gamma}) + 2N_2(\dot{\gamma}) = \text{const} \tag{4.3}$$

from which follows the normal stress $\sigma_{\vartheta\vartheta}$, the negative value of which gives the pressure on the plate and the cone respectively,

$$\sigma_{\vartheta\vartheta}(R) = \sigma_{\vartheta\vartheta}(R_0) + [N_1(\dot{\gamma}) + 2N_2(\dot{\gamma})] \ln \frac{R}{R_0} \tag{4.4}$$

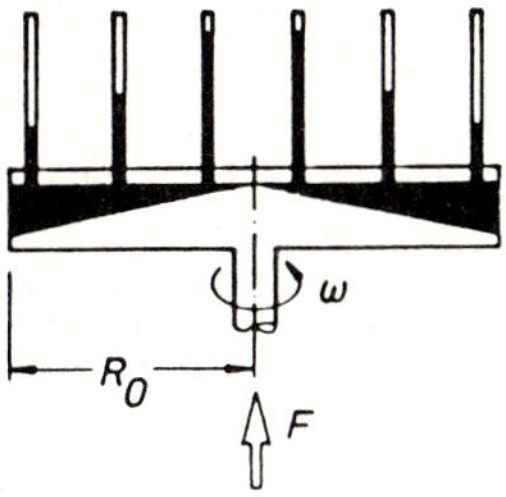

Fig. 4.1 Illustration of the pressure distribution in a cone-and-plate flow

The normal stress properties of non-Newtonian fluids accordingly can be expressed as a position dependent pressure distribution along the plate. Because the quantity $N_1 + 2N_2$ is generally positive, the pressure $-\sigma_{\vartheta\vartheta}$ increases towards the axis of rotation. Fig. 4.1 shows this normal stress effect. From equation (4.4) the combination $N_1 + 2N_2$ of the normal stress functions for discrete shear rates can be determined experimentally as follows. One measures at constant rate of rotation the pressure on the plate at various positions R, records the data in the form log R/R_0, reads off the slope of the resulting straight line and thus obtains a value for $N_1 + 2N_2$ for given shear rate $\dot{\gamma} = \omega/\beta$. In this way the function $N_1(\dot{\gamma}) + 2N_2(\dot{\gamma})$ can be determined step by step if the shear rate is varied by changing the speed of rotation.

In order to determine with the cone-and-plate device both functions separately and not only their sum, an independent statement about the normal stress functions has to be added to the information in equation (4.4). One obtains such a relationship by applying the following limiting conditions. If the test fluid exactly occupies the space between the cone and the plate ($R \leqq R_0$) and on the outside it is immediately bounded by the surrounding atmosphere, then one can by neglecting capillary forces at the free surface use the ambient pressure for the normal stress in the radial direction. Because this can be assumed to be zero without loss of the generality, then there would apply $-p + \tau_{RR} = 0$ for $R = R_0$, or

$$\sigma_{\vartheta\vartheta}(R_0) = N_2 \tag{4.5}$$

i.e. the wall pressure at $R = R_0$ would immediately give the property $-N_2$. Against this method of determining N_2 there is the fact that in the vicinity of the free surface possible distorting end effects can arise, which affect the flow there and hence cause false readings of the stress $\sigma_{\vartheta\vartheta}(R_0)$. One therefore conveniently uses the resultant force

$$F := -2\pi \int_0^{R_0} R\,\sigma_{\vartheta\vartheta}(R)\,dR = \frac{\pi}{2} R_0^2(N_1 + 2N_2) - \pi R_0^2\,\sigma_{\vartheta\vartheta}(R_0) \tag{4.6}$$

which is comparatively little affected by disturbances at the boundary, as an independent measurement. The right-hand side follows from using equation (4.4) alone. With the additional assumption (4.5) one can write

$$\frac{2F}{\pi R_0^2} = N_1 \tag{4.7}$$

The force with which the plate must be retained so that it does not move away from the cone is accordingly a direct measurement of the first normal stress function. If the rotation speed and hence the shear rate $\dot{\gamma} = \omega/\beta$ is varied, then the value of the force naturally varies too. It is therefore possible to determine the first normal stress function $N_1(\dot{\gamma})$ by measuring a force at different values of rotation speed. With a cone-and-plate device one can therefore in principle find both normal stress functions. One should of course remember that equations (4.5) and (4.7) are based on a special assumption about the stresses at the external boundary, but equation (4.4) is quite independent of these boundary conditions. The method based on equation (4.4) of finding the sum $N_1(\dot{\gamma}) + 2N_2(\dot{\gamma})$ should therefore be considered as more reliable than the determination of $N_1(\dot{\gamma})$ or $N_2(\dot{\gamma})$ by using equations (4.7) and (4.5).

One can totally dispense with the relationships (4.5) and (4.7) if one infers the still absent information about the two normal stress functions of another kind of viscometric flow. Thus a torsional flow is suitable in particular where the test fluid is contained between two parallel discs, one of which rotates (Fig. 1.11). In this case too the pressure on the walls ($-\sigma_{zz}$) increases towards the axis of rotation, and of course has a finite value there, in contrast to the cone-and-plate arrangement. The reader is left to satisfy himself that for the normal stress $\sigma_{zz}(r)$ the following formula applies:

$$r\frac{d\sigma_{zz}}{dr} = \dot{\gamma}\,\frac{dN_2(\dot{\gamma})}{d\dot{\gamma}} + N_1(\dot{\gamma}) + N_2(\dot{\gamma}) \tag{4.8}$$

In the derivation of this relationship the inertia of the fluid and the force of gravity have once again been neglected (horizontal discs). Hence the local shear rate $\dot{\gamma}$ depends in a simple way on the difference in angular velocity $\Delta\omega$ of the two discs, on their distance apart h and the distance from the axis r: $\dot{\gamma} = r\Delta\omega/h$ (cf. Section 1.6.5). Equation (4.8) gives the relationship between the pressure on the discs and the two normal stress functions. It permits on the one hand the calculation of the pressure distribution for known normal stress properties. On the other hand it gives information about the functions $N_1(\dot{\gamma})$ and $N_2(\dot{\gamma})$ when the pressure distribution $-\sigma_{zz}(r)$ is taken as a measured quantity. One can transform equation (4.8) into

$$N_2(\dot{\gamma}) = r\int_0^r \frac{1}{r'}\,\frac{d\sigma_{zz}(r')}{dr'}\,dr' - \dot{\gamma}\int_0^{\dot{\gamma}} \frac{N_1(\kappa) + 2N_2(\kappa)}{\kappa^2}\,d\kappa \tag{4.9}$$

If one has determined the function $N_1(\dot{\gamma}) + 2N_2(\dot{\gamma})$ as described above by the use of a cone-and-plate device and has recorded the pressure distribution $-\sigma_{zz}(r)$ on the walls in the parallel disc device, then the second normal stress function can obviously be calculated from equation (4.9). Note that here $\dot{\gamma} = r\Delta\omega/h$. Therefore the two devices form a suitable rheometer for determining both the normal stress functions. Because no assumptions about the stresses at the edge of the discs (for $r = r_0$) in equations (4.8) and (4.9) are involved, it is not necessary to limit the fluid to the space between the discs. In order to reduce troublesome boundary effects it is of course advantageous to keep the length ratio h/r_0 as small as possible.

4.2 Weissenberg effect

We now turn to a phenomenon which occurs on the free surface of a fluid. This is the situation in the simplest case: the fluid is in a wide container between a vertical cylinder and the coaxial container wall, but only occupies the space up to a certain height, and above that there is air at pressure p_0. In the state of rest the free fluid surface is plane if the effect of capillary forces is neglected. But if the cylindrical rod is rotated at constant angular velocity about its axis then the fluid starts to move and the surface no longer remains plane. For Newtonian fluids the surface of the fluid sinks on the inside and climbs outwards. With non-Newtonian fluids one observes exactly the opposite in certain cases: the fluid reaches its highest level at the stirrer, and the surface falls outwards (Fig. 4.2).

This is called the *Weissenberg effect* and is caused by the normal stress differences. In order to make this clear we imagine in passing that the fluid is to have no free surface, but is to fill completely the annular space between the rotating rod and the stationary container. In these conditions a Couette flow will develop, in which all the material points move round the axis of rotation in circles. The angular velocity ω thus depends only on the distance from the axis r, and for the shear rate $\dot{\gamma} = r d\omega/dr$ applies (cf. Section 1.6.3). Thus on the one hand there is the inertia force $\rho r \omega^2$ acting radially outwards on the fluid particles; on the other hand because this is a viscometric flow two normal stress differences occur, namely $\tau_{\varphi\varphi} - \tau_{rr} = N_1(\dot{\gamma})$, and $\tau_{rr} - \tau_{zz} = N_2(\dot{\gamma})$. If we identify the negative normal stress in the z-direction as the pressure p, make the associated extra-stress equal to zero ($\tau_{zz} = 0$), then the two normal stress differences on an element of volume of size $r\,\Delta\varphi\,\Delta r\,\Delta z$ can be viewed as forces that are to be superposed on the forces caused by the pressure. The sum $N_1 + N_2$ corresponds to a tensile force in the circumferential direction, N_2 to a tension in the radial direction. Therefore four additional normal forces act on the element of volume besides the pressure forces (Fig. 4.3). They can be combined into one radial force. Dividing by the volume of the element one obtains a radial volume force

$$\frac{1}{r}\left[-(N_1+N_2)+\frac{d}{dr}(rN_2)\right]\mathbf{e}_r,$$

which adds to the centrifugal force $\rho r\omega^2\mathbf{e}_r$.

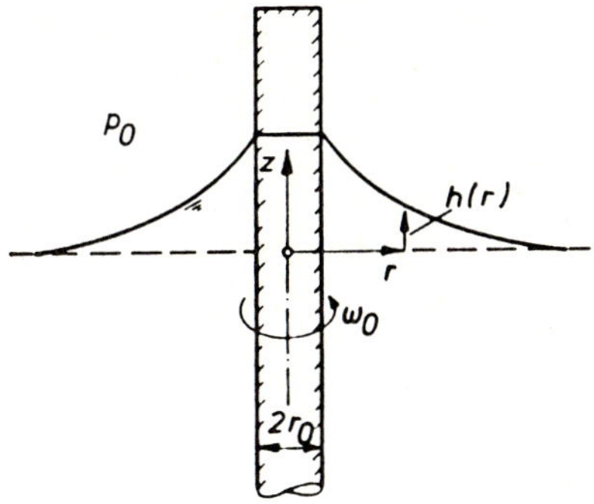

Fig. 4.2 Weissenberg effect at a rotating bar

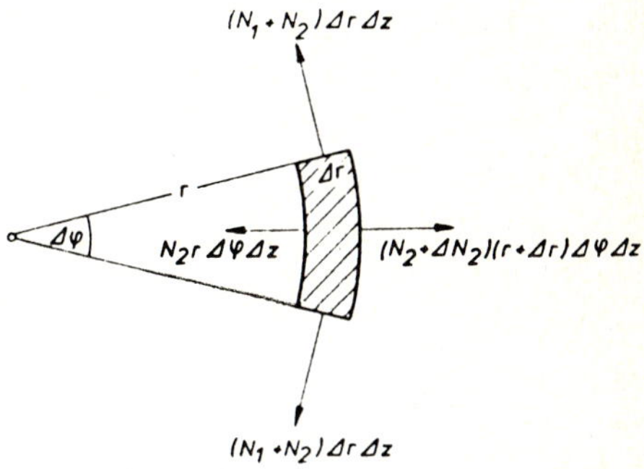

Fig. 4.3 Normal stress differences at an element

Because the fluid particles move neither in the radial, nor in the axial direction, these volume force fields are compensated together with the force of gravity in the negative z-direction by a pressure gradient:

$$\text{grad } p = \left[\rho r\omega^2 - \frac{N_1}{r} + \frac{dN_2}{dr}\right]\mathbf{e}_r - \rho g\mathbf{e}_z \tag{4.10}$$

Hence the pressure varies in radial direction, and it is therefore clear that a plane z = constant cannot be a free surface on which the pressure must be constant. When therefore the annular space is only partly filled with fluid, then a curved surface is immediately established. Thereby the flow is altered in the vicinity of the surface, i.e. on the viscometric undisturbed flow there is superposed an additional motion.

Now it is a peculiarity of the cylindrical device considered here that this perturbation motion does not become noticeable at a low angular velocity ω_0 of the rotating rod. To be more accurate: in the framework of a second order theory in which only the linear and quadratic terms in ω_0 are considered, there arises a displacement of the surface of magnitude $O(\omega_0^2)$, but the movement of the fluid is the same shear flow as for the completely filled space. Hence for slow rotary movement the equation (4.10) also applies when the fluid is bounded by a free surface. Because in this approximation the flow law is approximated by a linear relationship ($\tau = \eta_0\dot{\gamma}$), this is the flow field of a Newtonian fluid. If the external boundary is at infinity, then the angular velocity in particular depends in a simple way on the axial distance r and the two constants ω_0 and r_0 (but not on η_0): $\omega = \omega_0 r_0^2/r^2$ (potential vortex). If one now replaces the normal stress functions $N_1(\dot{\gamma})$ and $N_2(\dot{\gamma})$ by the quadratic expressions $\nu_{10}\dot{\gamma}^2$ and $\nu_{20}\dot{\gamma}^2$ respectively (equations (2.28) - (2.30)) and in this replaces $\dot{\gamma}$ by $r d\omega/dr = -2\omega_0^2 r_0^2/r^2$, then equation (4.10) transforms into

$$\operatorname{grad} p = \left[\rho\omega_0^2 \frac{r_0^4}{r^3} - 4(\nu_{10} + 4\nu_{20})\omega_0^2 \frac{r_0^4}{r^5}\right] \mathbf{e}_r - \rho g \mathbf{e}_z \tag{4.11}$$

Because the pressure on the surface of the fluid is constant the vector grad p must be normal to the surface and therefore parallel to the surface normals $\mathbf{n} \sim \mathbf{e}_z - (dh/dr)\mathbf{e}_r$. Here h(r) is the rise in the surface of the fluid above the undisturbed level; cf. Fig. 4.2. Hence it follows that the conditional equation

$$\frac{dh}{dr} = \frac{\omega_0^2}{\rho g}\left[\rho \frac{r_0^4}{r^3} - 4(\nu_{10} + 4\nu_{20}) \frac{r_0^4}{r^5}\right] \tag{4.12}$$

describes the shape of the free surface. One obtains by integrating

$$\frac{h(r)}{r_0} = \frac{r_0\omega_0^2}{g}\left[-\frac{1}{2} + \frac{\nu_{10} + 4\nu_{20}}{\rho r^2}\right]\left(\frac{r_0}{r}\right)^2 \tag{4.13}$$

The different signs of the two terms indicate that the inertia and normal stresses are acting in opposite directions. For a sufficiently thin rod the second term is predominant on the inside. The fluid therefore climbs against the force of gravity at the rotating rod and reaches its highest position there as long as $\nu_{10} + 4\nu_{20} > 0$ (Weissenberg effect). The critical rod radius $r_* = \sqrt{2(\nu_{10} + 4\nu_{20})/\rho}$, which results from equating to zero the terms within square brackets in equation (4.13) depends remarkably only on the constitutive constants, and turns out to be the greater the greater the normal stress coefficients are. Equation (4.13) shows that the Weissenberg effect occurs only for comparatively thin rods ($r_0 < r_*$). If $r_0 > r_*$, then the surface of the fluid qualitatively behaves like a fluid that does not have normal stress properties. It lies totally below the reference level and increases monotonically for $r > \sqrt{2}r_*$.

The analytical result (4.13) can in principle serve to determine the constant coefficient of the second order $\nu_{10} + 4\nu_{20}$. The two normal stress coefficients ν_{10} and ν_{20} cannot in fact be distinguished from each other in this way. The measurement would be of the displacement of the surface of the fluid, particularly the height of the fluid $h(r_0)$ immediately adjacent to the rod for given values of r_0 and ω_0, or of the position of the zero intercept r_*. The formula then obviously allows the calculation of the quantity $(\nu_{10} + 4\nu_{20})/\rho$.

The rise of the free surface of a liquid because of normal stress differences is naturally not only observed for long circular section cylindrical stirrers, and it is also not necessary for the rotating boundary to project out of the fluid. Thus one can also observe in non-Newtonian fluids for example an 'upwelling' near the axis of rotation when the stirrer is reduced to a disc lying on the bottom of the container, and rotated say by electromagnetic remote control (Fig. 4.4).

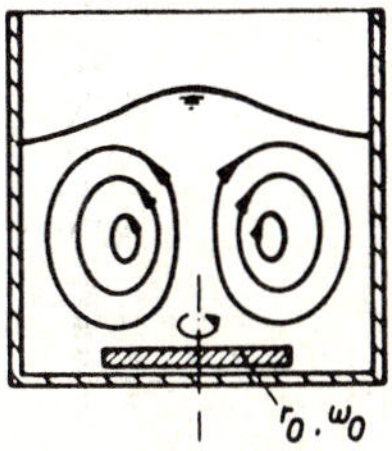

Fig. 4.4 'Upwelling effect'

The principal difference from the previously described situation consists of the fact that already within a second order theory a 'secondary flow' occurs in the meridian plane in the form of one or more vortices which have an important effect on the shape of the free surface. Because of this, any accurate calculation of the shape of the free surface is considerably more difficult. For a characteristic rise of the surface of the fluid above the undisturbed level, particularly for the fluid level h(0) at the axis of rotation the corresponding however applies as previously: h(0) increases for slow movement as the square of the rotation speed of the stirrer and is composed of the sum of three terms, two of which contain the normal stress coefficients, and one describes the effect of the inertia. Altogether, h(0) is affected by the quantities ν_{10}, ν_{20}, ρ, g, ω_0, and three lengths, namely the radius of the container, the height of the undisturbed fluid level, and the radius r_0 of the rotating disc, but not the zero-shear viscosity η_0. The following relationship results from dimensional analysis and the previous statements:

$$\frac{h(0)}{r_0} = \frac{r_0\omega_0^2}{g}\left[-a + \frac{b_1\nu_{10} + b_2\nu_{20}}{\rho r_0^2}\right] \qquad (4.14)$$

The three quantities a, b_1 and b_2 depend on the two length ratios which are made up from the height of the fluid level, the radius of the container, and the radius of the stirrer, but cannot be determined analytically here in contrast to the situation described at the beginning.

4.3 Die-swell

This is another normal stress effect which occurs when a pseudoplastic fluid emerges as a free stream from a die or from the nozzle of a cylindrical pipe. The phenomenon appears thus: the stream of a non-Newtonian fluid emerging downwards from a vertical nozzle under certain conditions first expands considerably before it contracts again under the influence of gravity. Essentially two patterns are observed: at relatively low flow rates the swell begins immediately below the nozzle, so that the jet boundary stream lines at the nozzle exhibit a distinct kink (Fig. 4.5a). Because the free-falling fluid is accelerated by the force of gravity and the cross-section of the stream therefore later contracts, a situation develops which has a form rather reminiscent of an onion. One can therefore appropriately refer to this phenomenon as 'onion formation' in the free jet. For a relatively large flow rate the fluid flows smoothly out of the pipe and the 'onion' does not develop until some distance beyond the nozzle of the pipe (Fig. 4.5b).

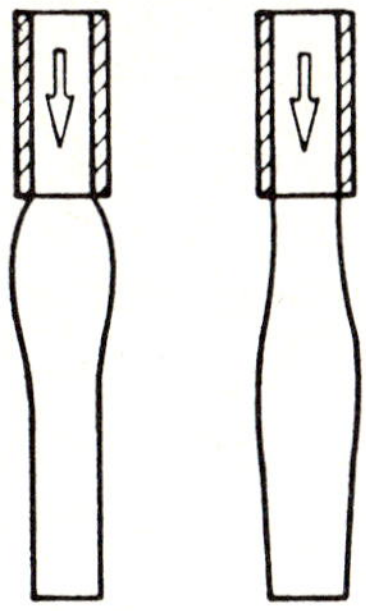

Fig. 4.5 Die-swell: (a) attached swell, (b) detached swell

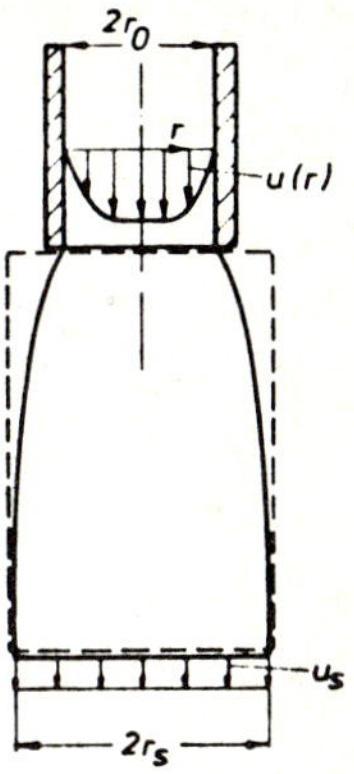

Fig. 4.6 Control volume and notations

Because the force of gravity is certainly not the cause of the expansion of the jet, it will be totally disregarded from the start, i.e. an idealised process will be considered, by which the jet after emerging from the pipe widens to a certain diameter which then stays constant downstream, and at the same time the velocity differences over the cross-section diminish. The aim of the following statements is to combine the flow widening with the normal stress differences in the pipe, and thus explain the phenomenon quantitatively. To this end the law of conservation of mass and momentum are formulated for the control volume sketched in Fig. 4.6. Because the flow is steady and the fluid is regarded as having constant density, the conservation of mass leads to

$$\pi r_0^2 \bar{u} = \pi r_s^2 u_s \tag{4.15}$$

In this $\bar{u}$ indicates the mean value over the cross-section of the nozzle, hence in the present case the mean value of the axial velocity in the pipe:

$$\bar{u} := \frac{1}{\pi r_0^2} 2\pi \int_0^{r_0} r\, u(r)\, dr \tag{4.16}$$

The meaning of the other symbols is shown in Fig. 4.6. According to the law of momentum the difference between the momentum flux emerging from the control volume and the entering momentum flux is equal to the resultant force on the surfaces of the control volume. Because the fluid after the flow widening flows without velocity differences (like a rigid body), no extra-stresses occur there. In addition the free boundary of a jet emerging into a gas of low viscosity, say air, is free of stress. The constant ambient pressure can be equated to zero without limitation of the generality. If finally capillary forces are neglected, then obviously only the axial stresses in the jet cross-section contribute to the resultant force on the control volume, thus:

$$\rho\pi r_s^2 u_s^2 - \rho\pi r_0^2 \overline{u^2} = -\pi r_0^2 \overline{\sigma_{zz}} \tag{4.17}$$

(σ_{zz} is the axial normal stress and therefore is an abbreviation for $-p + \tau_{zz}$). If one eliminates the jet velocity u_s from the equations (4.15) and (4.17), then one obtains the rigorous relationship:

$$\frac{r_0^2}{r_s^2}\bar{u}^2 - \overline{u^2} = -\frac{\overline{\sigma_{zz}}}{\rho} \tag{4.18}$$

We shall now evaluate this expression, assuming that Poiseuille flow is present in the nozzle cross-section. This certainly presumes that the pipe from which the fluid emerges is sufficiently long, so that inlet effects are absent and the flow can as a whole develop fully. However, even with a long pipe it cannot be expected that a viscometric flow will occur in the outlet cross-section. In effect the changed conditions behind the nozzle have a backward effect in the capillary, and the flow there will change more or less. Because however the fluid up to the nozzle adheres to the wall and the flow velocity ought to increase monotonically to the axis, the velocity profile is not very much different from that which occurs in the fully developed pipe flow. To that extent the above assumption can be accepted. It implies that the axial normal stress is connected with the two normal stress functions $N_1(\dot{\gamma})$ and $N_2(\dot{\gamma})$, when in the present case du/dr must be written for $\dot{\gamma}$. The equation of motion in the radial direction thus reduces for a Poiseuille flow to

$$\frac{\partial(-p + \tau_{rr})}{\partial r} + \frac{1}{r}(\tau_{rr} - \tau_{\varphi\varphi}) = 0 \tag{4.19}$$

(see Formula Appendix). Hence the difference $\tau_{rr} - \tau_{\varphi\varphi}$ corresponds to the second normal stress difference $N_2(\dot{\gamma})$. It is expedient here to identify the negative normal stress in the radial direction with the pressure p, hence writing $\tau_{rr} = 0$. One immediately obtains by integrating equation (4.19)

$$p(r) = p_w + \int_{r_0}^{r} \frac{N_2}{r}\, dr \tag{4.20}$$

The symbol p_w denotes the pressure on the pipe wall immediately in front of the nozzle, which is indicated via a pressure transducer located flush within the wall. If one now uses the expression $\tau_{zz} - \tau_{rr} = N_1$, which can also be written in the form $\sigma_{zz} + p = N_1$, then one obtains the following relationship between the axial normal stress σ_{zz} and the functions N_1 and N_2:

$$\sigma_{zz}(r) = -p_w + N_1 - \int_{r_0}^{r} \frac{N_2}{r}\, dr \tag{4.21}$$

From this follows the normal stress in the axial direction averaged over the outlet cross-section

$$\overline{\sigma_{zz}} = -p_w + \overline{N_1} + \frac{1}{2}\overline{N_2} \tag{4.22}$$

Hence equation (4.18) transforms to

$$\frac{r_0^2}{r_s^2} = \frac{\overline{u^2}}{\bar{u}^2} - \frac{\overline{N_1} + \frac{1}{2}\overline{N_2} - p_w}{\rho\, \bar{u}^2} \tag{4.23}$$

According to this relationship the widening ratio r_s^2/r_0^2 of the free jet depends on the velocity profile u(r) of the developed Poiseuille flow and hence on the viscometric flow properties of the fluid, but also on the two other viscometric functions $N_1(\dot{\gamma})$ and $N_2(\dot{\gamma})$. Note that $\dot{\gamma} = du/dr$ holds. In addition, there enter into the result the density of the fluid and the wall pressure at the nozzle of the pipe.

In the case of the detached swell for a smoothly flowing jet (Fig. 4.5b) one can equate the pressure on the wall with the external pressure, so one can write $p_w = 0$. Hence flow widening $(r_s > r_0)$ occurs according to equation (4.23) only when $\overline{N_1} + \frac{1}{2}\overline{N_2} > \rho\,(\overline{u^2} - \bar{u}^2) = \rho\,\overline{(u - \bar{u})^2} > 0$, i.e. when the mean value of the controlling normal stress function $N_1 + \frac{1}{2}N_2$ exceeds a certain positive value. It is instructive to see what this signifies on the basis of a hypothetical fluid with Newtonian flow properties and quadratic normal stress properties ($N_1 = \nu_{10}\dot{\gamma}^2$, $N_2 = \nu_{20}\dot{\gamma}^2$). The velocity profile of the Poiseuille flow would of course be parabolic in this case, $u(r) = 2\bar{u}(1 - r^2/r_0^2)$. From this follows $\overline{u^2} = \frac{4}{3}\bar{u}^2$, and $\overline{\dot{\gamma}^2} = \overline{(du/dr)^2} = 8\bar{u}^2/r_0^2$. Hence equation (4.23) can be written as

$$\frac{r_0^2}{r_s^2} = \frac{4}{3} - \frac{4(2\nu_{10} + \nu_{20})}{\rho\, r_0^2} \tag{4.24}$$

This expression shows that the effect of the die-swell only appears with sufficiently narrow capillaries. To that extent a relationship exists with the Weissenberg effect, which for its part only occurs for sufficiently thin rods. The critical radius $r_* = \sqrt{12(2\nu_{10} + \nu_{20})/\rho}$ depends once more only on the material constants ν_{10}/ρ and ν_{20}/ρ.

For an adjacent swell with a strongly kinked wall stream line (Fig. 4.5a) it would be doubtful to assume once again that $\tau_w = 0$ holds. Because in this case the velocities are relatively small, there is reason to neglect in the momentum equation the terms multiplied by the square of the flow velocity, therefore to replace equation (4.17) by $\overline{\sigma_{zz}} = 0$ (creeping flow). If we further assume that in the nozzle cross-section there is also a developed Poiseuille flow, then we come to the following relationship

$$p_w = \overline{N_1} + \frac{1}{2}\overline{N_2} = \frac{2}{r_0^2}\int_0^{r_0} (N_1 + \frac{1}{2}N_2)\, r\, dr \tag{4.25}$$

Because here quantities defined in the nozzle cross-section only occur and there is a lack of any information about the expanded cross-section, the swelling behaviour of the cross-section area cannot of course be obtained in this way. The question therefore arises about what can be done with this relation. If one replaces the coordinates r by the local shear stress τ, by means of the relationship $\tau = \tau_w r/r_0$ which applies to a fully developed steady pipe flow, then equation (4.25) can be written as

$$p_w = \frac{2}{\tau_w^2}\int_0^{\tau_w} (N_1 + \frac{1}{2}N_2)\, \tau\, d\tau \tag{4.26}$$

Thus one should note that N_1 and N_2 strictly speaking depend on $\dot{\gamma}$, but can just as well be looked on as functions of τ if one takes into account the relationship between $\dot{\gamma}$ and τ, i.e. the flow law.

Equation (4.26) states that the wall pressure p_w at the pipe nozzle is related to the wall shear stress τ_w, where in this relation also the normal stress functions of the fluid enter. Equation (4.26) therefore allows one, by measurement of the wall pressure at given wall stress (i.e. at given pressure drop and given radius of the capillary, cf. equation (3.114)), to arrive at the quantities $N_1 + \frac{1}{2}N_2$. To that extent the events at the outlet of a capillary are also of interest from the rheometric point of view. In the evaluation one records p_w under various conditions for τ_w (variable pressure drop in the pipe). If one takes τ_w as the variable, then equation (4.26) can be solved for the quantity

$$\left(N_1 + \frac{1}{2} N_2\right)_w = p_w + \frac{1}{2} \tau_w \frac{dp_w}{d\tau_w} \tag{4.27}$$

One therefore records the measured quantities p_w against τ_w, obtains from this graph the derivative $dp_w/d\tau_w$ and then calculates using (4.27) the value of the normal stress difference $N_1 + \frac{1}{2}N_2$ belonging to τ_w. With the help of the flow function of the fluid, which is assumed to be known, one obtains the value of $\dot{\gamma}_w$ pertaining to τ_w, and thus obtains the sum of the normal stress functions in the desired form as a function of the shear rate.

An analogous consideration for a slit rheometer leads to the formula

$$N_{1,w} = p_w + \tau_w \frac{dp_w}{d\tau_w} \tag{4.28}$$

(cf. Problem 4.2). The evaluation of the wall pressure measurements accordingly leads to the first normal stress function N_1. The phenomenon of die-swell can become troublesome during the manufacture of synthetic fibres or in die-casting. The flow extruded from a rectangular die does not in fact have a rectangular shape, but has a pin-cushion cross-section shape. If the cross-section of the extrudate is to be as desired, then the nozzle of the die must be correspondingly smaller and made of the correct shape to compensate for this die-swell effect. In the case of a rectangular section extrudate a die with inward curved sides would be required (Fig. 4.7).

Fig. 4.7 Nozzle section for rectangular extrusion

4.4 Axial shear flow

When a Newtonian fluid performs a laminar flow through a straight pipe of arbitrary cross-section under the action of an axial pressure gradient, then the pressure is constant across the plane of the cross-section, and each material point moves parallel to the axis of the pipe (in the x-direction), so that the velocity depends on the position of the point in the plane of the cross-section. We now ask whether or under what conditions the steady flow of an incompressible fluid arises within a cylindrical pipe of arbitrary cross-section in the same way. We therefore consider as a basis the velocity field

$$\mathbf{v} = u(y, z)\mathbf{e}_x \tag{4.29}$$

in which y and z are coordinates in the plane of the cross-section; $\mathbf{e}_x$ is the unit vector in the axial direction. The continuity equation div $\mathbf{v}$ = 0 for constant density fluids is therefore obeyed. The constant velocity surfaces u(y,z) = constant form a set of cylindrical layers, which slide like the tubes of a telescope in the axial direction. One therefore calls it an axial shear flow. It is viscometric, because any chosen fluid particle experiences at all times a shear with constant shear rate

$$\dot{\gamma} = |\operatorname{grad} u| = \sqrt{\left(\frac{\partial u}{\partial y}\right)^2 + \left(\frac{\partial u}{\partial z}\right)^2} \tag{4.30}$$

With reference to the localised 'natural' basis shown in Fig. 4.8 with the unit vectors $\mathbf{e}_1 = \mathbf{e}_x$, $\mathbf{e}_2 = \operatorname{grad} u/|\operatorname{grad} u|$, $\mathbf{e}_3 = \mathbf{e}_1 \times \mathbf{e}_2$ the stress tensor therefore has the form (2.34). Because Cartesian co-ordinates will be used, we more conveniently use the coordinate invariant form (2.36) of the general equation for viscometric flows, and in this way obtain the following matrix form of the extra-stress tensor referred to the basis with the fixed constant unit vectors $\mathbf{e}_x$, $\mathbf{e}_y$ and

e_z:

$$\mathbf{T} = \begin{bmatrix} N_1(\dot\gamma) + N_2(\dot\gamma) & \eta(\dot\gamma)\dfrac{\partial u}{\partial y} & \eta(\dot\gamma)\dfrac{\partial u}{\partial z} \\ \eta(\dot\gamma)\dfrac{\partial u}{\partial y} & \nu_2(\dot\gamma)\left(\dfrac{\partial u}{\partial y}\right)^2 & \nu_2(\dot\gamma)\dfrac{\partial u}{\partial y}\dfrac{\partial u}{\partial z} \\ \eta(\dot\gamma)\dfrac{\partial u}{\partial z} & \nu_2(\dot\gamma)\dfrac{\partial u}{\partial y}\dfrac{\partial u}{\partial z} & \nu_2(\dot\gamma)\left(\dfrac{\partial u}{\partial z}\right)^2 \end{bmatrix} \tag{4.31}$$

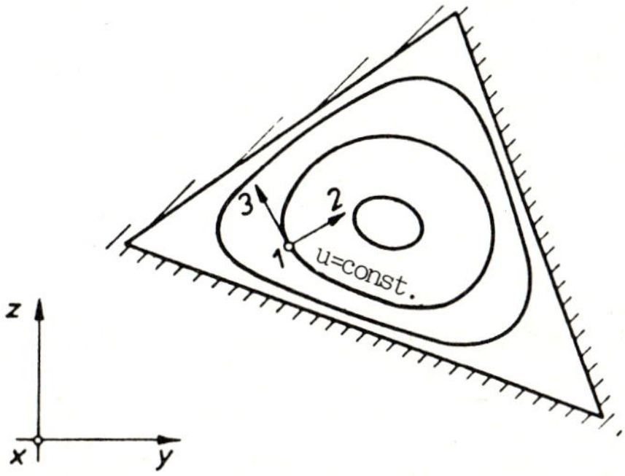

Fig. 4.8 Axial shear flow; Cartesian and local natural basis

One sees that the extra-stresses, like the flow rate, depend only on y and z. Because this is a motion without acceleration the equations of motion reduce to div **T** = grad p (cf. equation (1.116)) if volume forces are neglected. It immediately follows from this that the pressure p can vary only linearly with the coordinate in the axial direction

$$p = -kx + p'(y, z) \tag{4.32}$$

Thus the constant $k[N/m^3]$ defines the axial pressure gradient in the pipe (impressed from outside). Now the three equations of motion in full are as follows:

$$\frac{\partial}{\partial y}\left[\eta\frac{\partial u}{\partial y}\right] + \frac{\partial}{\partial z}\left[\eta\frac{\partial u}{\partial z}\right] = -k \tag{4.33}$$

$$\frac{\partial}{\partial y}\left[\nu_2\left(\frac{\partial u}{\partial y}\right)^2\right] + \frac{\partial}{\partial z}\left[\nu_2\frac{\partial u}{\partial y}\frac{\partial u}{\partial z}\right] = \frac{\partial p'}{\partial y} \tag{4.34}$$

$$\frac{\partial}{\partial y}\left[\nu_2\frac{\partial u}{\partial y}\frac{\partial u}{\partial z}\right] + \frac{\partial}{\partial z}\left[\nu_2\left(\frac{\partial u}{\partial z}\right)^2\right] = \frac{\partial p'}{\partial z} \tag{4.35}$$

Note that the first normal stress function $N_1(\dot{\gamma})$ has been completely eliminated. For a given force density k (and meaningful given boundary conditions) equation (4.33) expresses the conditional equation for the velocity field u(y,z). It is a non-homogeneous second order partial differential equation of the elliptical type. For Newtonian fluids (η = constant) the left-hand side is linear in the required function u ('Poisson equation'). For non-Newtonian flow properties there is a non-linear differential equation, because η depends on $\dot{\gamma} = \sqrt{(\partial u/\partial y)^2 + (\partial u/\partial z)^2}$. The velocity field u(y,z) is therefore affected by the flow properties, but not by the normal stress properties of the fluid.

If one eliminates the pressure from the two other equations of motion one then obtains

$$\left(\frac{\partial^2}{\partial y^2} - \frac{\partial^2}{\partial z^2}\right)\left[\nu_2 \frac{\partial u}{\partial y}\frac{\partial u}{\partial z}\right] - \frac{\partial^2}{\partial y \partial z}\left[\nu_2\left(\frac{\partial u}{\partial y} - \frac{\partial u}{\partial z}\right)\left(\frac{\partial u}{\partial y} + \frac{\partial u}{\partial z}\right)\right] = 0 \qquad (4.36)$$

The assumed flow form can obviously only come about when the solution of equation (4.33) satisfies this condition. Otherwise a uniaxial shear flow would not be possible. The secondary condition (4.36) is obviously trivial for $\nu_2 = 0$, hence for fluids with a vanishing second normal stress function and in fact independent of the properties of the two other viscometric functions. The condition is in consequence satisfied for any $\nu_2(\dot{\gamma})$ only by certain cross-section forms, particularly by pipes of circular or annular cross-section. For an arbitrary cross-section shape equations (4.33) and (4.36) are generally in disagreement, unless the two material functions $\eta(\dot{\gamma})$ and $\nu_2(\dot{\gamma})$ are proportional to each other. One can of course by use of equation (4.33) alone obtain the relationship analogous to the secondary conditions

$$\left(\frac{\partial^2}{\partial y^2} - \frac{\partial^2}{\partial z^2}\right)\left[\eta \frac{\partial u}{\partial y}\frac{\partial u}{\partial z}\right] - \frac{\partial^2}{\partial y \partial z}\left[\eta\left(\frac{\partial u}{\partial y} - \frac{\partial u}{\partial z}\right)\left(\frac{\partial u}{\partial y} + \frac{\partial u}{\partial z}\right)\right] = 0 \qquad (4.37)$$

To derive this first multiply both sides of equation (4.33) by the vector grad u and then add the expression $(\eta/2)$ grad $\dot{\gamma}^2$, which one can express by use of the potential $\Omega(\dot{\gamma}) := \int_0^{\dot{\gamma}} \eta(\dot{\gamma})\dot{\gamma}\, d\dot{\gamma}$, also in the form grad Ω. If one now performs the rotation, the right-hand side vanishes and equation (4.37) results. For fluids that have the property

$$\nu_2(\dot{\gamma}) = c\,\eta(\dot{\gamma}) \tag{4.38}$$

the equations (4.36) and (4.37) are obviously identical, hence the secondary condition is fulfilled. Hence under these circumstances for any shape of pipe cross-section an axial shear flow does in fact occur. The truth of this statement rests most of all on the direct consequence that with sufficiently small axial pressure gradients in any simple fluid an axial shear flow will establish itself. For a slow flow the shear rate $\dot{\gamma}$ of course remains low in such a way that the functions $\eta(\dot{\gamma})$ and $\nu_2(\dot{\gamma})$ can be approximated by the first (absolute) terms of their Taylor expansions for $\dot{\gamma} = 0$, which are the zero values η_0 and ν_{20} defined by equations (2.6) and (2.30) respectively. If the temperature of the fluid is constant everywhere, then η_0 and ν_{20} are constant quantities, and the condition (4.38) is obviously fulfilled.

For a large pressure gradient not only the constant initial values but the wider properties of the two functions are involved. Because in general the necessary relationship in equation (4.38) does not exist, an axial shear flow is no longer possible, and additional to the motion in the axial direction there occurs a secondary flow in the plane of the cross-section. However, such a secondary flow can also occur in a second order fluid (when η and ν_2 are not dependent on $\dot{\gamma}$), when marked temperature differences exist in the fluid.

The temperature dependence of the quantities η_0 and ν_{20} are related to each other by equation (2.33), so that the condition (4.38) is once more infringed. In Section 9.4 we consider these secondary flows in more detail. Figs. 9.12 and 9.14 show flow line patterns of such secondary flows. Here the main intention has been to examine why such phenomena must necessarily occur because of normal stress differences.

Problems

4.1 There is an incompressible simple fluid which has the viscometric functions $F(\dot{\gamma})$, $N_1(\dot{\gamma})$, $N_2(\dot{\gamma})$ located between a fixed and a similar parallel disc (radius r_0, distance apart h) which rotates round the vertical axis with an angular velocity Ω.

Prove that, starting from the equations of motion in cylindrical coordinates and neglecting the inertia (creeping flow) between the discs, a torsional flow can occur which has the velocity field (1.91). First find $\omega(z)$. Then calculate the components of the stress tensor $-p + \tau_{rr}$, $\tau_{r\varphi}$, etc. as functions of r and z under the condition that the normal stress in a radial direction vanishes at the boundary, i.e. $p - \tau_{rr} = 0$ for $r = r_0$. With what torque must the rotating disc be driven in order to maintain the motion? How much axial force must be applied to the discs so that they do not separate from each other? How do the general results simplify in the case of a fluid of constant viscosity and constant normal stress coefficients ($F(\dot{\gamma}) = \eta_0\dot{\gamma}$, $N_1(\dot{\gamma}) = \nu_{10}\dot{\gamma}^2$, $N_2(\dot{\gamma}) = \nu_{20}\dot{\gamma}^2$)?

4.2 Consider the exit of a non-Newtonian fluid from a plane channel (slit nozzle). Assuming that in the channel up to the outlet a plane shear flow exists, that the momentum flux of the free jet is negligibly small and that gravity and surface tension forces play no part, deduce a relationship between the wall pressure at the outlet, the wall shear stress, and the first normal stress difference. Discuss how the relationship can be applied to the determination of $N_1(\dot{\gamma})$.

4.3 Consider the Couette flow between two coaxial cylinders of radius r_1 and r_2 ($>r_1$), for which the inner cylinder is stationary and the outer one rotates at an angular velocity ω_2. The gap is narrow, i.e. $r_2 - r_1 \ll r_1$, and one can therefore assume that the angular velocity of the fluid in the gap is a linear function of the distance from the axis. Show that the first normal stress function of the fluid is related to the normal stresses at the cylinders by the expression

$$N_1(\dot{\gamma}) = \frac{\sigma_{rr}(r_2) - \sigma_{rr}(r_1)}{r_2 - r_1}\, r_1 + \frac{1}{3}\rho\, r_1^2\omega_2^2$$

in which $\dot{\gamma} = r_1\omega_2/(r_2 - r_1)$. By measuring the pressure difference at the two cylinders one can therefore determine N_1.

5 SIMPLE UNSTEADY FLOWS

We now consider unsteady flows of incompressible fluids, and firstly restrict ourselves to the simplest case of a plane shear flow with the velocity field $\mathbf{v} = u(y,t)\mathbf{e}_x$. For a simple fluid the extra-stresses on a particle at time t depend, as has already been explained in the introduction to Section 2, on the shear rate at which the particle was deformed at an earlier time t - s ($s \geqq 0$). Because the particle is at all times located in a plane y = constant, the statement leads in particular to the already known relationship

$$\tau(y,t) = \mathop{\mathrm{F}}_{s=0}^{\infty} [\dot{\gamma}(y, t-s)] \tag{2.3}$$

for the shear stress between the layers in the direction of movement. Because with an isochoric shear flow two more extra-stresses, namely the normal stress differences $\tau_{xx} - \tau_{yy}$ and $\tau_{yy} - \tau_{zz}$ occur, two constitutive equations arise:

$$\tau_{xx}(y,t) - \tau_{yy}(y,t) = \mathop{\mathrm{N}_1}_{s=0}^{\infty} [\dot{\gamma}(y, t-s)] \tag{5.1}$$

$$\tau_{yy}(y,t) - \tau_{zz}(y,t) = \mathop{\mathrm{N}_2}_{s=0}^{\infty} [\dot{\gamma}(y, t-s)] \tag{5.2}$$

These three expressions describe the friction behaviour of a simple fluid under unsteady shear flow conditions.

The three *functionals* on the right-hand sides are at first unknown for an actual fluid. Experimental rheology aims to determine these properties over wide areas of application by measuring the extra-stresses in suitable unsteady shear flows. One should of course clearly understand that on principle there should be an unlimited number of experiments in order to determine completely the three functionals for a given fluid. Thus for motions of various kinds, say with a purely oscillatory or a monotonically increasing deformation, obviously different extra-stresses and hence different properties of the functionals become apparent.

Also for motions which in the present proceed in the same way, but which in the past differed from each other, a material with a memory reacts with different stresses. Consider, say, two shear flows, one of which was always steady, but the other was not 'switched on' until a certain time. To that extent every imaginable shear flow can only provide a part of that information which the three functionals as a whole contain for a certain material. However the functionals may appear in the concrete case, they always have the property

$$\overset{\infty}{\underset{s=0}{F}}[0] = \overset{\infty}{\underset{s=0}{N_1}}[0] = \overset{\infty}{\underset{s=0}{N_2}}[0] = 0 \tag{5.3}$$

because the extra-stresses vanish in a fluid which has never been sheared ($\dot{\gamma} \equiv 0$). In addition, the following symmetry conditions apply because for flow reversal ($u \to -u$ and hence $\dot{\gamma} \to -\dot{\gamma}$) the shear stress obviously changes its sign, but the normal stresses remain unaltered:

$$\overset{\infty}{\underset{s=0}{F}}[-\dot{\gamma}(t-s)] = -\overset{\infty}{\underset{s=0}{F}}[\dot{\gamma}(t-s)] \tag{5.4}$$

$$\overset{\infty}{\underset{s=0}{N_{1,2}}}[-\dot{\gamma}(t-s)] = \overset{\infty}{\underset{s=0}{N_{1,2}}}[\dot{\gamma}(t-s)] \tag{5.5}$$

We shall incidentally suppress the unimportant argument y here.

5.1 Linear viscoelasticity

If in an unsteady shear flow the shear rate at every place and at all times remains small, then the right-hand sides of equations (2.3), (5.1) and (5.2) can be approximated by linear expressions in $\dot{\gamma}(y,t-s)$. Absolute terms are then absent because of the property expressed by (5.3). In the case of the normal stress differences the linear term too is absent because of the symmetry requirements (5.5), so that within the framework of a linear theory the normal stresses are of the same magnitude in all three directions, $\tau_{xx} = \tau_{yy} = \tau_{zz}$. The occurrence of different normal stresses is therefore as for steady shear flows a non-linear phenomenon and is not considered initially. Therefore there remains only the linear version of the constitutive law (2.3) for the shear stress. Such a linear functional under the conditions which for actual substances are usually realised, can be expressed as an integral:

$$\tau(y,t) = \int_0^\infty G(s)\,\dot{\gamma}(y,t-s)\,ds \tag{5.6}$$

Therefore the material properties of the fluid reduce to a single function G(s), the kernel of the integral. It obviously evaluates the influence of the shear at the past time (t - s) on the shear stress at the present time. It is feasible, and is confirmed by experimental tests on actual substances, that this influence function can only take on positive values inclusive of zero. It is moreover clear that the influence of a past event on the present stress will in general be the less the further this event lies in the past. In fact for actual fluids G(s) generally decreases monotonically with increasing values of s. One calls this *'fading memory'*.

In the special case of a steady shear flow $\dot{\gamma}$ in equation (5.6) is constant with respect to s, and the expression reduces to $\tau = \eta_0 \dot{\gamma}$. One learns from this that the integral over the influence function is in agreement with the zero-shear viscosity of the fluid:

$$\int_0^\infty G(s)\, ds = \eta_0 \qquad (5.7)$$

According to the preceding statements on qualitative properties it is conceivable that G(s) decreases exponentially with s and can be expressed with the help of the two constants G(0) and λ in the form:

$$G(s) = G(0)\, e^{-s/\lambda} \qquad (5.8)$$

Hence from equation (5.7) and under these conditions $G(0)\lambda = \eta_0$. On the basis of this example it can be easily seen that fluids with a memory have not only viscous but also elastic properties. One therefore refers to them as *viscoelastic substances*. From

$$\tau(t) = G(0) \int_0^\infty e^{-s/\lambda}\, \dot{\gamma}(t-s)\, ds \qquad (5.9)$$

there follows for the time derivative of the stress

$$\dot{\tau}(t) = -\,G(0) \int_0^\infty e^{-s/\lambda}\, \frac{\partial \dot{\gamma}(t-s)}{\partial s}\, ds \qquad (5.10)$$

After one partial integration the expression (5.9) reappears on the right-hand side, and one obtains

$$\dot{\tau}(t) = G(0)\, \dot{\gamma}(t) - \frac{1}{\lambda}\, \tau(t) \qquad (5.11)$$

This expression can also be written in the form

$$\tau(t) + \lambda\, \dot{\tau}(t) = \eta_0\, \dot{\gamma}(t) \qquad (5.12)$$

Note that here only those quantities for the actual time occur, besides τ and $\dot{\gamma}$ also the time derivative of the stress. The integral constitutive equation (5.9) can therefore just as well be replaced by the differential form (5.12). The two relationships are completely equivalent. Equation (5.12) can be interpreted as a stress-strain relationship of a *Maxwell body*, which according to Fig. 5.1 consists of a purely viscous component of viscosity η_0 and a purely elastic component with a modulus of elasticity G(0). From the statement that the elongations of the two components add up to the total elongation

$$\gamma = \gamma_1 + \gamma_2 \tag{5.13}$$

and that the stress is related to the elongations according to

$$\tau = \eta_0 \, \dot{\gamma}_1 = G(0) \, \gamma_2 \tag{5.14}$$

expression (5.12) follows after elimination of γ_1 and γ_2.

The function G(s) describing a linear viscoelastic substance is immediately evident in the *relaxation test*. In this the substance is under no load up to time t = 0, and at rest ($\tau = \gamma = 0$ for $t < 0$). At time t = 0, and from then, a shear deformation is applied with a constantly maintained angle γ_0 (Fig. 5.2). For this to happen in a Maxwell fluid the initial shear stress $G(0)\gamma_0$ is necessary, because only the spring starts to expand immediately; the fluid damper remains 'frozen' at first. After that the stress decreases with time, because

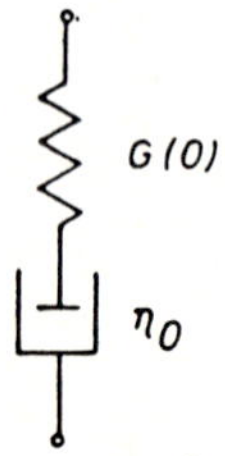

Fig. 5.1 Maxwell model

the fluid damper yields when under load, and the material 'relaxes'. Equation (5.6) shows that the relaxation curve $\tau(t)$ plotted from it, up to a factor γ_0, corresponds to the influence function G(t), which is therefore called the *relaxation function*:

$$\tau(t) = G(t) \, \gamma_0 \quad \text{for } t > 0 \tag{5.15}$$

for the relaxation test. For a Maxwell fluid the relaxation curve is described by a decreasing exponential function. The time constant λ

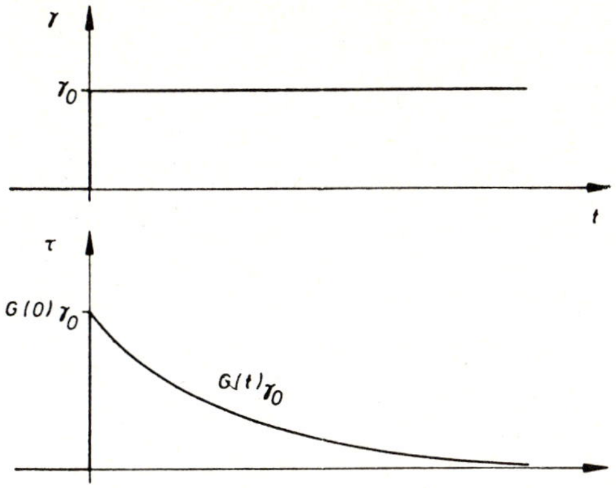

Fig. 5.2 Stress relaxation after sudden shear

of the material, during which time the stress has fallen to 1/e times the original value, is called the *relaxation time*. For real viscoelastic fluids the relaxation function can be expressed if not by a single function, then mostly by the sum of several exponential functions:

$$G(s) = \sum_{i=1}^{n} G_i(0)\, e^{-s/\lambda_i} \tag{5.16}$$

This property can be illustrated by means of a mechanical model, which consists of n Maxwell elements, which are arranged parallel to each other (Fig. 5.3). Because in the relaxation test all elements experience the same 'elongation' γ_0, and the stresses on the individual elements add up to the total stress τ, the relaxation function of this body appears from equation (5.15) as the sum of the contributions of the individual elements, so that equation (5.16) relates to the model shown in Fig. 5.3.

Now it is extremely difficult to make conclusions from observations made during the relaxation test, and hence for a known function G(t), about the number of analogue models and their relaxation times. One therefore sometimes starts from the assumption that in the material not only analogues with discrete relaxation times occur, but a whole spectrum comes in. This leads to an integral expression for the relaxation function, which will be described briefly. For this purpose we consider a chain model as shown in Fig. 5.3, but with an infinite

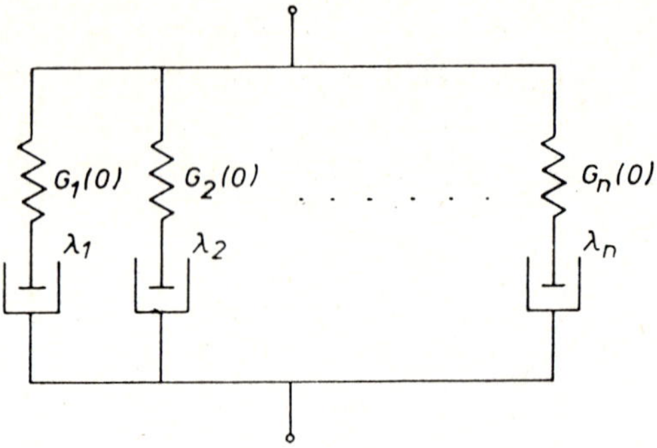

Fig. 5.3 Generalised Maxwell model

number of elements ($n \to \infty$), for which the relaxation times for two adjoining elements vary about the fixed value $\Delta\lambda$. The smaller $\Delta\lambda$ is chosen, the smaller must be the product $G_i(0)\lambda_i$ too, because their sum corresponds, according to equation (5.7), to the zero-shear viscosity of the fluid, and therefore has a quite definite value. Therefore $G_i(0)$ is proportional to $\Delta\lambda$ and is inversely proportional to λ_i, and this is expressed by the following equation:

$$G_i(0) = \frac{H_i}{\lambda_i} \Delta\lambda \tag{5.17}$$

One now makes $\Delta\lambda$ notionally smaller and smaller and proceeds from the discrete to the continuous distribution, thus obtaining for the relaxation function instead of equation (5.16) the general expression

$$G(t) = \int_0^\infty \frac{H(\lambda)}{\lambda} e^{-t/\lambda} d\lambda \tag{5.18}$$

The function $H(\lambda)$ which occurs here is called the *relaxation spectrum* of the viscoelastic fluid.

The relaxation test described above requires a constant angle of shear and hence for fluids in contrast to solid bodies cannot be off-hand realised. It is therefore advisable in the case of fluids to look to other test methods which lead directly to the relaxation function G(s). There is a discussion about suitable tests in the three following sections.

5.1.1 Sudden change in shear rate

In the case of fluids it is considerably easier to bring about a sudden change in strain rate than a change in the deformation itself. This is done for example in a Couette device, in that the movable cylinder is moved suddenly at constant angular velocity, or alternatively is braked suddenly. The curvature of the stream lines need not be considered if the gap between the cylinders is small in comparison with the radii of the cylinders. If in addition the dimensionless coefficient $\rho h^2/\eta_0\lambda$, formed from the gap h, the density ρ, the viscosity η_0 and the relaxation time λ of the fluid, is small compared with unity, then the inertia of the fluid plays no part, i.e. one can start from the assumption that the fluid particles at every point in the gap deform in the same way and are under the same mechanical stress. Under these conditions the shear rate and the shear stress are not dependent on position. We shall establish this statement more precisely in Section 5.1.7.

By sudden acceleration of the movable cylinder at time t = 0 at constant angular velocity it is brought about that the fluid is from this time deformed at constant shear rate $\dot{\gamma}_0$, the amount of which is found from equation (1.85). The following expression holds for the shear rate behaviour with respect to time:

$$\dot{\gamma}(t) = \begin{cases} 0 & \text{for } t < 0 \\ \dot{\gamma}_0 & \text{for } t > 0 \end{cases} \tag{5.19}$$

One can conveniently call a process of this kind a *'stress growth experiment'*.

From equation (5.6) the stress response of the substance is

$$\tau(t) = \begin{cases} 0 & \text{for } t < 0 \\ \int_0^t G(s)\,ds \cdot \dot{\gamma}_0 & \text{for } t > 0 \end{cases} \tag{5.20}$$

The shear stress accordingly increases monotonically with time and asymptotically reaches the steady value $\eta_0\dot{\gamma}_0$ as $t \to \infty$ (Fig. 5.4).

It has already been mentioned that the relaxation function G(s) only takes on positive values. In practice the stress growth process after a period of time, which corresponds to a multiple of the greatest relaxation time, is regarded as terminated.

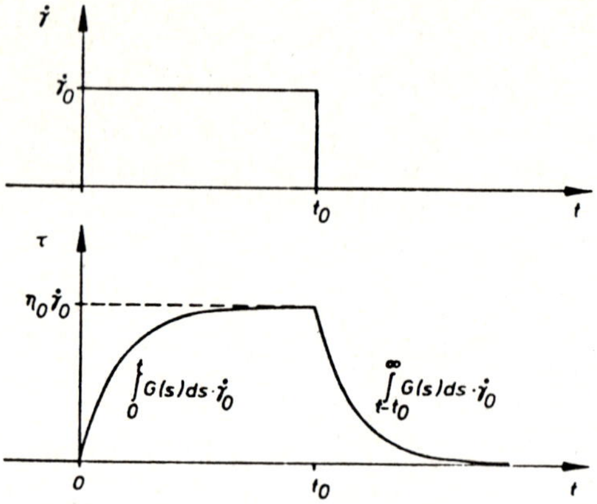

Fig. 5.4 Stress increase and stress relaxation at the start and end of a steady shear flow

If conversely at time $t = t_0$ a steady shear flow is cut off

$$\dot{\gamma}(t) = \begin{cases} \dot{\gamma}_0 & \text{for } t < t_0 \\ 0 & \text{for } t > t_0 \end{cases} \tag{5.21}$$

the shear stress does not fall immediately to the rest state zero, but dies away gradually (Fig. 5.4). Equation (5.6) gives the time law of this stress relaxation:

$$\tau(t) = \begin{cases} \eta_0\dot{\gamma}_0 & \text{for } t < t_0 \\ \int\limits_{t-t_0}^{\infty} G(s)\,ds \cdot \dot{\gamma}_0 & \text{for } t > t_0 \end{cases} \tag{5.22}$$

If one records the stress history for one of the two processes in a fluid of unknown properties, one can easily find the relaxation function G(t) from the test results by using equations (5.20) and (5.22). In fact the stress processes recorded do not correspond directly with the relaxation function, but their time derivatives do. Further, it follows from equations (5.20) and (5.22) that the stress values of both processes taking place at corresponding times add up to the same constant value of $\eta_0\dot{\gamma}_0$. Therefore the two stress processes shown in Fig. 5.4 follow each other by reflection across the t-axis and with corresponding displacement. Stress growth and stress relaxation therefore follow exactly similar courses. This reciprocity no longer exists if the material is so strongly sheared that non-linear properties make themselves felt.

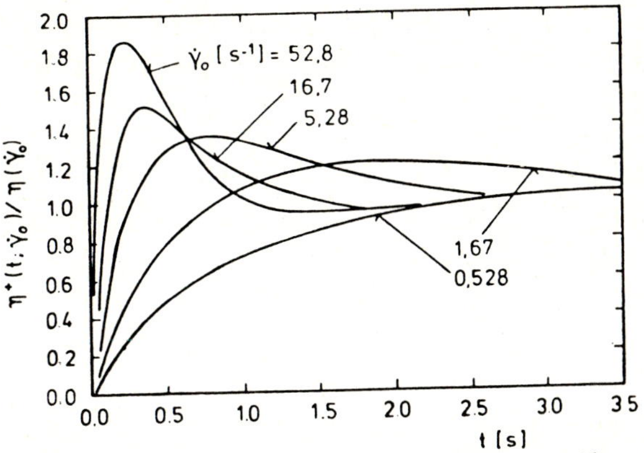

Fig. 5.5 Normalised viscosity in the stress growth experiment of a polymer solution (4% polystyrene in chlorodiphenol at 25°C) (after Carreau)

If one carries out the two standard tests in which the shear rate defined by equations (5.19) and (5.21) is changed in jumps for various values $\dot{\gamma}_0$, then one observes a monotonic rise in stress, and the reciprocity just mentioned only occurs at relatively small values. The more strongly the fluid is sheared the more do the departures from linear viscoelastic properties become noticeable. In the 'stress growth experiment' for example the shear stress reaches in the initial period under certain conditions considerably greater values (stress-overshoot) than under steady conditions for large time values. The stress-time graphs therefore have for sufficiently large shear rates a well defined maximum. The curves for various values of $\dot{\gamma}_0$ can best be compared if the actual shear stress is referred to the steady value (Fig. 5.5). This dimensionless quantity can also be interpreted as the ratio of the time dependent *'viscosity in the stress growth experiment'* $\eta^+(t;\dot{\gamma}_0)$ to the steady value $\eta(\dot{\gamma}_0)$ Hence the viscosity in the stress growth experiment is defined as the quotient of the actual shear stress and the constant shear rate (the + symbol indicates that this is a growth process),

$$\eta^+(t;\dot\gamma_0) := \frac{\tau(t;\dot\gamma_0)}{\dot\gamma_0}$$

Within the framework of a linear theory η^+ does not depend on $\dot\gamma_0$ (cf. equation (5.20)). The dependence shown in Fig. 5.5 of the growth curves on the shear rate therefore reflects non-linear properties, which were not considered in the previously developed theory.

Correspondingly the non-linear behaviour brings it about that after the cut-off of a steady shear flow the shear stress relaxation depends on the value of the original shear rate $\dot\gamma_0$. Fig. 5.6 shows that the relaxation process dies away the faster the greater the shear rate. The stress in contrast to the growth process always dies away monotonically. The two processes are therefore no longer in a reciprocal relationship to one another in the non-linear region. The recorded stress processes can again be just as well regarded as viscosity curves, because in the cut-off process the time dependent quantity defined by

$$\eta^-(t;\dot\gamma_0) := \frac{\tau(t;\dot\gamma_0)}{\dot\gamma_0}$$

is measured with the dimensions of viscosity.

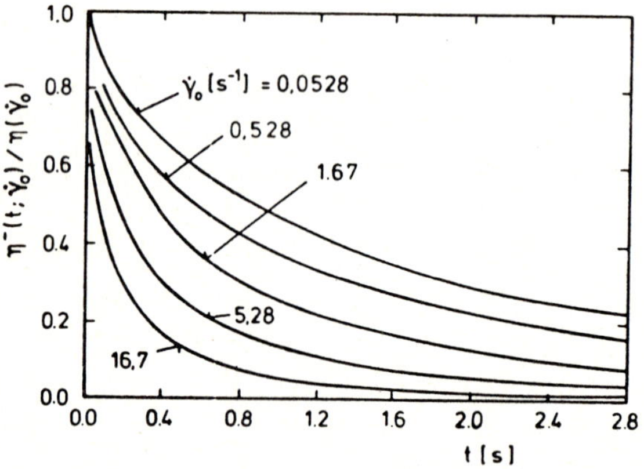

Fig. 5.6 Shear stress relaxation function $\eta^-(t;\dot\gamma_0)/\eta(\dot\gamma_0)$ of a polymer solution (the same fluid as shown in Fig. 5.5) (after Carreau)

5.1.2 Creep test and creep recovery

In the *creep test* one subjects an initially unstressed and at rest material to a suddenly applied constant shear stress τ_0 at the time $t = 0$, and records the shear strain $\gamma(t)$. One therefore measures the *creep compliance* $J(t)$ as defined by

$$\gamma(t) = J(t)\,\tau_0 \text{ for } t > 0 \tag{5.23}$$

For a Newtonian fluid of viscosity η_0 this would be represented by a straight line in the t-γ diagram; for a viscoelastic fluid under the applied load not only the viscous but also the elastic parts elongate, so that a characteristic for $J(t)$ results, which lies above the straight line mentioned above (see the left-hand part of Fig. 5.7).

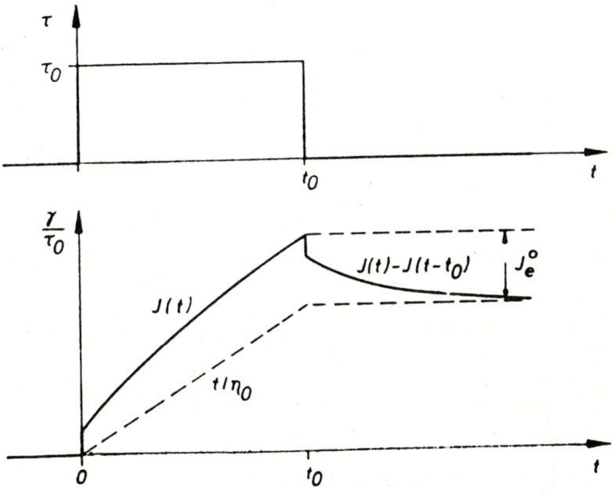

Fig. 5.7 Creep test and creep recovery

At a time after the shear stress τ_0 has been applied, more accurately: for $t \gg \lambda$, in which λ is a characteristic relaxation time of the material (e.g. the maximum relaxation time), the elastic components have elongated about a fixed value $J_e^0 \tau_0$ appertaining to this stress.

One calls the elastic property J_e^0 the equilibrium compliance of the fluid. The graph $J(t)$ then runs parallel to the characteristic straight lines t/η_0 for the viscous behaviour, but is displaced by the constant J_e^0 at greater values:

$$J(t) \simeq \frac{t}{\eta_0} + J_e^0 \quad \text{for} \gg \lambda \tag{5.24}$$

It is clear that the property $J(t)$ shown in the creep test is connected with the relaxation function $G(t)$. If one inserts equation (5.23) into equation (5.6), then one obtains

$$1 = \int_0^t G(s)\,\dot{J}(t-s)\,ds + G(t)\,J(0) \tag{5.25}$$

By integrating with respect to t a double integral arises on the right-hand side. If one interchanges the integration sequence and then calculates the inner integral, one finds that the connection between G(t) and J(t) can also be written in the form

$$\int_0^t G(s)\,J(t-s)\,ds = t \tag{5.26}$$

This relationship shows that both functions are reciprocal to each other in a certain way. It immediately follows from this that the Laplace transforms $\bar{G}(p)$ and $\bar{J}(p)$ are reciprocally proportional to each other, thus:

$$p\,\bar{G}(p) = \frac{1}{p\,\bar{J}(p)} \tag{5.27}$$

If one of the two functions is known and is described by an analytical expression, then the other one can accordingly be calculated. The evaluation of a creep test yields in particular the equilibrium compliance J_e^0. This constant can also be determined by a *creep recovery test* (see the right-hand side of Fig. 5.7). It is of course the same as the value of the 'reversible shear' relative to τ_0, to which the material springs back after removal of the load. By J_e^0 one essentially measures the weighted integral over the relaxation function $\int_0^\infty s\,G(s)\,ds$. If one inserts the asymptotic expression for J(t) into the integral relationship between G(t) and J(t), cancels a term t on both sides, and then lets t tend to infinity, and noting equation (5.7), one obtains

$$J_e^0 = \frac{1}{\eta_0^2}\int_0^\infty s\,G(s)\,ds \tag{5.28}$$

One can show that the integral is half the magnitude of the first zero normal stress coefficient ν_{10}, hence

$$J_e^0 = \frac{\nu_{10}}{2\eta_0^2} \tag{5.29}$$

By determining the reversible shear in the creep recovery test one therefore obtains information about the first normal stress coefficients for low shear rates.

5.1.3 Oscillatory stress and deformation

In order to learn about the material behaviour of a linear viscoelastic substance, besides the previously described standard tests, a pure sinusoidal motion can also be suitable. Such dynamic experiments require an instrument which creates in the test fluid a sinusoidal deformation with respect to time as the input, and the resulting stress is recorded as output. Devices which consist of a plate and an oscillating cone are customary, with two concentric cylinders, one of which oscillates, and with two parallel discs steadily rotating round different fixed axes, therefore not oscillating.

For oscillatory motion the fluid is subjected to a variable sinusoidal deformation of amplitude γ_0 and angular velocity ω.

$$\gamma^* = \frac{\gamma_0}{i} e^{i\omega t} \tag{5.30}$$

Complex quantities are denoted here with a star; only the real part has any physical significance.

In the case of linear behaviour the material responds likewise with a sinusoidal stress of the same frequency, but the stress and strain are not in phase (Fig. 5.8).

$$\tau^* = \frac{\tau_0}{i} e^{i(\omega t+\delta)} \tag{5.31}$$

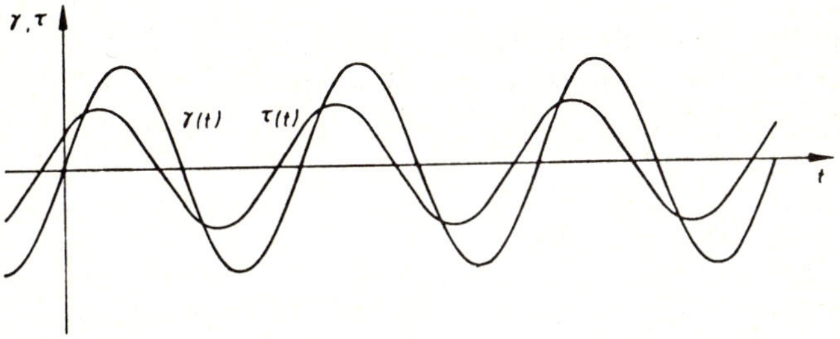

Fig. 5.8 Stress and shear for pure sinusoidal deformation

By analogy with purely elastic bodies which obey Hooke's law, for which $\tau(t) = G\gamma(t)$ and with a purely viscous (Newtonian) fluid for which $\tau(t) = \eta\dot{\gamma}(t)$ one now defines a frequency dependent *complex shear modulus* defined by the expression

$$G^*(\omega) = G'(\omega) + i\,G''(\omega) := \frac{\tau^*}{\gamma^*} \tag{5.32}$$

and a frequency dependent *complex viscosity* defined by

$$\eta^*(\omega) = \eta'(\omega) - i\,\eta''(\omega) := \frac{\tau^*}{\dot\gamma^*} = \frac{\tau^*}{i\omega\gamma^*} \tag{5.33}$$

Note the minus sign in the imaginary part of the complex viscosity. Obviously $G^* = i\omega\eta^*$, hence

$$G'(\omega) = \omega\,\eta''(\omega), \quad G''(\omega) = \omega\,\eta'(\omega) \tag{5.34}$$

If η' and G'' were zero, then the stress and the shear would be in phase, i.e. this would be a purely elastic body. Thus G' represents the effective elastic part of the real fluid under oscillatory load. On the other hand η' describes the viscous part, because this quantity is linked with a dissipation of energy. One can easily prove that the mechanical energy dissipated per unit time and unit volume $\mathrm{Re}\tau^* \cdot \mathrm{Re}\dot\gamma^*$ has the value $\frac{1}{2}\,\eta' \cdot (\omega\gamma_0)^2$ averaged over a period of time, which for a given shear rate amplitude $\omega\gamma_0$ is only affected by η', but not by G'. The real part η' of the complex viscosity is incidentally called the dynamic viscosity; G' is called the dynamic shear modulus.

In the test one generally measures the amplitude τ_0 of the response and the phase angle δ for given input values ω and γ_0. Hence because $G^* = \tau^*/\gamma^* = (\tau_0/\gamma_0)e^{i\delta}$:

$$G'(\omega) = \frac{\tau_0}{\gamma_0}\cos\delta; \quad \eta'(\omega) = \frac{\tau_0}{\gamma_0\omega}\sin\delta \tag{5.35}$$

Because they have the amplitude ratio τ_0/γ_0 and phase angle δ, in general G' and η' also depend on the frequency. Fig. 5.9 shows typical graphs of the functions $G'(\omega)$ and $\eta'(\omega)$ for a polymer solution and a polymer melt. The characteristic properties $\eta(\dot\gamma)$ and $N_1(\dot\gamma)$ for steady shear flows are also included for comparison. These show that on the one hand the graphs of the dynamic viscosity $\eta'(\omega)$ and the shear viscosity $\eta(\dot\gamma)$, and on the other hand the graphs of the dynamic shear modulus $G'(\omega)$ and the first normal stress function $N_1(\dot\gamma)$ have similar properties. One can show that the following relationships between these functions apply:

$$\lim_{\omega\to 0} \eta'(\omega) = \lim_{\dot\gamma\to 0} \eta(\dot\gamma) \equiv \eta_0 \tag{5.36}$$

$$\lim_{\omega \to 0} \frac{G'(\omega)}{\omega^2} = \frac{1}{2} \lim_{\dot{\gamma} \to 0} \frac{N_1(\dot{\gamma})}{\dot{\gamma}^2} \equiv \frac{1}{2} \nu_{10} \qquad (5.37)$$

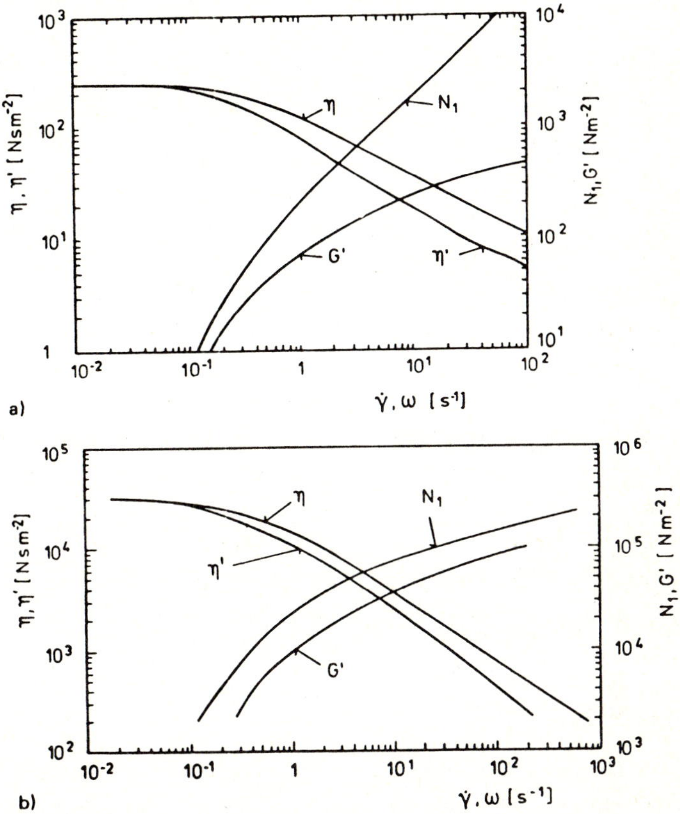

Fig. 5.9 The properties $\eta'(\omega)$, $G'(\omega)$, $\eta(\dot{\gamma})$ and $N_1(\dot{\gamma})$ of a polymer solution and a polymer melt

(a) 4% polystyrene in chlorodiphenol at 25°C (after Ashare)

(b) Polystyrene at 200°C (after Han et al.)

The first relationship can be easily understood if one remembers that a very low frequency periodic deformation of small amplitude can be considered as a nearly steady slow flow, and therefore in the limit as $\omega \to 0$ and $\dot{\gamma} \to 0$ the zero-shear viscosity must result. One also observes with certain polymer systems that the absolute value of the complex viscosity $|\eta^*(\omega)|$ and the shear viscosity $\eta(\dot{\gamma})$ almost coincide when $\omega = \dot{\gamma}(\neq 0)$. Sometimes a relationship is also found between the shear viscosity and the stress growth viscosity recorded for low amplitude movements, say $\eta^{+}(t,0) = \eta(\dot{\gamma})$ for $t\dot{\gamma}$ = constant. But so far no basis has been conclusive for these noteworthy relationships between the linear viscoelastic and the non-linear viscous properties. We cannot yet explain the expression (5.37) here, but only take note of it. It states that the normal stress coefficient ν_{10} can be determined by low frequency dynamic shear stress measurements.

The connection between the physical properties $\eta'(\omega)$ and $G'(\omega)$ when measured under oscillatory load and the relaxation function $G(s)$ of the fluid, starting from the constitutive equation gives

$$\tau^*(t) = \int_0^\infty G(s)\dot{\gamma}^*(t-s)\,ds \tag{5.6}$$

Hence by use of equation (5.30) there follows for the complex viscosity

$$\eta^* = \frac{\tau^*(t)}{\dot{\gamma}^*(t)} = \int_0^\infty G(s)\frac{\dot{\gamma}^*(t-s)}{\dot{\gamma}^*(t)}\,ds = \int_0^\infty G(s)\,e^{-i\omega s}\,ds$$

If one separates both sides of the equation into real and imaginary parts, then one obtains the two real relationships

$$\eta'(\omega) = \int_0^\infty G(s)\cos\omega s\,ds, \quad \eta''(\omega) = \int_0^\infty G(s)\sin\omega s\,ds \tag{5.38}$$

Therefore for $\eta' + i\eta''$ (the conjugate complex number for η^*) this is the same as the Fourier transform of the relaxation function $G(s)$.

Concerning the dependence of the linear viscoelastic properties on the external parameters like temperature Θ, the mean molecular weight $\bar{M}$ of the fluid, etc., the following similarity laws should apply by analogy to the properties of the viscometric functions described in Chapter 2 (cf. equations (2.8) and (2.31)), which lead to universal expressions for the data:

$$\frac{\eta'(\omega;\Theta,\bar{M})}{\eta_0(\Theta,\bar{M})} = H'\left(\frac{\eta_0(\Theta,\bar{M})\omega}{\tau_*}\right), \quad \frac{G'(\omega,\Theta,\bar{M})}{\tau_*} = \tilde{G}'\left(\frac{\eta_0(\Theta,\bar{M})\,\omega}{\tau_*}\right) \tag{5.39}$$

Hence one can also derive the corresponding expressions for η'' and G''.

The factorisation

$$\frac{G(s;\Theta,\bar{M})}{\tau_*} = \tilde{G}\left(\frac{\tau_* s}{\eta_0(\Theta,\bar{M})}\right) \qquad (5.40)$$

would then hold for the relaxation function $G(s;\Theta,\bar{M})$. Note that the zero-shear viscosity appears here in the denominator of the argument of the universal function G. Accordingly the characteristic memory time of a viscoelastic fluid is proportional to the zero-shear viscosity $\eta_0(\Theta,\bar{M})$. The cooler the fluid the higher its average molecular weight, the greater therefore is η_0; the more slowly in consequence does the stress decay during a relaxation test.

5.1.4 Tuning a shock absorber

Fig. 5.10 shows the model of a shock absorber. The plane wall at y=0 represents the vibrating part of the machine. The vibration of the generator has an angular velocity ω and velocity amplitude U_1. Without limiting the generality we can therefore write its velocity in the form $u(0,t) = U_1\cos\omega t$. As in the preceding chapter we shall however prefer a complex notation in which the complex quantities will again be denoted by a star, and shall agree that the real part of quantities thus marked has a physical significance. Thus the motion of the vibration generator is described by

$$u^*(0,t) = U_1\, e^{i\omega t} \qquad (5.41)$$

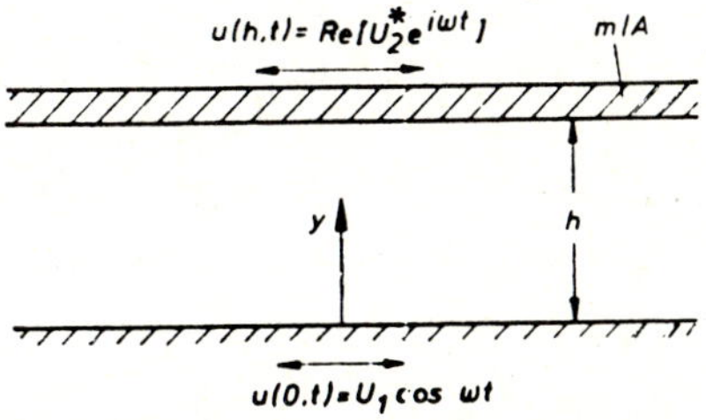

Fig. 5.10 Model of a shock absorber

There is a freely moving plate of mass m, to which the vibration is transferred, separated from the vibration generator by a linear viscoelastic fluid layer of thickness h. Because the mechanically coupled mass follows the generator with a phase lag, we can represent its velocity in the form

$$u^*(h, t) = U_2^* e^{i\omega t} \tag{5.42}$$

in which U_2^* represents a complex constant, the magnitude of which describes the velocity amplitude of the coupled mass.

In order to find the relationship between the vibration amplitudes U_2^* and U_1, one must study the motions of the coupled plate and the fluid layer. Because apart from the shear stress exerted from the fluid on the underside of the plate no further restoring force acts on the coupled plate, its equation of motion is

$$m \frac{du^*(h, t)}{dt} = - A\, \tau^*(h, t) \tag{5.43}$$

in which A is the wetted surface area. The shear stress on the right-hand side depends on the motion of the layer of fluid and its constitutive properties. We assume for the sake of simplicity that the friction forces are dominant in the fluid and that the inertia forces on the contrary can be neglected. Later we shall show that this assumption is valid if the mass of the enclosed fluid is small compared with the mass of the coupled plate. This assumption leads to the conclusion that the fluid at any moment flows in such a way as would the two bounding walls be moved at constant velocities. The resulting drag flow is characterised by a spatially constant shear stress field and a variable velocity profile linear in y.

Hence the shear rate $\dot{\gamma}^* = \partial u^*/\partial y$ according to equations (5.41) and (5.42) works out as

$$\dot{\gamma}^* = \frac{U_2^* - U_1}{h} e^{i\omega t} \tag{5.44}$$

For such an oscillatory motion the constitutive law of a linear viscoelastic fluid can be written in the form $\tau^* = \eta^* \dot{\gamma}^*$ (cf. equation (5.33)). In the present case therefore

$$\tau^* = (\eta' - i\eta'') \frac{U_2^* - U_1}{h} e^{i\omega t} \tag{5.45}$$

If one enters this result on the right-hand side of equation (5.43) and takes into account on the left-hand side the expression (5.42), then a connection arises between the two vibration amplitudes U_2^* and

U_1, in which besides the angular velocity ω, the physical properties $\eta'(\omega)$ and $\eta''(\omega)$ as well as the constants m, A and h are involved. By use of the quantity χ of dimensions length/mass defined by the equation

$$\chi := \frac{A}{m\,h} \tag{5.46}$$

this relationship can be expressed in the form

$$U_2^* - U_1 = \frac{-i\omega U_1}{\chi\eta'(\omega) + i\omega\left(1 - \chi\frac{\eta''(\omega)}{\omega}\right)} \tag{5.47}$$

Hence U_2^* can be eliminated from equation (5.45), so that the shear stress as a function of time and dependent on the frequency and amplitude of the vibration generator is found to be

$$\tau^* = -\frac{\eta'' + i\eta'}{\chi\frac{\eta'}{\omega} + i\left(1 - \chi\frac{\eta''}{\omega}\right)} \cdot \frac{U_1}{h}\,e^{i\omega t} \tag{5.48}$$

In order to be able to estimate the effectiveness of the shock absorber we calculate the energy W which is dissipated during each period of oscillation. It corresponds to that energy which is fed to the fluid across the walls during each period. The coupled plate retains its average energy for the period, therefore it transmits no energy to the fluid. Thus W turns out to be that work which is expended against the shear stress during the motion of the generator

$$W = -U_1 A \int_0^{2\pi/\omega} \cos\omega t \cdot \mathrm{Re}\,\tau^*\,dt \tag{5.49}$$

By use of equation (5.48) after a short calculation the expression follows:

$$\frac{W}{2\pi E_0} = \frac{\chi\frac{\eta'(\omega)}{\omega}}{\left(\chi\frac{\eta'(\omega)}{\omega}\right)^2 + \left(1 - \chi\frac{\eta''(\omega)}{\omega}\right)^2} \tag{5.50}$$

The kinetic energy generated by the oscillating mass and the amplitude of the generator thus appears on the left-hand side as a reference quantity.

$$E_0 := \frac{m}{2} U_1^2$$

The right-hand side of equation (5.50) contains two independent parameters, namely $\chi\eta'/\omega$ and $\chi\eta''/\omega$. The graphic representation of the result leads to the single parameter set of graphs shown in Fig. 5.11. The graph denoted by the parameter value 0 describes the damping behaviour when a Newtonian fluid is used (for which η' = constant and $\eta'' = 0$). In this case one achieves optimum damping when $\chi\eta'/\omega = 1$, i.e. at a quite definite frequency, the value of which depends on the properties of the shock absorber, including the viscosity of the Newtonian fluid. By the use of a viscoelastic oil larger damping values may obviously be obtained, but smaller damping values are also possible.

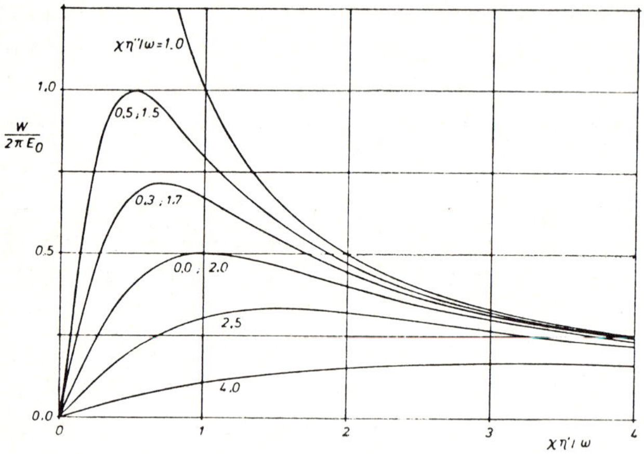

Fig. 5.11 Energy dissipated per period of oscillation

The practical utility of this knowledge is especially clear when we limit ourselves to fluids for which the physical function $\eta''(\omega)$ increases linearly with ω.

$$\eta''(\omega) = \alpha\,\omega \tag{5.51}$$

The parameter of the graphs $\chi\eta''/\omega = \chi\alpha$ in Fig. 5.11 then does not depend in any way on the vibration frequency; ω enters only into the abscissa value $\chi\eta'/\omega$. Because for fixed abscissa values the greatest possible damping is reached when $\chi\alpha = 1$ (Fig. 5.11), it is possible to obtain a frequency independent optimum 'tuning' of the shock absorber, because the matching of the coupled plate and the geometrical dimensions of the damper, represented by the quantity χ, with the physical coefficient α, from $\chi\alpha = 1$, independently of the frequency leads to the greatest possible energy dissipation. The value of the maximum dissipated energy W_{max} per cycle of oscillation gives according to equation (5.50) the expression $W_{max}/2\pi E_0 = \alpha\omega/\eta'(\omega) = \eta''(\omega)/\eta'(\omega)$. For optimum tuning moreover the coupled mass vibrates with greater amplitude than the generator. From equation (5.47) there follows for $\chi\alpha = 1$: $|U_2^*|/U_1 = \sqrt{1 + [\eta''(\omega)/\eta'(\omega)]^2} > 1$. As a comparison we consider the situation by use of a Newtonian fluid (η' = constant and $\eta'' = 0$). Hence equation (5.47) leads after a short calculation to $|U_2^*|/U_1 = 1/\sqrt{1 + (\omega/\chi\eta')^2} < 1$, i.e. the amplitude of the coupled plate vibration is always smaller than the amplitude of the generator.

From the above statements the question arises about the existence of fluids which have the physical property given by (5.51), for which a frequency independent optimum tuning only is possible. Here one can consider that the practical requirements of a shock absorber in general do not involve suppressing any high frequency vibrations, but much more must one damp vibrations of an upwards limited band of frequencies. Hence this concept plays a part, because for a sufficiently low vibration, i.e. when the period of vibration is large compared with the relaxation time of the fluid, the behaviour of every real fluid can be approximated by the relationship (5.51) (cf. equations (5.37) and (5.34)). Hence in general $\eta'(\omega)$ can at the same time be replaced by the constant zero-shear viscosity, which of course is not essential here. To that extent any linear viscoelastic fluids, preferably those with the shortest possible relaxation time, are relevant here, for example silicone oil ($\bar{M}$ = 20,000) with a relaxation time of about 1 ms. The less the relaxation time the higher will be the vibration frequencies which are within the range of optimum tuning.

We shall now prove the basic assumption that in the fluid the inertia forces can be neglected compared with the friction forces. The inertia force per unit volume is given by the product of the density ρ of the fluid and the local acceleration $\partial u/\partial t$.

For an estimate of the order of magnitude the time derivative can be replaced by a factor ω and the velocity by a characteristic value, e.g. U_1. Over the whole volume hA filled with fluid the inertia forces accordingly add up to an expression of magnitude $\rho\omega U_1 hA$. Opposing this are the friction forces of magnitude $|\tau^*|A$. The condition $\rho\omega U_1 hA << |\tau^*|A$ can be transformed by use of the expressions (5.48) for the shear stress and (5.46) for the quantity χ in the case of optimum tuning (hence for $\chi\eta''/\omega = 1$) in the relationship $\rho hA << m\sqrt{1 + (\eta''/\eta')^2}$.

Therefore it is valid to disregard the inertia when the mass of the damper fluid is small compared with the mass of the coupled plate.

Note that the main result, equation (5.50) and Fig. 5.11, was deduced on the basis of a simplified model of a shock absorber, but it also remains valid for more practical applications. For the torsional shock absorber shown for example in Fig. 5.22, there results with a suitably altered definition of the reference energy E_0 the same expression (5.50) for the energy W dissipated per cycle. The only difference is that the constant χ there is composed of the moment of inertia of the coupled rotating mass and the various characteristic length scales of the instrument (cf. Problem 5.5).

5.1.5 Flow in the vicinity of a vibrating wall

We consider below an unsteady process in which besides the mechanical behaviour discussed above, the inertia of the fluid also has a controlling effect. The fluid occupies the half space $y > 0$, and is bounded by a plane wall (for $y = 0$), which in accordance with the law

$$u(0, t) = U \cos \omega t \tag{5.52}$$

performs sinusoidal oscillations in the x-direction. U is the maximum velocity of the wall, ω the angular velocity of the oscillation. At a large distance from the wall the fluid is at rest at all times,

$$u(\infty, t) = 0 \tag{5.53}$$

An unsteady shear flow will develop in the x-direction with an initially unknown velocity field $u(y,t)$. Thus the shear stress τ between layers also is naturally dependent on time and position; other extra-stresses do not occur. We shall not consider volume forces, particularly the force of gravity.

Because the fluid remote from the wall is at rest there is no pressure gradient. The equation of motion for the direction parallel to the wall, equation (1.113), relates for these conditions the acceleration of the fluid particles to the spatial derivative of the shear stress. Because the convective components in the acceleration vanish (see the first of equations (1.13)), the equation of motion reduces to

$$\rho \frac{\partial u}{\partial t} = \frac{\partial \tau}{\partial y} \tag{5.54}$$

In the oscillatory state u and τ are sinusoidal functions of time with angular velocity ω. It is expedient once more to use complex quantities u^* and τ^*, whose real parts are u and τ.

For oscillatory load, according to equation (5.33) $\tau^* = \eta^*(\omega)\partial u^*/\partial y$, so that the equation of motion (5.54) transforms into the heat conduction equation

$$\rho \frac{\partial u^*}{\partial t} = \eta^*(\omega) \frac{\partial^2 u^*}{\partial y^2} \tag{5.55}$$

The product trial function

$$u^* = \varphi^*(y)\, e^{i\omega t} \tag{5.56}$$

reduces it to the ordinary differential equation $i\omega\rho\varphi^* = \eta^* d^2\varphi^*/dy^2$. The integral which satisfies the conditions (5.52) and (5.53) is as follows:

$$\varphi^*(y) = U\, e^{-(\alpha+i\beta)y} \tag{5.57}$$

The positive constants α and β thus introduced depend on the density ρ of the fluid, the angular velocity ω and the related quantities $\eta'(\omega)$ and $\eta''(\omega)$. Note that $\eta^* = \eta' - i\eta''$, cf. equation (5.33):

$$\alpha = \sqrt{\frac{\rho\omega}{2} \cdot \frac{\sqrt{\eta'^2 + \eta''^2} - \eta''}{\eta'^2 + \eta''^2}}\,; \qquad \beta = \sqrt{\frac{\rho\omega}{2} \cdot \frac{\sqrt{\eta'^2 + \eta''^2} + \eta''}{\eta'^2 + \eta''^2}} \tag{5.58}$$

Hence one obtains for the velocity field

$$u = U\, e^{-\alpha y} \cos(\omega t - \beta y) \tag{5.59}$$

This is a damped *transverse wave* travelling in the y-direction at a velocity ω/β. Note that its phase velocity ω/β varies as the angular velocity ω changes (*dispersion*). Every layer oscillates at the angular velocity ω, but the amplitude of the oscillation decreases exponentially outwards; the phase shift relative to the motion of the wall increases linearly. The fluid is obviously only noticeably disturbed in the vicinity of the oscillating wall (Fig. 5.12).

If one defines the depth of penetration of the wave as that distance from the wall where the velocity amplitude is only 1% of the amplitude at the wall, then one obtains for this *'boundary layer thickness'* the value $4.6/\alpha$. The expression (5.58) for α shows that with a Newtonian fluid for which $\eta'' = 0$ and η' has a constant value independent of ω, the boundary layer thickness decreases with increasing frequency as $\omega^{-1/2}$. For viscoelastic fluids this applies only to low frequencies. At higher frequencies the boundary layer thickness can decrease much more rapidly.

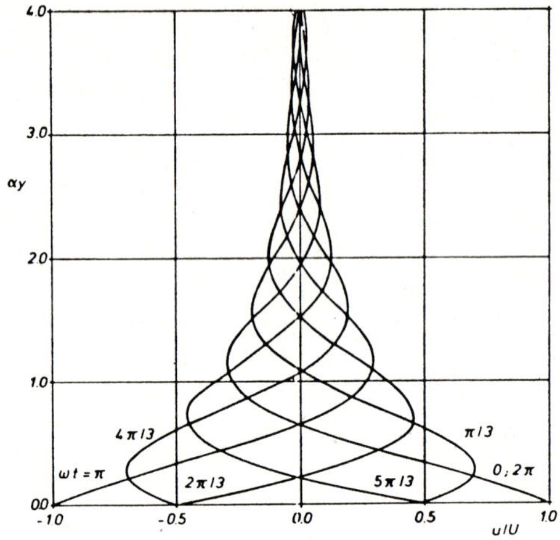

Fig. 5.12 Fluid velocity in the vicinity of an oscillating wall; $\eta''/\eta' = 1$

For the polymer solution, whose properties for example are shown in Fig. 5.9a the ratio η''/η' for $\omega > 1$ remains almost constant (the numerical value is about unity), so that there, because of the decrease in the dynamic viscosity with the frequency, the decay factor α in accordance with equation (5.58) increases much faster than $\omega^{1/2}$. Hence the boundary layer thickness decreases correspondingly faster than in a Newtonian fluid.

5.1.6 Rayleigh problem for a Maxwell fluid

This concerns the following unsteady process. The fluid once again occupies the half space $y > 0$, but is at rest until time $t = 0$, so that the initial conditions

$$u(y, 0) = 0; \quad \tau(y, 0) = 0 \tag{5.60}$$

apply. At time $t = 0$ the boundary wall is suddenly accelerated to a constant velocity and immediately moves with this velocity in the x-direction:

$$u(0, t) = U \text{ for } t > 0 \tag{5.61}$$

A pressure drop does not exist, but there is a pure drag flow.

Hence the fluid is at rest at all times remote from the wall, so that here too the boundary condition (5.53) has to be taken into account. The unsteady shear flow which arises is again described by the equation of motion (5.54). We assume that the fluid has linear viscoelastic properties. For the sake of simplicity we take as a basis the constitutive equation of a Maxwell fluid:

$$\tau + \lambda \frac{\partial \tau}{\partial t} = \eta_0 \frac{\partial u}{\partial y} \tag{5.62}$$

If one eliminates the stress from the expressions (5.54) and (5.62), then the following wave equation results for the velocity field, which contains a damping term:

$$\rho \frac{\partial u}{\partial t} + \rho\lambda \frac{\partial^2 u}{\partial t^2} = \eta_0 \frac{\partial^2 u}{\partial y^2} \tag{5.63}$$

If u were eliminated then one would obtain the same wave equation for the stress τ. One finds from the coefficients of the terms that have the second derivatives that

$$c := \sqrt{\frac{\eta_0}{\rho\lambda}} \tag{5.64}$$

expresses the *wave speed*. It is at this speed that the disturbance which the wall generates at the start of its motion in the fluid spreads outwards into the fluid. Hence it follows that at time $t > 0$ the fluid at position $y > ct$ is completely at rest. Therefore there is, unlike a Newtonian fluid for which there is a diffusion process, a sharp *wavefront*, which separates layers still at rest from those layers already accelerated, and which moves outwards into the fluid at constant speed c from the wall, a transverse shock front.

The mathematical treatment of the phenomenon is simplified if one uses the following dimensionless quantities:

$$\xi := \frac{y}{c\lambda}, \qquad \theta := \frac{t}{\lambda} \tag{5.65}$$

$$\Omega := \frac{u}{U}, \qquad T := \frac{\tau}{\rho c U} \tag{5.66}$$

The set of differential equations (5.54) and (5.62) is then

$$\frac{\partial \Omega}{\partial \theta} = \frac{\partial T}{\partial \xi} \tag{5.67}$$

$$T + \frac{\partial T}{\partial \theta} = \frac{\partial \Omega}{\partial \xi} \tag{5.68}$$

The initial conditions (5.60) and the boundary conditions (5.61), (5.53) transform into

$$\Omega(\xi, 0) = 0; \quad T(\xi, 0) = 0 \tag{5.69}$$

$$\Omega(0, \theta) = 1 \quad \text{for } \theta > 0 \tag{5.70}$$

$$\Omega(\infty, \theta) = 0; \quad T(\infty, \theta) = 0 \tag{5.71}$$

This normalised linear initial boundary value problem can be solved by using the Laplace transform (relative to the dimensionless time θ). One thus uses instead of $\Omega(\xi,\theta)$ the Laplace transform

$$\bar{\Omega}(\xi, p) := \int_0^\infty e^{-p\theta}\, \Omega(\xi, \theta)\, d\theta \tag{5.72}$$

Hence one associates a function $\bar{\Omega}(\xi, p)$ with the function $\Omega(\xi, \theta)$ according to equation (5.72), which is expressed symbolically as follows

$$\Omega(\xi, \theta) \circ\!\!-\!\!\!-\!\!\bullet\ \bar{\Omega}(\xi, p)$$

In the same way we convert the function $T(\xi,\theta)$ into its Laplace transform

$$T(\xi, \theta) \circ\!\!-\!\!\!-\!\!\bullet\ \bar{T}(\xi, p)$$

There are corresponding integrals for the time derivatives, concerning which one can easily prove by partial integration:

$$\frac{\partial \Omega(\xi, \theta)}{\partial \theta} \circ\!\!-\!\!\!-\!\!\bullet\ p\bar{\Omega}(\xi, p) - \Omega(\xi, \theta = 0)$$

$$\frac{\partial T(\xi, \theta)}{\partial \theta} \circ\!\!-\!\!\!-\!\!\bullet\ p\bar{T}(\xi, p) - T(\xi, \theta = 0)$$

If the homogeneous initial conditions (5.69) are taken into account the last term on the right-hand side vanishes. Hence the set of differential equations (5.67) and (5.68) can be converted to

$$p\bar{\Omega} = \frac{\partial \bar{T}}{\partial \xi} \tag{5.73}$$

$$\dot{\bar{T}} + p\bar{T} = \frac{\partial \bar{\Omega}}{\partial \xi} \tag{5.74}$$

By elimination of $\bar{T}$ it follows that

$$p(1 + p)\,\bar{\Omega} = \frac{\partial^2 \bar{\Omega}}{\partial \xi^2} \tag{5.75}$$

The solution of this ordinary differential equation which satisfies the transformed boundary conditions $\bar{\Omega}(0,p) = 1/p$ and $\bar{\Omega}(\infty,p) = 0$ is as follows

$$\bar{\Omega}(\xi, p) = \frac{1}{p}\, e^{-\sqrt{p+p^2}\,\xi} \tag{5.76}$$

Because the expression on the right-hand side cannot be immediately transformed backwards, we look at $\bar{T}(\xi,p)$, for which by equation (5.74):

$$\bar{T}(\xi, p) = -\frac{1}{\sqrt{p + p^2}}\, e^{-\sqrt{p+p^2}\,\xi} \tag{5.77}$$

One can find this expression in the tables of Laplace transforms[1]. The appropriate original function is

$$T(\xi, \theta) = \begin{cases} 0 & \text{for } \xi > \theta \\ -\,e^{-\theta/2}\, I_0\left(\frac{1}{2}\sqrt{\theta^2 - \xi^2}\right) & \text{for } \xi < \theta \end{cases} \tag{5.78}$$

In this $I_0(x)$ denotes the modified Bessel function of zero order and first type, i.e. the solution of the modified Bessel differential equation which is finite at $x = 0$ and normalised to unity for $n = 0$.

$$x^2 y'' + xy' - (x^2 + n^2)\, y = 0 \tag{5.79}$$

The shear stress is found by using equation (5.78) as a function of position and time. The minus sign obviously shows that the fluid velocity decreases from the wall outwards, and therefore the shear stress by the usual convention correspondingly works out as negative. The normalising undertaken causes the quantity T at the wall to jump to the value unity initially, $T(0, +0) = 1$. Thereafter the absolute value of wall shear stress for $\xi = 0$ decreases monotonically with time.

---

1 e.g. in Abramovitz, M. and Stegun, I.A. (Eds.): Handbook of Mathematical Functions. New York: Dover 1970, p. 1027.

On the basis of equation (5.79) one can easily prove from this that $dI_0(x)/dx = I_1(x)$ holds, i.e. the derivative of the modified Bessel function of zero order yields the Bessel function of the first order. We need this information to determine the velocity field $\Omega(\xi,\theta)$. Hence we return to the expression (5.67), from which

$$\frac{\partial\Omega}{\partial\theta} = \begin{cases} 0 & \text{for } \xi > \theta \\ e^{-\theta/2}\, I_1\left(\frac{1}{2}\sqrt{\theta^2-\xi^2}\right)\cdot\frac{\xi}{2\sqrt{\theta^2-\xi^2}} & \text{for } \xi < \theta \end{cases} \tag{5.80}$$

follows. By integrating we obtain

$$\Omega(\xi,\theta) = \begin{cases} 0 & \text{for } \xi > \theta \\ \Omega^+(\xi) + \frac{\xi}{2}\int_\xi^\theta e^{-\sigma/2}\,\frac{I_1\left(\frac{1}{2}\sqrt{\sigma^2-\xi^2}\right)}{\sqrt{\sigma^2-\xi^2}}\,d\sigma & \text{for } \xi < \theta \end{cases} \tag{5.81}$$

The quantity $\Omega^+(\xi)$ describes the normalised velocity immediately behind the wavefront (for $\theta = \xi + 0$; cf. Fig. 5.13). We find this quantity in the following way. An observer who is moving towards the wavefront at a velocity $d\xi/d\theta = -1$ records as a time derivative of the fluid velocity

$$\frac{d^-\Omega}{d\theta} = \frac{\partial\Omega}{\partial\theta} - \frac{\partial\Omega}{\partial\xi} = \frac{\partial T}{\partial\xi} - T - \frac{\partial T}{\partial\theta} = -T - \frac{d^-T}{d\theta}\ .$$

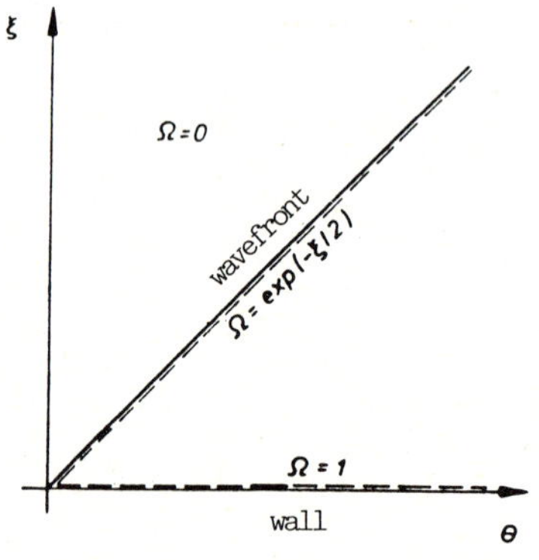

Fig. 5.13 Wavefront in a θ-ξ-diagram

The expressions (5.67) and (5.68) have been used in the transformation. By integrating this expression over the wavefront one obtains by bringing in equation (5.69) the following relationship between the flow velocity and the shear stress immediately behind the shock front $\Omega^+(\xi) = -T^+(\xi)$. Because at the wavefront $\theta = \xi$ holds, either ξ or θ can be used as the argument. An observer who is travelling with the wavefront, hence at the velocity $d\xi/d\theta = +1$, records the time derivative

$$\frac{d^+\Omega}{d\theta} = \frac{\partial\Omega}{\partial\theta} + \frac{\partial\Omega}{\partial\xi} = \frac{\partial T}{\partial\xi} + T + \frac{\partial T}{\partial\theta} = T + \frac{d^+T}{d\theta}\,.$$

By use of the already established *jump relationship* $T^+ = -\Omega^+$ it follows that $2d\Omega^+/d\theta = -\Omega^+$. By integrating and taking into account the initial condition (5.70) one obtains

$$\Omega^+(\xi) = e^{-\xi/2} \tag{5.82}$$

The velocity immediately behind the wavefront therefore decreases exponentially with increasing distance from the wall. The numerical evaluation of the expression (5.81) leads to the graphs in Figs. 5.14 and 5.15. They show the fluid velocity as a function of position at certain times, and as a function of time for fixed distance from the wall respectively, as experienced by individual fluid particles.

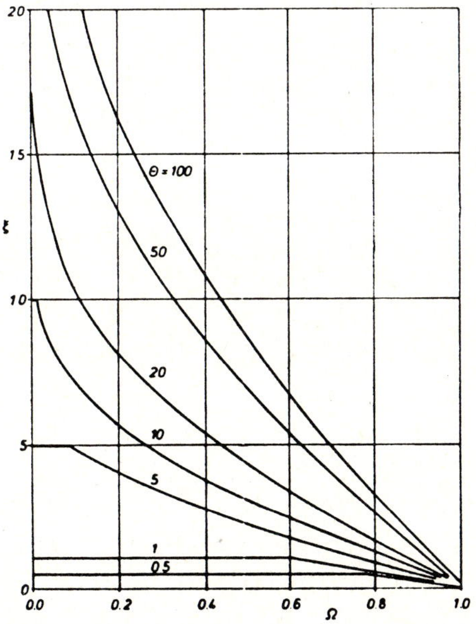

Fig. 5.14 Velocity profiles of a Maxwell fluid at different times

Fig. 5.14 clearly shows how the wavefront spreads out at a constant velocity into the fluid with a progressive decrease in amplitude. After time $\theta = 15$ there is practically no velocity jump at the wavefront, and the velocity behaviour is effectively the same as that of a Newtonian fluid, which has no sharp wavefront and which is characterised by the asymptotic decay for $\xi \to \infty$.

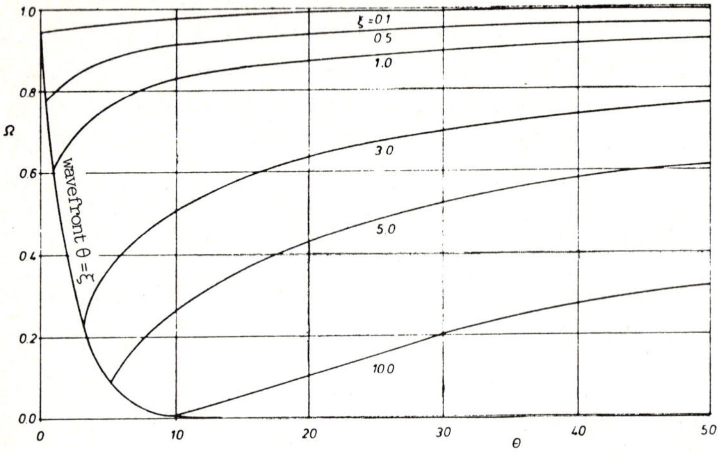

Fig. 5.15 Velocity as a function of time for different layers

If one defines the boundary layer thickness as that distance from the wall up to which marked velocities are recorded (say $\Omega \geqq 0.01$), then this position at first is situated in the wavefront, so that the boundary layer thickness at first increases strictly linearly with time. For longer periods of time compared with the relaxation time this is no longer true, because obviously that position at which the velocity has practically fallen to zero remains behind the wavefront (the graph for $\theta = 20$ in Fig. 5.14). One could obtain the increase in the boundary layer for greater values of time from an asymptotic representation of the exact result (5.81). We use however for the sake of simplicity an engineering approximation procedure, which is based on integral relationships. To this end we integrate equation (5.68) with respect to position and find by taking into account the boundary conditions at the wall and at infinity

$$\left(1 + \frac{d}{d\theta}\right) \int_0^\infty T(\xi, \theta)\, d\xi = -1 \qquad (5.83)$$

This relationship can be integrated with respect to time by taking into account the homogeneous initial condition (5.69) for the stress:

$$\int_0^\infty T(\xi,\theta)\,d\xi = -1 + e^{-\theta} \tag{5.84}$$

We shall now turn to the other field equation (5.67). If one were to act with it in the same way the shear stress at the wall $T(0,\theta)$ would appear in the integrated equation as an unknown quantity which should not occur in the formulas. We therefore multiply equation (5.67) by ξ and then integrate. In this way we obtain after partial integration and use of the second boundary condition (5.71)

$$\frac{d}{d\theta}\int_0^\infty \xi\,\Omega(\xi,\theta)\,d\xi = -\int_0^\infty T(\xi,\theta)\,d\xi \tag{5.85}$$

If one enters the explicit expression (5.84) on the right-hand side and integrates, taking into account the initial condition (5.69) for the velocity, then one obtains

$$\int_0^\infty \xi\,\Omega(\xi,\theta)\,d\xi = \theta + e^{-\theta} - 1 \tag{5.86}$$

We now introduce as a suitable measure for the thickness of the boundary layer the dimensionless *displacement thickness* Δ defined below (note that $\Omega(0,\theta) = 1$).

$$\Delta(\theta) := \int_0^\infty \Omega(\xi,\theta)\,d\xi \tag{5.87}$$

Without restricting the generality one can put

$$\Omega(\xi,\theta) = \varphi\left(\frac{\xi}{\Delta(\theta)},\theta\right) \tag{5.88}$$

The function on the right-hand side has because of (5.70) and (5.87) the properties

$$\varphi(0,\theta) = 1; \qquad \int_0^\infty \varphi(\zeta,\theta)\,d\zeta = 1 \tag{5.89}$$

If one inserts the expression (5.88) into equation (5.86) one then obtains for the displacement thickness the still quite exact result

$$\Delta^2(\theta) = \frac{\theta + e^{-\theta} - 1}{\int_0^\infty \zeta\,\varphi(\zeta,\theta)\,d\zeta} \tag{5.90}$$

Hence the displacement thickness can be approximately calculated as a function of time if one, as is usual for the approximation method of boundary layer theory, introduces an a-priori trial function $\varphi(\zeta,\theta)$, which satisfies the conditions (5.89). In the simplest case one makes a 'similarity assumption', i.e. one disregards the explicit dependence of the function φ on θ. Such a trial function, which satisfies the condition (5.89), is $\varphi(\zeta,\theta) = e^{-\zeta}$. With this function the denominator on the right-hand side of equation (5.90) gives the numerical value 1, so that the expression in the numerator is approximately the square of the displacement thickness. Hence there results particularly

$$\Delta \simeq \frac{\theta}{\sqrt{2}} + O(\theta^2) \quad \text{for } \theta \ll 1 \tag{5.91}$$

$$\Delta \simeq \sqrt{\theta} + O\left(\frac{1}{\sqrt{\theta}}\right) \quad \text{for } \theta \gg 1 \tag{5.92}$$

Initially therefore, the boundary layer thickness as previously mentioned increases linearly with time; after that however (for long periods of time as compared with the relaxation time) it increases only as the square root of the time. If one represents the asymptotic result $\Delta \simeq \sqrt{\theta}$ in terms of physical quantities, then the relaxation time does not enter into this relationship, and there is the same relationship as for a Newtonian fluid $\delta \simeq \sqrt{\eta_0 t/\rho}$. For a sufficient period of time therefore only the viscous but not the elastic properties of the fluid are involved.

5.1.7 Unsteady Couette flow

We now modify the previously discussed matter in a way in which we consider a channel of width h instead of a half space filled with fluid. The Maxwell fluid is therefore to be located between an unsteadily moved wall and a fixed wall parallel to it, as is the case in an unloaded journal bearing, or in Couette type viscometers. We shall look into the initial events of such a Couette flow when the moving wall at time $t = 0$ is suddenly brought to a constant velocity. With the channel width h a new dimensionless expression enters

$$\epsilon := \sqrt{\frac{\rho h^2}{\eta_0 \lambda}} \tag{5.93}$$

This parameter can on the one hand be interpreted as the ratio of channel depth h to the path $c\lambda$ traversed by the wave within a relaxation time; but on the other hand also as the root of the ratio of the two periods of time, namely that time $\rho h^2/\eta_0$ by which the motion of the wall in consequence of the viscous properties of the fluid at a distance h is taken into account, and the relaxation time λ.

Because the fluid adheres to the fixed wall ($y = h$, or $\xi = \varepsilon$), one has to replace the boundary condition (5.71) by

$$\Omega(\epsilon, \theta) = 0 \tag{5.94}$$

It is clear that the solution obtained in Section 5.1.6 is also valid here as long as the wavefront leaving the moving wall has not yet reached the fixed wall, i.e. for $\theta < \varepsilon$. At the time $\theta = \varepsilon$ the wavefront has just reached the fixed wall and is reflected because it cannot pass through it. Hence there arises a second wave travelling in the opposite direction, whose amplitude is determined by the condition (5.94). One can imagine this in such a way that at time $\theta = 0$ at the point $\xi = 2\varepsilon$, where there is in fact no fluid at all, a wave will be transmitted in the opposite direction, of exactly the same kind, except for the sign. Both extinguish themselves at the fixed wall at all times and moreover are additively superposed on each other because the phenomenon is linear. We denote the function defined by equation (5.81) by $\Omega_\infty(\xi, \theta)$, in which the subscript ∞ indicates that this is a solution of the problem for $h \to \infty$.

Thus in accordance with the above statements in the case of finite channel depth the expression $\Omega(\xi,\theta) = \Omega_\infty(\xi,\theta) - \Omega_\infty(2\varepsilon - \xi, \theta)$ applies until the front of the reflected wave has reached the moved wall, hence for $\theta < 2\varepsilon$. Accordingly it is reflected there again, so that a third elementary wave arises whose amplitude results from the condition (5.70), and which one can think of as starting at time $\theta = 0$ from the point $\xi = -2\varepsilon$. Thus each reflection of the wavefront brings in a new additional contribution. The complete solution for the velocity field can be expressed according to this statement in the following way by the function Ω_∞ in equation (5.81):

$$\Omega(\xi, \theta) = \Omega_\infty(\xi, \theta) - \sum_{n=1}^{\infty} \Omega_\infty(2n\epsilon - \xi, \theta) + \sum_{n=1}^{\infty} \Omega_\infty(2n\epsilon + \xi, \theta) \tag{5.95}$$

The same applies to the shear stress field. If one denotes the expression in equation (5.78) by $T_\infty(\xi,\theta)$, then the result is

$$T(\xi,\theta) = T_\infty(\xi,\theta) + \sum_{n=1}^{\infty} T_\infty(2n\epsilon - \xi,\theta) + \sum_{n=1}^{\infty} T_\infty(2n\epsilon + \xi,\theta) \tag{5.96}$$

The different sign in the second term is explained by the fact that between Ω and T, also between Ω_∞ and T_∞ the expression (5.67) applies. Note that at a certain time only a finite number of terms contribute to the solution, because the functions Ω_∞ and T_∞ vanish, if the first argument is greater than the second (the time).

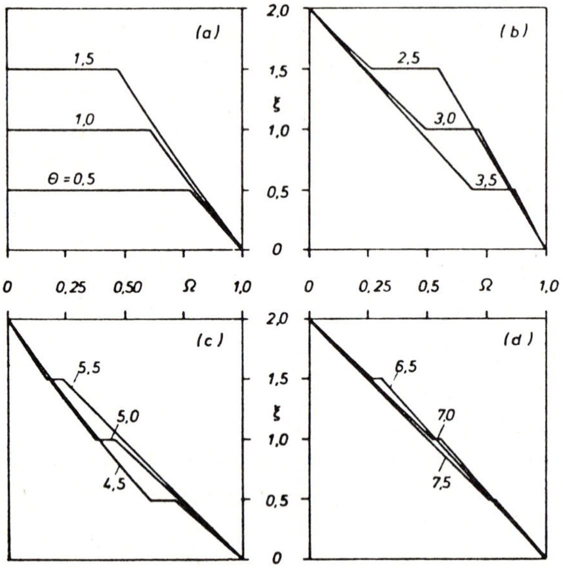

Fig. 5.16 Velocity profiles for various times for $\varepsilon=2$; (a) $0<\theta<2$, (b) $2<\theta<4$, (c) $4<\theta<6$, (d) $6<\theta<8$

Fig. 5.16 illustrates the result (5.95). During the time interval $0 < \theta < \varepsilon$ only the primary wave $\Omega_\infty(\xi,\theta)$ is present, which has a sharp wavefront. For $\varepsilon < \theta < 2\varepsilon$ the first reflected wave occurs. The jump in the velocity profile thus moves in the opposite direction as before, its amplitude decreases further and the propagation speed continues to be constant. During the time interval $2\varepsilon < \theta < 3\varepsilon$ the shock moves once more in the original direction and gives rise to a second reflected wave. With increasing time more and more terms in equation (5.95) gain influence, corresponding to the increasing number of reflections, and the velocity profile approaches the straight line which is characteristic of a steady drag flow in a channel.

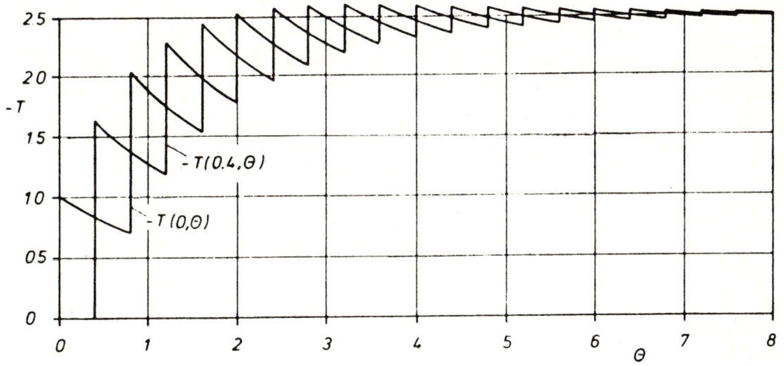

Fig. 5.17 Shear stresses at moved and stationary walls; $\varepsilon = 0.4$

The reflections of the shear shock at the two walls are especially clear if one expresses the wall shear stresses as functions of time (Fig. 5.17). Thus the values of the stresses increase suddenly if the disturbance has just reached the wall concerned. At the moving wall this is the case at the times 0, 2ε, 4ε, etc.; at the fixed wall at the times ε, 3ε, 5ε, etc. Hence the shear stress can rise during the intervening time well above the equilibrium value $T = 1/\varepsilon$ (correspondingly $\tau = \eta_0 U/h$), which occurs under steady conditions after a long time. This 'overshoot' of the shear stress in a fluid which has linear properties is an inertia effect and should not be confused with a similar phenomenon described in Section 5.1.1, which reflects non-linear behaviour.

The number of the steps and the step height of the 'sawtooth' in the stress behaviour depends on the magnitude of the coefficient ε. The smaller ε is the more 'teeth' appear and the smaller is their step height. The stress curves at both walls thus approach each other and finally converge in the limiting case $\varepsilon \to 0$ (Fig. 5.18). The result for $\varepsilon \to 0$, in the above case the exponential function

$$-\epsilon T = 1 - e^{-\theta} \tag{5.97}$$

thereby arises from theory without consideration of the inertia forces.

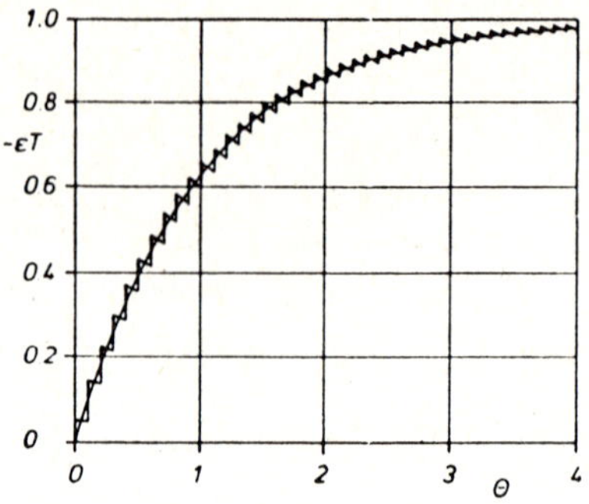

Fig. 5.18 Shear stress as a function of time for $\varepsilon = 0.05$ and $\varepsilon \to 0$

If one neglects the left-hand side in the momentum equation (5.54), then $\partial\tau/\partial y = 0$, i.e. the shear stress in the channel depends only on time. In consequence the right-hand side of the constitutive equation (5.62) can also only depend on time. The velocity profile is therefore linear at all times and corresponds to the function $u = U(1 - y/h)$ because of the boundary conditions at the walls. Hence the right-hand side of (5.62) becomes $-\eta_0 U/h$, and the integration leads to the result $\tau = -(\eta_0 U/h)\cdot(1 - e^{-t/\lambda})$, which is expressed in dimensionless form by equation (5.97). Hence a theory that does not take into account the inertia forces describes the velocity field in the channel in such a way as if the flow were steady under the momentarily acting boundary conditions. This assumption which is often taken as a basis of the unsteady flow test technique is valid according to the previous considerations if the coefficient defined in equation (5.93) $\varepsilon \ll 1$.

5.2 Non-linear effects in unsteady pipe flow

During the movement of viscous fluids through circular pipes steady conditions will generally apply. In the case of a Newtonian fluid and with linear-viscoelastic fluids, unsteady periodic components which are superposed on a steady pressure flow, would only lead to increased dissipation. The mean volume flux would be the same as for steady flow under constant mean pressure drop. For viscoelastic fluids that have non-linear properties on the contrary unsteady behaviour like the periodic movement of the pipe wall or an oscillatory pressure drop in certain cases increases the discharge.

We consider the kinematically simplest case of an unsteady flow through a circular pipe of radius r_0 in which the fluid moves parallel to the pipe axis throughout in the x-direction, and the velocity field u(r,t) depends, apart from being dependent on the time t, only on the axial distance r. Inlet processes are therefore not considered, and an axial shear flow occurs. A constant pressure drop $-\partial p/\partial x = k$ prevails along the axis of the pipe, and the wall of the pipe oscillates in the axial direction at an angular velocity ω and a maximum velocity U:

$$u(r_0, t) = U \cos \omega t \tag{5.98}$$

Only fully developed oscillatory motion with a periodic flow field is of interest. In the case of pseudoplastic fluids the oscillation of the wall leads to an increase in the discharge, which can be explained qualitatively as follows. Because of the motion of the wall there is an increase in the relative velocities of adjacent layers and hence there is an increase in the shear rate averaged over time. Because the viscosity of a pseudoplastic fluid decreases with shear rate, the fluid becomes thinner averaged over time because of the motion of the wall, so that for a given pressure drop the volume flux increases. This argument on a basis of the steady flow properties of a pseudoplastic fluid is of course only definite for sufficiently slowly varying flow. In the case of a markedly unsteady flow not the viscous properties alone, but also the memory of the fluid are involved. For an unsteady pipe flow the volume flux passing through the pipe cross-section naturally depends on time. However, only the mean value is of interest, which we shall denote by $\bar{\dot{V}}$, in which the bar will indicate the mean value with respect to time. $\bar{\dot{V}}$ is obtained by integration of the averaged velocity field $\bar{u}(r)$ over the cross-section:

$$\bar{\dot{V}} := 2\pi \int_0^{r_0} r\,\bar{u}(r)\,dr = -\pi \int_0^{r_0} r^2 \frac{d\bar{u}}{dr}\,dr \tag{5.99}$$

The second equality follows after a partial integration, taking into account the fact that according to equation (5.98) the mean velocity of the fluid particles vanishes at the wall. We shall determine with the help of this expression the mean discharge as soon as the time averaged velocity field is known.

In order to be able to analyse the unsteady pressure-drag flow we must formulate the equation of motion in the x-direction. It connects the acceleration $\partial u/\partial t$ of the fluid particles at the specific pressure force k/ρ and the specific friction force, which has its origin in the shear stress τ between the cylindrical layers. The equation reads:

$$\rho \frac{\partial u}{\partial t} = k + \frac{1}{r} \frac{\partial}{\partial r} (r\tau) \tag{5.100}$$

(cf. the set of formulas for cylindrical coordinates in the Appendix).

For the average over time of this expression the left-hand side vanishes and after one integration with respect to r it follows that

$$\bar{\tau} = -\frac{k}{2} r \tag{5.101}$$

Hence the shear stress averaged over time, like the shear stress of a steady pipe flow, increases independently of the physical properties, proportionally to the distance from the pipe axis, so that the amount of the increase depends on the axial pressure gradient k (cf. equation (3.113)).

It is expedient to consider first the velocity field u_L of linear viscoelastic fluids. Because of the linearity of the equation of motion (5.100) and the physical law (5.6) involved with it, the flow field for such fluids is composed additively of two components. One is proportional to the pressure gradient k and describes the conditions for a stationary wall, i.e. a steady pure pressure flow. Because a linear viscoelastic fluid under steady shear flow behaves like a Newtonian fluid in which the zero-shear viscosity η_0 is critical (cf. equation (5.7)), there is for this pressure flow a Poiseuille flow with a parabolic velocity profile. The second term is proportional to the amplitude U of the wall motion, and describes the motion without pressure gradient, that is a pure drag flow oscillating at the angular velocity ω. Hence the linear viscoelastic behaviour can be expressed in terms of two constants η' and η'' dependent on frequency, the real and the imaginary parts of the complex velocity (cf. Section 5.1.3). The velocity field can therefore be expressed as follows:

$$u_L(r, t) = \frac{kr_0^2}{4\eta_0}\left(1 - \frac{r^2}{r_0^2}\right) + U\,g\left(\frac{r}{r_0}, \omega t\right) \tag{5.102}$$

The dimensionless function $g(\sigma,\zeta)$ can be considered to be the real part of a complex quantity $g^*(\sigma,\zeta)$, for which according to equation (5.100) in combination with the constitutive law $\tau^* = \eta^*\dot{\gamma}^*$ (equation (5.33)) yields the equation

$$\rho\omega\frac{\partial g^*}{\partial \zeta} = \frac{\eta' - i\eta''}{r_0^2}\cdot\frac{1}{\sigma}\frac{\partial}{\partial\sigma}\left(\sigma\frac{\partial g^*}{\partial\sigma}\right) \tag{5.103}$$

Its solution, which satisfies the boundary condition (5.98), i.e. the expression $g^*(1,\zeta) = e^{i\zeta}$ and which is finite on the axis (for $\sigma \to 0$) can be found by using a trial function of product form and can be represented by a Bessel function of zero order with complex argument. Equation (5.103) shows that g^* depends on two real numerical parameters, which are the dimensionless quantities

$$St := \frac{\rho\omega r_0^2}{\eta'}, \qquad V := \frac{\eta''}{\eta'} \tag{5.104}$$

We shall denote the *Stokes number* by St; the viscoelastic parameter V is a pure material quantity. By use of these abbreviations the analytical expression for the second term in equation (5.102) is as follows:

$$g(\sigma, \zeta) = \mathrm{Re}\left\{\frac{J_0\left(\sigma\sqrt{\frac{-iSt}{1-iV}}\right)}{J_0\left(\sqrt{\frac{-iSt}{1-iV}}\right)}\,e^{i\zeta}\right\} \tag{5.105}$$

Because the time averaged value of this periodic drag flow vanishes over the total cross-section, only the steady pressure flow contributes to the mean volume flux. The first term in equation (5.102) yields in accordance with equation (5.99)

$$\bar{\dot{V}}_L = \frac{\pi k r_0^4}{8\eta_0} \tag{5.106}$$

For a linear viscoelastic fluid therefore the oscillatory motion exerts no effect on the mean volume flux. A change in the flow rate in consequence of this kind of unsteady action can only occur in fluids that have non-linear properties.

We now turn therefore to fluids that have non-linear properties. Because this concerns an unsteady shear flow, and of the extra-stresses only the shear stress τ between the layers is involved, we can start from the constitutive law in the form (2.3). By separating all linear components of the functional the law can be expressed in the form

$$\tau(t) = \int_0^\infty G(s)\, \dot{\gamma}(t-s)\, ds + \underset{s=0}{\overset{\infty}{H}} [\dot{\gamma}(t-s)] \tag{5.107}$$

in which the second term describes the non-linear properties. For the sake of brevity only the time dependence of the field quantities is noted, but the argument r which describes the position dependence is suppressed. For $\dot{\gamma}$ we can write $\partial u/\partial r$ for the pipe flow. The time average of equation (5.107) gives

$$\bar{\tau} = \eta_0 \frac{d\bar{u}}{dr} + \overline{\underset{s=0}{\overset{\infty}{H}} \left[\frac{\partial u}{\partial r}(t-s) \right]} \tag{5.108}$$

Hence we have taken into account the fact that the integral on the relaxation function G(s) according to equation (5.7) agrees with the zero-shear η_0 of the fluid. According to equation (5.101) the left-hand side is a known function of r.

Hitherto the theory has been rigorously accurate; the rest depends on the special properties of the non-linear part $H[\dot{\gamma}]$ in the constitutive law. In order to obtain concrete results we shall assume that

$$\frac{1}{\eta_0} \underset{s=0}{\overset{\infty}{H}} [\dot{\gamma}(t-s)] = -\lambda_2^2\, \dot{\gamma}^3(t) + \lambda_4^3\, \dot{\gamma}^2(t) \frac{\partial \dot{\gamma}(t)}{\partial t} \tag{5.109}$$

We therefore only consider two cubic terms, which contain the shear rate and its time derivative at present time t; λ_2 and λ_4 are two constants with the dimension of time. In Section 5.2.1 one will see that these assumptions are not by chance, but for slow, and slowly varying processes represent an approximation (fourth order) of the real behaviour of any simple fluid (cf. the non-linear components in equation (5.118)). In this approximation the flow curve of the fluid is moreover described by the cubic expression $\tau = \eta_0(1 - \lambda_2^2\dot{\gamma}^2)\dot{\gamma}$ because in the limiting case of a steady shear flow the right-hand side of equation (5.109) reduces to the first term. The minus sign of the term which is multiplied by $\dot{\gamma}^3$ has been inserted because in this case only pseudoplastic materials are considered, whose shear viscosity decreases with increasing shear rate.

The second term on the right-hand side vanishes in the time average of equation (5.109). Hence equation (5.108) is rewritten as

$$\frac{\bar{\tau}}{\eta_0} = \frac{d\bar{u}}{dr} - \lambda_2^2 \overline{\left(\frac{\partial u}{\partial r}\right)^3} \qquad (5.110)$$

Now the cubic expression in the velocity field u remains small compared with the linear terms for slow and slowly varying flow. This allows us to consider it as a perturbation term and for the purpose of computation to replace the actual, but unknown, velocity field u(r,t) by that field $u_L(r,t)$ which has resulted from the sole consideration of the linear components in the constitutive equation (5.102). Its additive resolution into a term proportional to the pressure gradient k and a term proportional to the velocity amplitude U of the wall leads to quantities with the factors k^3, k^2U, kU^2 and U^3, because the sum is taken to the third power. The terms that are linear in U depend on the dimensionless time ζ just as the function $g(\sigma,\zeta)$ in equation (5.105), therefore they contain either $\sin\zeta$ or $\cos\zeta$ as a factor and vanish when averaged over time. The terms that are multiplied by U^3 belong to a pure periodic flow without a pressure gradient (k = 0). Their mean values therefore vanish likewise. Hence it follows from $\overline{(\partial u_L/\partial r)^3}$ that there are only contributions proportional to k^3 and kU^2. By use of the expression (5.102) for the field $u_L(r,t)$ and taking into account the result (5.101) for the time averaged shear stress, equation (5.110) transforms into

$$-\frac{d\bar{u}}{dr} = \frac{kr}{2\eta_0} + \frac{\lambda_2^2 k^3 r^3}{8\eta_0^3} + \frac{3\lambda_2^2 kU^2 r}{2\eta_0 r_0^2}\overline{\left(\frac{\partial g}{\partial \sigma}\right)^2} \qquad (5.111)$$

By integrating this expression and taking into account equation (5.99) the time average value of the volume flux $\bar{\dot{V}}$ follows. The first two terms on the right-hand side are independent of U, and lead therefore to that discharge which arises without oscillation of the wall (U = 0), i.e. with steady pressure flow, and which will be accordingly denoted by $\dot{V}_{st}$. Because the fluid has shear thinning properties $\dot{V}_{st}$ is naturally larger than the volume flux of the associated linear viscoelastic fluid (having $\lambda_2 = 0$). The last term on the right-hand side of equation (5.111) leads to a flow rate increase due to the oscillating wall. The result can be expressed by use of the abbreviation (5.106) thus:

$$\bar{\dot{V}} = \dot{V}_{st} + \bar{\dot{V}}_L \left(\frac{\lambda_2 U}{r_0}\right)^2 \cdot \Lambda(St, V) \tag{5.112}$$

This shows that for the unsteady flow the additionally transported volume flux increases as the square of the amplitude of the wall velocity U. The dimensionless factor Λ is given by

$$\Lambda(St, V) := 12 \int_0^1 \sigma^3 \, \overline{\left(\frac{\partial g}{\partial \sigma}\right)^2} \, d\sigma \tag{5.113}$$

from that velocity field $g(\sigma,\zeta)$ which describes the pure drag flow of the linear viscoelastic reference fluid compared. It depends on the two characteristic quantities St and V as defined by equation (5.104) and is always positive. Therefore for a pseudoplastic fluid that has a memory the oscillatory motion of the pipe wall leads to an increase in the discharge. By use of the explicit result (5.105) and taking into account the fact that the functions $\cos^2\zeta$ and $\cos\zeta \sin\zeta$ have the same mean values 1/2 and 0 respectively, one finds the dependence on St and V illustrated in Fig. 5.19. This shows that the discharge increases with the frequency of the motion of the wall. If the frequency is increased at constant amplitude of wall displacement, then the drag velocity U, the Stokes number St, in general also the viscoelastic parameter V, and the second term in equation (5.112) collectively increase. A suitable means of enhancing the discharge would therefore increase the frequency of the wall motion.

In the limit of a large Stokes number the drag effect of the oscillating wall is restricted to a thin boundary layer near the wall. The function $g(\sigma,\zeta)$, which describes the drag flow of the linear viscoelastic reference fluid, is therefore different from zero only in the immediate vicinity of the wall. One can therefore, calculating it to a first approximation, omit the curvature of the wall. The pipe wall can be replaced by an oscillating plane wall. The geometrically simple problem which thus arises has been dealt with in Section 5.1.5, the appropriate velocity field has been given in equation (5.59). Thus

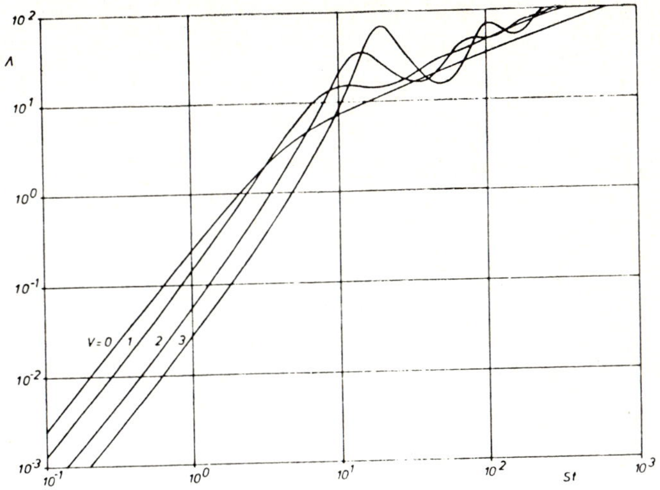

Fig. 5.19 The function $\Lambda(St,V)$ describing the discharge increase for an oscillating wall

u/U corresponds to the function g used here, and $r_0(1 - \sigma)$ would be substituted for the wall distance y. If one forms with this expression the integral in equation (5.113), then one obtains after a short calculation the asymptotic expression $\Lambda(St,V) = 3\sqrt{2St(V + \sqrt{1 + V^2})} + O(1)$ for $St \gg 1$. Therefore for a large Stokes number the function $\Lambda(St,V)$ represented in Fig. 5.19 increases as $\sqrt{St}$, and the factor depends in a simple way on the viscoelastic parameter V. For a small Stokes number, $St \ll 1$, on the other hand $\Lambda(St,V) = St^2/4(1 + V^2) + O(St^3)$, which however will not be proved in more detail here.

One should mention that the result (5.112) not only describes the situation considered so far, which is characterised by constant pressure gradient and an oscillating pipe wall, but also applies in the case of a sinusoidally oscillating pressure gradient and stationary wall. Indeed an observer moving with the oscillating wall records the same time averaged volume flux as a stationary observer, because his directional velocity vanishes in the average over time. For him the wall is stationary, but there arises per unit volume a time dependent inertia force $-\rho\partial(U\cos\omega t)/\partial t$, which is equivalent to an additional axial pressure gradient of the form $\rho U\omega\sin\omega t$. Such a sinusoidally oscillating pressure gradient superposed on the constant value k therefore leads to the same increase in discharge as the oscillating wall. Therefore if there is altogether a pressure gradient $-\partial p/\partial x = k + k'\sin\omega t$ and the pipe wall is at rest, then with the same assumptions as before equation (5.112) still holds. There would only be a substitution of $k'/\rho\omega$ for U.

5.2.1 Constitutive equation for slow and slowly varying processes

The following statements are mainly intended to serve the purpose of justifying the expression (5.109) for the non-linear components in the constitutive law for the shear stress. They point however at the same time to an alternative to the integral constitutive equations used in Section 5.1.

The constitutive equation (5.6) which describes linear viscoelastic behaviour with wide-ranging memory can be regarded as an approximation of the general law for shear flows with a small shear rate at all times. For processes whose amplitude is not small enough in regard to linearisation, but which however, measured over the range of the memory capability of the fluid, change slowly, another approximation of the general law can be introduced. We explain this for unsteady shear flows in which the shear stress $\tau(t)$ between two layers depends on the history of the shear rate $\dot{\gamma}(t - s)$, $s \geqq 0$ (cf. equation (2.3)). For the sake of simplicity consider a simple shear flow. The dependence on position of the quantities τ and $\dot{\gamma}$ is not involved in this connection and is therefore not considered.

We start from a certain shear flow of deformation rate $\dot{\tilde{\gamma}}(t - s)$ and also consider a set of motions with reduced amplitude, and at the same time with an extended past history, for which the shear rate at time t - s is given by the expression

$$\dot{\gamma}(t, s) = \epsilon\,\dot{\tilde{\gamma}}(t - \epsilon s); \quad 0 < \epsilon \leqslant 1 \tag{5.114}$$

The smaller the numerical parameter ε is chosen, the slower does the fluid flow at the actual time and the slower has the shear rate changed with time in the past (Fig. 5.20). We now approximate the past histories of motions shown in Fig. 5.20, and hence the right-hand side of equation (5.114) by the first n members of a Taylor series expansion for s = 0.

$$\dot{\gamma}(t, s) = \epsilon \sum_{k=0}^{n-1} \frac{(-\epsilon s)^k}{k!} \cdot \frac{\partial^k \tilde{\dot{\gamma}}(t-\epsilon s)}{\partial t^k} \Bigg|_{s=0} \tag{5.115}$$

We write partial derivatives because there is possibly also a dependence on position. The smaller the parameter ε, the greater is the time interval in which equation (5.115) is 'valid' for the given number of terms, and the fewer are the terms sufficing for the approximation in a given time interval.

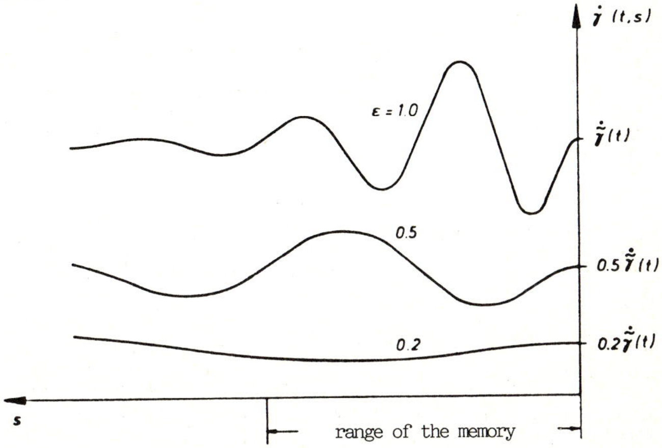

Fig. 5.20 Illustration of slow and slowly varying processes

Because the motion far back in the past, on account of fading memory, has only an unimportant effect on the stress at the present time, the approximation formula does not need to be valid for an infinite range of s, but it is adequate if the approximation extends to a certain time in the past. The length of the time interval of interest here is determined by the *range of the memory*, say the greatest relaxation time of the fluid. For a sufficiently small value of ε, i.e. for sufficiently slow and slowly varying processes, equation (2.3) can therefore be replaced by

$$\tau(t) = \underset{s=0}{\overset{\infty}{\mathrm{F}}} \left[\sum_{k=0}^{n-1} \frac{\epsilon^{k+1}}{k!} \cdot \frac{\partial^k \tilde{\dot{\gamma}}(t)}{\partial t^k} (-s)^k \right] \tag{5.116}$$

Because the retardation s occurs explicitly, one can consider the functional to be evaluated with respect to s. Hence the right-hand side reduces to a function of the coefficients depending on the present time t, but not on the retardation s:

$$\tau(t) = \Phi\left(\epsilon\dot{\tilde{\gamma}}(t), \epsilon^2\frac{\partial\dot{\tilde{\gamma}}(t)}{\partial t}, \ldots, \epsilon^n\frac{\partial^{n-1}\dot{\tilde{\gamma}}(t)}{\partial t^{n-1}}\right) \tag{5.117}$$

The memory of the fluid expresses itself here in the occurrence of time derivatives of the velocity field at the present time. One can in general proceed from the assumption that the constitutive function on the right-hand side has a regular series expansion. In the course of a rigorous expansion in powers of ε, Φ can be replaced by the first members of this expansion up to the order $O(\varepsilon^n)$ inclusive. An absolute term is lacking in this because the shear stress vanishes in the state of rest. Thus there results, dependent on the degree n of the approximation, an expression with a greater or less number of terms,which approximately describes the shear stress behaviour for slow and slowly varying flow. Thus the special properties of the fluid come out in constant expansion coefficients independent of the degree of the approximation.

We shall only use here an expansion up to terms of the order ε^4 inclusive. The individual terms result from various products, which can be formed with the four arguments in equation (5.117). There arise: $\varepsilon\dot{\tilde{\gamma}}$ as a single term of the first order in ε ; $\varepsilon^2\dot{\tilde{\gamma}}^2$, and $\varepsilon^2\partial\dot{\tilde{\gamma}}/\partial t$ as second order terms; $\varepsilon^3\dot{\tilde{\gamma}}^3$, $\varepsilon^3\dot{\tilde{\gamma}}\partial\dot{\tilde{\gamma}}/\partial t$ and $\varepsilon^3\partial^2\dot{\tilde{\gamma}}/\partial t^2$ as third order terms; $\varepsilon^4\dot{\tilde{\gamma}}^4$, $\varepsilon^4\dot{\tilde{\gamma}}^2\partial\dot{\tilde{\gamma}}/\partial t$, $\varepsilon^4\dot{\tilde{\gamma}}\partial^2\dot{\tilde{\gamma}}/\partial t^2$ and $\varepsilon^4\partial^3\dot{\tilde{\gamma}}/\partial t^3$ as fourth order terms. The components that are even in $\dot{\tilde{\gamma}}$, for example $\varepsilon^3\dot{\tilde{\gamma}}\,\partial\dot{\tilde{\gamma}}/\partial t$, cannot of course occur in the law for the shear stress. For a flow reversal ($\dot{\tilde{\gamma}} \to -\dot{\tilde{\gamma}}$) these terms would remain unaltered, whilst the shear stress changes its sign ($\tau \to -\tau$). This symmetry property excludes four of the previously quoted terms. The other expressions multiplied by corresponding constants yield in the sum the approximation of the fourth order for the shear stress for slow and slowly varying unsteady shear flows. After the derivation of this expression has been explained we can immediately identify the reference flow $\dot{\tilde{\gamma}}(t)$ with the actual motion $\dot{\gamma}(t)$, i.e. put $\varepsilon = 1$ and express the constitutive equation as follows:

$$\tau = \eta_0 \left\{ \underbrace{\dot{\gamma}}_{\text{1st.}} \underbrace{- \lambda_1 \frac{\partial \dot{\gamma}}{\partial t}}_{\text{2nd.}} \underbrace{- \lambda_2^2 \dot{\gamma}^3 + \lambda_3^2 \frac{\partial^2 \dot{\gamma}}{\partial t^2}}_{\text{3rd.}} \underbrace{+ \lambda_4^3 \dot{\gamma}^2 \frac{\partial \dot{\gamma}}{\partial t} - \lambda_5^3 \frac{\partial^3 \dot{\gamma}}{\partial t^3}}_{\text{4th. order}} \right\} \tag{5.118}$$

In this η_0 denotes the zero-shear viscosity of the fluid, and $\lambda_i (i = 1,\ldots,5)$ are constants which have the dimension of time. For a steady shear flow equation (5.118) reduces to $\tau = \eta_0(\dot{\gamma} - \lambda_2^2\dot{\gamma}^3)$, i.e. the constants η_0 and λ_2 are found from the flow curve of the fluid. The minus sign in the cubic term relates to shear thinning materials. Some other constants in equation (5.118), λ_1, λ_3 and λ_5 can be derived from the relaxation function G(s) of the material, critical for low amplitude processes. The linear components of the general law are of course, as has been established earlier, described by the expression (5.6). If one replaces in that $\dot{\gamma}(t - s)$ by the expression (5.115) and again puts $\varepsilon = 1$, then instead of the history of the shear rate there appears a sum of time derivatives $\partial^k \dot{\gamma}(t)/\partial t^k$, taken at the present time t. Thus certain integrals over the relaxation function appear as coefficients. A comparison with equation (5.118) yields in addition to the earlier used expression (5.7) the following relationships:

$$\eta_0 \lambda_1 = \int_0^\infty s\, G(s)\, ds \tag{5.119}$$

$$\eta_0 \lambda_3^2 = \frac{1}{2} \int_0^\infty s^2\, G(s)\, ds \tag{5.120}$$

$$\eta_0 \lambda_5^3 = \frac{1}{6} \int_0^\infty s^3\, G(s)\, ds \tag{5.121}$$

The determination of the constant λ_4 is made neither under steady, nor under linear viscoelastic conditions, but requires an experiment with an unsteady motion of finite amplitude. In Section 5.2 the constitutive equation of the fourth order was used, in order to explain the increase in the discharge for periodic pipe flow (equation (5.109) and equation (5.118)). Hence information about the quantity λ_4 was not required; because of the two non-linear quantities in the law the term multiplied by λ_4^3 did not have any effect because of the periodicity of the flow.

Note that the approximations of lower order are included in the fourth order constitutive equation as special cases, and arise from it, whilst the higher order terms enclosed in brackets in equation (5.118) are omitted. In the first approximation only the first term $\eta_0\dot{\gamma}(t)$ remains on the right-hand side, i.e. every simple fluid has for slow and slowly varying flow Newtonian flow behaviour to a first approximation. In the second order approximation the term $-\eta_0\lambda_1\partial\dot{\gamma}(t)/\partial t$ is added, which takes into account the memory of the fluid, but the linear flow behaviour does not change. The negative sign of these terms is explained by the fact that the fluid always remembers the past motion and does not in fact foresee the future motion. Within a second order approximation the constitutive equation of a simple fluid can also be written in the form

$$\tau(t) = \eta_0\,\dot{\gamma}(t-\lambda_1) \tag{5.122}$$

According to this the deformation rate at the moment going back by λ_1 in time is critical for the value of the shear stress. The constant λ_1 here plays the part of a dead time. In the third approximation a non-linear term appears for the first time in the constitutive equation, so that the graph of the flow function is approximated by a cubic parabola. With increasing order in the approximation for slow and slowly varying processes the memory of the fluid expresses itself in higher and higher time derivatives of the shear rate at the present time.

Problems

5.1 In the creep test an initially at rest and unloaded material at time $t = 0$ is subjected to the shear stress τ_0, which is then kept constant (Fig. 5.7). Calculate and sketch how a Maxwell fluid which is described by the constitutive equation (5.12), and a material deforms, whose properties correspond to the model shown in Fig. 5.21 (*Lethersich fluid*). Determine the time dependence of the creep compliance $\gamma(t)/\tau_0$.

In the second case the differential form of the constitutive equation must first be found ($\ddot{\gamma}$ also occurs).

5.2 The creep compliance $J(t)$ for a linear viscoelastic fluid was measured. The result can be described by the two parameter analytical function

$$J(t) = \frac{1}{\eta_0}\left[t + \frac{3}{2}\lambda - \frac{1}{2}\lambda\, e^{-t/\lambda} \right]$$

Determine from this the relaxation function G(t) of the fluid. It is advisable to use the Laplace transforms $\bar{J}(p)$ and $\bar{G}(p)$, for which the expression (5.27) holds.

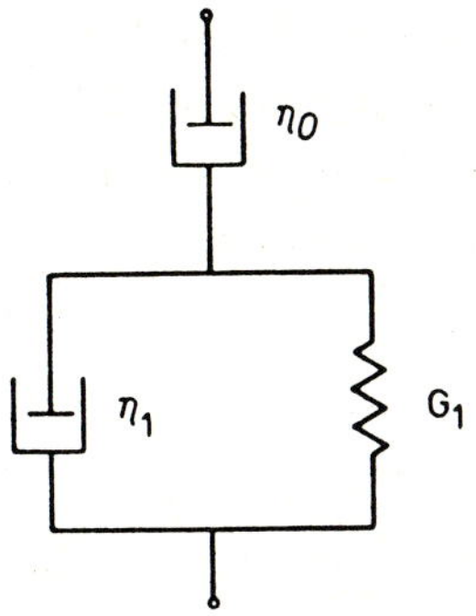

Fig. 5.21 Lethersich model

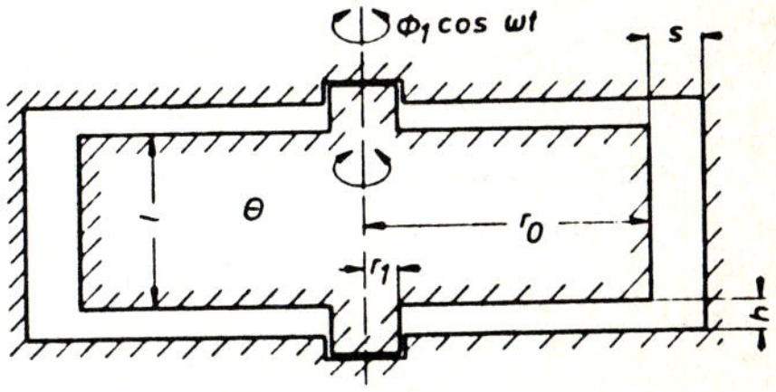

Fig. 5.22 Torsional shock absorber

5.3 Calculate the two components $\eta'(\omega)$ and $\eta''(\omega)$ of the complex viscosity for a Maxwell fluid which has the relaxation function $G(s) = (\eta_0/\lambda)e^{-s/\lambda}$, and compare the analytical results qualitatively with the real data shown in Fig. 5.9.

5.4 Determine for a Maxwell fluid which has the constants ρ = 970 kg/m^3, η_0 = 1.27 Ns/m^2 and λ = 10^{-4}s. (The data refer to a silicone oil of an average molecular weight 1000 at 24°C):

(a) the depth of penetration of a shear wave of frequency 10^{2}s^{-1}, 1 s^{-1} and 10^{-2} s^{-1} on the basis of the analytical results in Section 5.1.5;

(b) the displacement thickness for a suddenly accelerated wall after 10^{-2} s, 1 s and 10^2 s on the basis of equation (5.90).

5.5 The torsional shock absorber shown in Fig. 5.22 consists of a circular section cylindrical housing which performs rotational oscillations of given amplitude Φ_1 and given angular velocity ω, and a mass with moment of inertia Θ which is set into oscillation at the same frequency by a linear viscoelastic oil of complex viscosity $\eta'(\omega)-i\eta''(\omega)$ in the gap. The gap dimensions h and s are small compared with the external dimensions $r_0 - r_1$ and ℓ. Neglecting inertia forces in the fluid there is a Couette flow between the outer surfaces of the cylinders and there is a torsional flow between the end surfaces with a periodic shear rate, and boundary effects in the corners and near the axis can be disregarded. Prove that for the energy W dissipated per cycle for a suitably chosen definition of a reference energy E_0 equation (5.50) and the associated Fig. 5.11 still apply. Determine how the constant χ of the instrument depends on Θ, h, s, r_0, r_1 and ℓ.

5.6 On the basis of a Maxwell fluid model show that the propagation velocity of the wavefront in the Rayleigh problem agrees with the phase speed of high frequency sinusoidal transverse waves. By use of a concept used in gas dynamics one can denote this high frequency limiting value as a 'frozen' wave speed. What is 'frozen' in a Maxwell model under these conditions? What emerges for the low frequency limiting value, the 'equilibrium value' of the sinusoidal wave speed?

5.7 With reference to the statements made in Section 5.1.6 consider for an incompressible Maxwell fluid the following unsteady initial process: the fluid occupies the half space $y > 0$ and is at rest until time $t = 0$. From this time onwards there is a constant shear stress τ_0 acting on the plane boundary $y = 0$; the pressure in the fluid is spatially constant. The resulting unsteady plane shear flow is described by the differential equations (5.54) and (5.62) with appropriate initial and boundary conditions.

After the introduction of suitable dimensionless quantities the determination of the shear stress leads to the same initial boundary problem which describes the velocity field in the Rayleigh problem. The normalised shear stress field is accordingly described by the function defined in equation (5.81). Determine from that in particular the velocity field at the 'surface' and the velocity immediately behind the wavefront as a function of time. Which results for the displacement thickness and for the velocity at the surface provide an integral approximation procedure?

5.8 A non-Newtonian fluid of density ρ flows under the effect of a constant pressure gradient k in the x-direction along a channel between two parallel plane walls separated by a distance h. One wall (at y = h) is stationary, the other (at y = 0) oscillates at an angular velocity ω and the velocity amplitude U in its own plane, so that the displacement direction is independent of time and forms an angle δ with the x-direction. There arises a periodic plane shear flow which has two velocity components u(y,t) and w(y,t).

(a) Determine first the velocity field for a general linear viscoelastic fluid which has the properties $\eta'(\omega)$ and $\eta''(\omega)$, assuming that the damped transverse wave leaving the oscillating wall no longer has any effect at the position of the stationary wall (sufficiently large Stokes number $\rho\omega h^2/\eta'$).

(b) Now consider a weak non-linear viscoelastic fluid whose non-linear properties are described by the cubic contribution $-\eta_0\lambda_2^2[(\partial u/\partial y)^2 + (\partial w/\partial y)^2]\cdot\partial u/\partial y$ to the shear stress τ_{xy}. It corresponds to a third order term in the sense of an approximation for slow and slowly varying processes; cf. equation (5.118). Assuming that this term can be taken as a perturbation term, it is sufficient to form it with the velocity field previously established in part (a). Calculate in this way the time averaged flux in the x-direction as a function of ρ, η', η'', η_0, λ_2, h, k, ω, U and δ. For the greatest possible discharge is the longitudinal ($\delta = 0$) or the transverse motion ($\delta = \pi/2$) of the wall more favourable?

6 NEARLY VISCOMETRIC FLOWS

6.1 Shear flows with a weak unsteady component

In order to explain the concept 'nearly viscometric' we turn once more to the class of unsteady plane shear flows. Thus the extra-stresses operating on a fluid particle at time t are influenced by the history of the shear rate of the particle $\dot{\gamma}(t - s)$. We consider in particular those motions for which $\dot{\gamma}(t - s)$ within a past period of time, which expresses the range of the memory, deviates only slightly from a fixed (independent of s) reference value $\dot{\gamma}_0$.

This is say a motion which results from superposing a steady shear flow of constant shear rate $\dot{\gamma}_0$ on an oscillatory shear flow of small amplitude. Because the motion deviates only slightly from a viscometric flow, we speak of a *nearly viscometric flow.* The reference value $\dot{\gamma}_0$ however does not have to be absolutely constant, but is allowed to vary with the present time t, and in general also varies from particle to particle. Later we shall identify $\dot{\gamma}_0$ in particular with the local shear rate $\dot{\gamma}(t)$ at the time t.

If now the difference $\dot{\gamma}(t - s) - \dot{\gamma}_0$ is small, one can simplify the law (2.3) accordingly. If one now substitutes $\dot{\gamma}_0 + [\dot{\gamma}(t - s) - \dot{\gamma}_0]$ for $\dot{\gamma}(t - s)$ and expands the functional with respect to the terms within the square brackets, then the viscometric flow function $F(\dot{\gamma}_0)$ naturally enters as the absolute term (cf. equation (2.4)). The difference history $\dot{\gamma}(t - s) - \dot{\gamma}_0$ appears to a first approximation as an additional linear component, which in general can be expressed as an integral. For a nearly viscometric shear flow we can correspondingly write the constitutive equation for the shear stress in the form

$$\tau = F(\dot{\gamma}_0) + \int_0^\infty K_{\parallel}(s, \dot{\gamma}_0)\,[\dot{\gamma}(t - s) - \dot{\gamma}_0]\,ds \tag{6.1}$$

The kernel of the integral, apart from being dependent on the retardation time s, is also dependent on the reference value $\dot{\gamma}_0$ of the shear rate. The symbol $\parallel$ indicates that the basic viscometric flow is disturbed by a parallel oriented non-viscometric motion. For $\dot{\gamma}_0 = 0$ equation (6.1) is the linear viscoelastic constitutive equation. Hence follows the relationship with the linear viscoelastic influence function:

$$K_{\|}(s, 0) = G(s) \tag{6.2}$$

We may also imagine that $\dot{\gamma}(t - s)$ itself expresses a viscometric history and is therefore quite independent of $(t - s)$. Under this condition the right-hand side of equation (6.1) must of course agree with $F(\dot{\gamma})$. For an approximation this value can be replaced by the sum $F(\dot{\gamma}_0) + \hat{\eta}(\dot{\gamma}_0)[\dot{\gamma} - \dot{\gamma}_0]$, in which $\hat{\eta}$ denotes the differential viscosity (cf. Section 2.1). A comparison of both expressions shows that the integral with respect to the influence function $K_{\|}$ agrees with the differential viscosity:

$$\int_0^\infty K_{\|}(s, \dot{\gamma}_0)\, ds = \hat{\eta}(\dot{\gamma}_0) \tag{6.3}$$

We can see equation (5.7) in this when $\dot{\gamma}_0 \to 0$. In passing it may be mentioned that one can find the function $K_{\|}(s, \dot{\gamma}_0)$ by using dynamic experiments. If one superposes a parallel oriented pure sinusoidal oscillation of small amplitude $\dot{\gamma}_1$ on a steady shear flow $(\dot{\gamma}_0)$, then there follows:

$$\dot{\gamma}^*(t) = \dot{\gamma}_0 + \dot{\gamma}_1 e^{i\omega t} \tag{6.4}$$

Here too the real part of the complex quantities denoted by a star is meant. Under these conditions the fluid responds with an oscillatory stress of the same frequency, whereby a phase difference generally arises between the shear stress and the shear rate, which corresponds to a complex viscosity:

$$\tau^* = F(\dot{\gamma}_0) + \eta_{\|}^*(\omega, \dot{\gamma}_0)\, \dot{\gamma}_1 e^{i\omega t} \tag{6.5}$$

Fig. 6.1 shows the results of an appropriate dynamic experiment. The property $\eta_{\|}^*$ in this is closely connected with the memory function $K_{\|}$. If one inserts equation (6.4) into the right-hand side of equation (6.1) and compares the result with equation (6.5) one obtains

$$\eta_{\|}^*(\omega, \dot{\gamma}_0) = \int_0^\infty K_{\|}(s, \dot{\gamma}_0)\, e^{-i\omega s} ds \tag{6.6}$$

The conjugate complex value of $\eta_{\|}^*$ therefore represents the Fourier transform of $K_{\|}$.

So far we have only considered the shear stress. It is clear that similar expressions hold for the normal stress differences. In particular the normal stress differences for a nearly viscometric shear flow of the type expressed by equation (6.1) can be written as the sum of the corresponding viscometric function $N_1(\dot{\gamma}_0)$ or $N_2(\dot{\gamma}_0)$ (instead of

$F(\dot{\gamma}_0)$), and a correction term linear in $[\dot{\gamma}(t - s) - \dot{\gamma}_0]$, whereby different influence functions occur in each integral. Because no use is made of these relationships in what follows, no details of them will be given here.

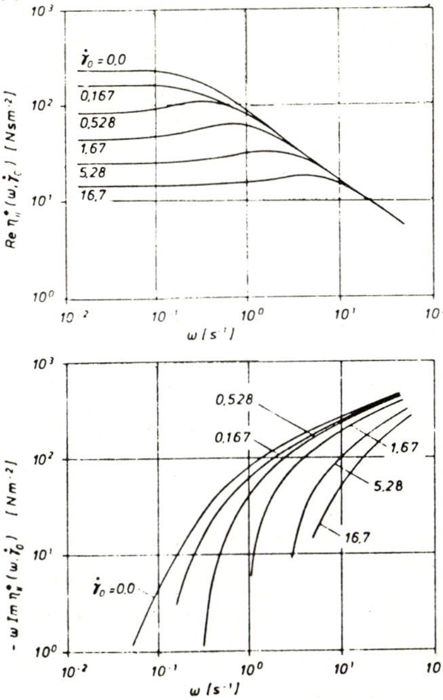

Fig. 6.1 The property $\eta^*_{\shortparallel}(\omega, \dot{\gamma}_0)$ of a polymer solution (4% polystyrene in chlorodiphenol at 25°C; after Macdonald)

In extension of the previous statements there now follows a short discussion of what happens when an unsteady weak perturbation $\dot{\gamma}_\perp(t)$ in the neutral direction (normal to the original shear plane) is superposed on a plane steady shear flow of shear rate $\dot{\gamma}_0$. There then naturally arises a shear stress $\tau_\perp$ in this direction, and similar arguments as before lead to the constitutive equation

$$\tau_\perp = \int_0^\infty K_\perp(s, \dot{\gamma}_0)\, \dot{\gamma}_\perp(t - s)\, ds \tag{6.7}$$

An absolute term is absent here, because without the transverse motion ($\dot{\gamma}_\perp = 0$), there is no shear stress $\tau_\perp$. If the base shear is absent ($\dot{\gamma}_0 = 0$), it is a matter once again of the law of the linear viscoelasticity, i.e. in addition the influence function $K_{\perp}(s,\dot{\gamma}_0)$ for the perturbations superposed transversely reduces in the limiting case of $\dot{\gamma}_0 \to 0$ to the linear viscoelastic relaxation function

$$K_{\perp}(s, 0) = G(s) \qquad (6.8)$$

If the base shear $\dot{\gamma}_0$ as well as the perturbation $\dot{\gamma}_{\perp}$ are constant with time, then we have got a viscometric flow, and for the stress $\tau_{\perp}$ the expression $\tau_{\perp} = \eta(\dot{\gamma}_0)\cdot\dot{\gamma}_{\perp}$ holds according to equation (2.15). By comparison with equation (6.7) there follows

$$\int_0^{\infty} K_{\perp}(s, \dot{\gamma}_0)\, ds = \eta(\dot{\gamma}_0) \qquad (6.9)$$

Note that the integral over the memory function $K_{\perp}$ agrees with the ordinary viscosity, whereas in the corresponding formula for $K_{\parallel}$ the differential viscosity occurs (equation (6.3)).

The laws (6.1) and (6.7) are based on the assumption that perturbations of a small amplitude are superposed on a viscometric base flow of arbitrary strength. No limiting assumptions at all have been made so far about the progress of the perturbation with respect to time. We shall now for the sake of simplicity restrict the discussion to slowly varying processes, for which the constitutive equations can be significantly simplified. We thus assume that the shear rate of a fluid particle within the past time interval, which corresponds to the range of the memory, has in fact changed, but the characteristic external time scale for the change is long compared with the memory time λ. One could think for instance that the shear rate has in the past fluctuated sinusoidally about a finite mean; then the external time would be identified with the reciprocal of the angular velocity $1/\omega$, and the criterion for the slow change would be $\omega\lambda \ll 1$. One sees from this that in the simplification aimed for not only the process itself but also the length of the memory is involved. The shorter the memory of the fluid is the quicker may the shear rate change without infringement of the condition for slow variation. In this context 'slow variation' also means 'sufficiently short memory', measured to the external time.

For a slowly varying motion the past history can be approximated by the present state and its change with respect to time, i.e. the process $\dot{\gamma}(t - s)$ occurring in the past time is replaced by the tangent to the function at the present time (Fig. 6.2):

$$\dot{\gamma}(t-s) \simeq \dot{\gamma}(t) - \frac{D\dot{\gamma}(t)}{Dt} \cdot s \tag{6.10}$$

Hence at times in the far distant past considerable differences possibly occurred between the approximation and the actual movement. Because of the fading memory the changed early history has only an insignificant effect on the present stress. We can therefore imagine that the memory has a finite range. Only within this time interval has the approximation to agree with the actual motion.

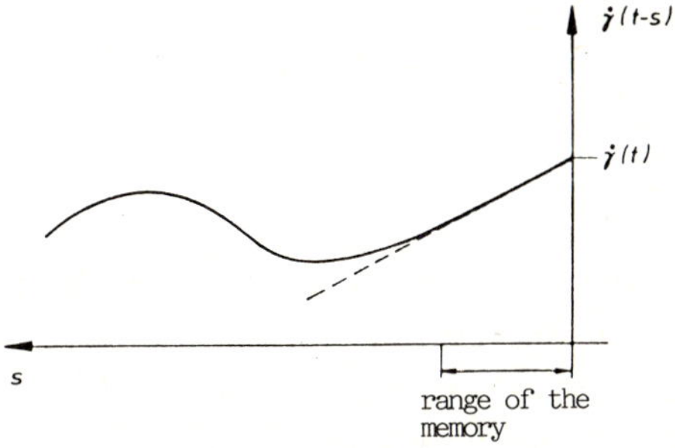

Fig. 6.2 Illustration of equation (6.10)

It is expedient in what follows to identify the reference value $\dot{\gamma}_0$ for the shear rate with its present value $\dot{\gamma}(t)$. By use of equation (6.10) the equation (6.1) then reduces to

$$\tau = F(\dot{\gamma}) - \beta(\dot{\gamma}) \frac{D\dot{\gamma}}{Dt} \tag{6.11}$$

(the argument t in $\dot{\gamma}(t)$ is henceforth suppressed). Hence the function $\beta(\dot{\gamma})$ obviously depends on the kernel $K_{\parallel}$, as follows:

$$\beta(\dot{\gamma}) = \int_0^{\infty} s\, K_{\parallel}(s, \dot{\gamma})\, ds \tag{6.12}$$

A more accurate expression would thus be $\beta_{\parallel}(\dot{\gamma})$. But because the integral formed with the influence function $K_{\perp}$ in the corresponding way below plays no part, we can dispense with the index $_{\parallel}$. In equation (6.11) a further function $\beta(\dot{\gamma})$ only dependent on $\dot{\gamma}$ occurs besides the viscometric flow function $F(\dot{\gamma})$. The term $-\beta D\dot{\gamma}/Dt$ deals with the memory of the fluid. For a simple fluid without any memory τ could only depend on the present value of $\dot{\gamma}$, but not on its change with respect to time $D\dot{\gamma}/Dt$, so that $\beta = 0$ would apply. For fluids that have a memory a new property $\beta(\dot{\gamma})$ in the constitutive equation for the shear stress occurs, which we shall call the *'second flow function'*. The difference from the original form (6.1) of the constitutive equation for nearly viscometric shear flows is that instead of the total past shear history in equation (6.11) only its time change at the present time comes into consideration, and hence the past is 'overcome'. We have been able to obtain this simplification only by the additional limitation of slow variation with respect to time, a fact which we must always remember.

We learned about another possibility in equation (5.118) in Section 5.2.1, of simplifying the general law for unsteady shear flows. For this the flow must, as we now repeat, be slowly varying. For this however the shear rate itself must be sufficiently small. In a comparison between the two approximations one should note that equation (5.118) is a fourth order approximation for slow and slowly varying processes, whereas equation (6.11) is a second order approximation for slowly varying processes. The present time shear rate $\dot{\gamma}$ ought therefore to be arbitrarily large. Naturally there are connections between both expressions. Thus obviously those terms of equation (5.118) which only contain $\dot{\gamma}$, namely $\eta_0(\dot{\gamma} - \lambda_2^2\dot{\gamma}^3)$ are merely the first members of a series expansion of the flow function $F(\dot{\gamma})$. Accordingly the expression $\eta_0(\lambda_1 - \lambda_4^3\dot{\gamma}^2)$, which appears in equation (5.118) multiplied by $\partial\dot{\gamma}/\partial t$, represents the start of a series expansion of the second flow function $\beta(\dot{\gamma})$.

As regards the experimental determination of the second flow function it is possibly significant to know that $\beta(\dot{\gamma})$ is connected with the imaginary part of the complex viscosity $\eta^*_{\parallel}(\omega,\dot{\gamma})$ defined by equation (6.5) and obtained by measurement.

$$\beta(\dot{\gamma}) = -\lim_{\omega \to 0} \frac{\operatorname{Im} \eta^*_{\parallel}(\omega, \dot{\gamma})}{\omega} \tag{6.13}$$

One obtains this expression by comparing (6.12) with (6.6). The 'second flow function' has a finite positive 'zero value' (for $\dot{\gamma} \to 0$), which will be denoted by β_0. From equations (6.2) and (6.12) it follows that

$$\beta_0 = \int_0^\infty s\,G(s)\,ds = \frac{\nu_{10}}{2} \tag{6.14}$$

The last equality results from equation (5.37) by use of equations (5.34) and (5.38). We therefore state that $2\beta_0$ is equal to the zero value of the first normal stress coefficient. However it would not be valid to conclude that the functions $2\beta(\dot{\gamma})$ and $\nu_1(\dot{\gamma})$ are always equal. Besides the connection given by (6.14) no generally valid relationships are known between the two functions. We must therefore look on the second flow function as a new and independent physical property. It is clear that with nearly viscometric flow non-viscometric functions will likewise occur in the constitutive equations for the normal stress differences. We note only the first normal stress difference:

$$\tau_{11} - \tau_{22} = N_1(\dot{\gamma}) - \kappa_1(\dot{\gamma})\frac{D\dot{\gamma}}{Dt} \tag{6.15}$$

The property $\kappa_1(\dot{\gamma})$ will play no part in the applications described in the following section, and therefore does not need to be considered in more detail.

6.2 Plane steady boundary layer flows

The purpose of the following remarks is firstly to develop adequate constitutive laws for boundary layer type flows. We consider in this context on the one hand lubricating films whose thickness varies slowly in the flow direction (Fig. 6.3); on the other hand the situation of real boundary layers in the vicinity of fixed walls. Both situations are characterised by the fact that the flow rate varies primarily in the direction normal to the wall, but insignificantly in the direction parallel to the wall. Furthermore in addition to the main velocity component parallel to the wall there occurs only a weak component normal to the wall. The stream lines are weakly curved as a result of this, and the motion of the liquid is as a whole a nearly viscometric flow.

In the derivation of adequate equations we shall assume, as in Section 6.1 that for each fluid particle the state of motion changes only slowly over the extent of the memory. In the one-dimensional case we could under this condition express the shear rate at earlier times in terms of its instantaneous value $\dot{\gamma}(t)$ and the variation of it with respect to time (equation (6.10)). In the three-dimensional case the first Rivlin-Ericksen tensor $\mathbf{A}_1$ takes the place of $\dot{\gamma}$, and the time change of the deformation rates is contained in the second Rivlin-Ericksen tensor $\mathbf{A}_2$. One can explain on the basis of equation (1.67) that for short memory $\mathbf{A}_2$ in addition to $\mathbf{A}_1$ is critical. We must therefore turn our attention to these two kinematic tensors.

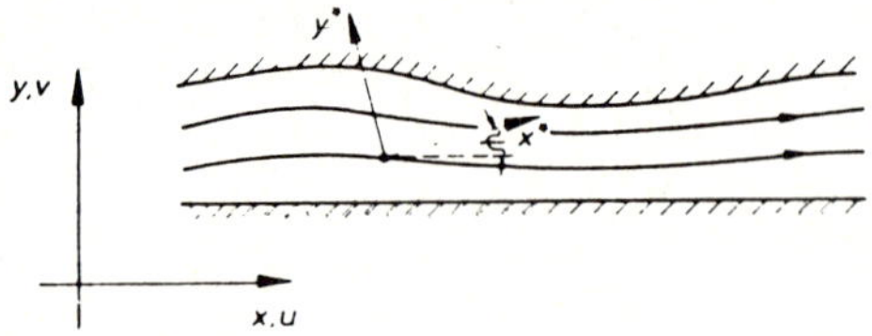

Fig. 6.3 Plane nearly viscometric flow

Referring to the Cartesian base (fixed wall x-y system) shown in Fig. 6.3, $\mathbf{A}_1$ is directly expressed by:

$$\mathbf{A}_1 = \begin{bmatrix} 2\dfrac{\partial u}{\partial x} & \dfrac{\partial u}{\partial y} + \dfrac{\partial v}{\partial x} \\ \dfrac{\partial u}{\partial y} + \dfrac{\partial v}{\partial x} & -2\dfrac{\partial u}{\partial x} \end{bmatrix} \tag{6.16}$$

Because only constant density fluids are considered, we can on the grounds of continuity here and in what follows substitute $-\partial u/\partial x$ for $\partial v/\partial y$. With reference to a suitably rotated local base the tensor takes on the form

$$\mathbf{A}_1^* = \begin{bmatrix} 0 & \dot{\gamma} \\ \dot{\gamma} & 0 \end{bmatrix} \tag{6.17}$$

Accordingly for a plane flow of a constant density fluid the instantaneous deformation of any fluid particle can be considered as a pure shear. The amount of the shear rate $\dot{\gamma}$ results from the fact that the matrices (6.16) and (6.17) have the same determinants, which are invariant for the rotation

$$\dot{\gamma}^2 = 4\left(\frac{\partial u}{\partial x}\right)^2 + \left(\frac{\partial u}{\partial y} + \frac{\partial v}{\partial x}\right)^2 \tag{6.18}$$

Now the following relationship exists between the two matrices that describe a symmetrical tensor **A** and the angle of rotation ψ:

$$\begin{bmatrix} A_{x^*x^*} & A_{x^*y^*} \\ A_{x^*y^*} & A_{y^*y^*} \end{bmatrix} = \begin{bmatrix} \cos\psi & \sin\psi \\ -\sin\psi & \cos\psi \end{bmatrix} \begin{bmatrix} A_{xx} & A_{xy} \\ A_{xy} & A_{yy} \end{bmatrix} \begin{bmatrix} \cos\psi & -\sin\psi \\ \sin\psi & \cos\psi \end{bmatrix}$$

There are three independent expressions, which we shall note in detail for further use:

$$A_{x^*y^*} = A_{xy} \cos 2\psi - \frac{1}{2}(A_{xx} - A_{yy}) \sin 2\psi \tag{6.19}$$

$$A_{x^*x^*} = \frac{1}{2}(A_{xx} + A_{yy}) + \frac{1}{2}(A_{xx} - A_{yy}) \cos 2\psi + A_{xy} \sin 2\psi \tag{6.20}$$

$$A_{y^*y^*} = \frac{1}{2}(A_{xx} + A_{yy}) - \frac{1}{2}(A_{xx} - A_{yy}) \cos 2\psi - A_{xy} \sin 2\psi \tag{6.21}$$

If one substitutes the expressions from equations (6.16) and (6.17) for the tensor components in equation (6.20), one obtains the angle ψ, about which the coordinate system must be rotated, so that the first Rivlin-Ericksen tensor takes on the form (6.17):

$$\tan 2\psi = -\frac{2\dfrac{\partial u}{\partial x}}{\dfrac{\partial u}{\partial y} + \dfrac{\partial v}{\partial x}} \tag{6.22}$$

For the second Rivlin-Ericksen tensor for a plane incompressible flow one uses equation (1.53) to obtain the following Cartesian matrix:

$$A_2 = \begin{bmatrix} 2\dfrac{\partial a_x}{\partial x} + 2\left(\dfrac{\partial u}{\partial x}\right)^2 + 2\left(\dfrac{\partial v}{\partial x}\right)^2 & \underline{\dfrac{\partial a_x}{\partial y}} + \dfrac{\partial a_y}{\partial x} + \underline{2\dfrac{\partial u}{\partial x}\dfrac{\partial u}{\partial y}} - 2\dfrac{\partial u}{\partial x}\dfrac{\partial v}{\partial x} \\ \underline{\dfrac{\partial a_x}{\partial y}} + \dfrac{\partial a_y}{\partial x} + \underline{2\dfrac{\partial u}{\partial x}\dfrac{\partial u}{\partial y}} - 2\dfrac{\partial u}{\partial x}\dfrac{\partial v}{\partial x} & 2\underline{\dfrac{\partial a_y}{\partial y}} + \underline{2\left(\dfrac{\partial u}{\partial y}\right)^2} + 2\left(\dfrac{\partial u}{\partial x}\right)^2 \end{bmatrix} \tag{6.23}$$

in which a_x and a_y are the components of the acceleration vector. We now make use of the fact that this is a boundary layer type flow. We start by assuming that the field quantities in the direction normal to the wall have a considerably greater variation than in the direction parallel to the wall. By using a nominal layer thickness h, a characteristic length ℓ parallel to the wall, a characteristic velocity U and the abbreviation q for U/h we can estimate the elements of the kinematic tensors as follows:

$$\frac{\partial u}{\partial y} \sim \frac{U}{h} =: q \qquad \frac{\partial u}{\partial x} \sim \frac{U}{\ell} = q\,\frac{h}{\ell} \sim \frac{\partial v}{\partial y}$$

$$v \sim q\,\frac{h^2}{\ell} \qquad \frac{\partial v}{\partial x} \sim q\left(\frac{h}{\ell}\right)^2$$

$$a_x \sim u\,\frac{\partial u}{\partial x} \sim q^2\,\frac{h^2}{\ell} \qquad \frac{\partial a_x}{\partial x} \sim q^2\left(\frac{h}{\ell}\right)^2 \qquad \frac{\partial a_x}{\partial y} \sim q^2\,\frac{h}{\ell}$$

$$a_y \sim u\,\frac{\partial v}{\partial x} \sim q^2\,\frac{h^3}{\ell^2} \qquad \frac{\partial a_y}{\partial x} \sim q^2\left(\frac{h}{\ell}\right)^3 \qquad \frac{\partial a_y}{\partial y} \sim q^2\left(\frac{h}{\ell}\right)^2$$

Because of the *slenderness condition* $h/\ell \ll 1$ only the components linear in h/ℓ are considered henceforth, apart from the absolute terms. These simplifications have the result that in equation (6.23) only the underlined elements remain. In equation (6.22) the term $\partial v/\partial x$ vanishes from the denominator and equation (6.18) can be replaced by

$$\dot{\gamma} = \frac{\partial u}{\partial y} \tag{6.24}$$

Because the right-hand side of (6.22) represents a quantity of the order of h/ℓ, one can write for the angle of rotation $\psi \sim h/\ell$, so that within the scope of the approximation $\cos 2\psi$ can be replaced by 1 and $\sin 2\psi$ by 2ψ. Thus equation (6.22) can be converted to

$$\frac{\partial u}{\partial x} + \psi\,\frac{\partial u}{\partial y} = 0 \tag{6.25}$$

Accordingly ψ indicates for boundary layer type flows that direction in which the velocity component u parallel to the wall does not vary (ψ does not describe the direction of the stream lines). We can now express the shortened form of the tensor $\mathbf{A}_2$ with the help of the transform formulas (6.19) to (6.21), and by the use of the result (6.25) for the angle of rotation ψ with reference to the rotated local basis. The result is:

$$\mathbf{A}_2^* = \begin{bmatrix} 0 & \dfrac{\partial a_x}{\partial y} \\ \dfrac{\partial a_x}{\partial y} & 2\dot{\gamma}^2 \end{bmatrix} \tag{6.26}$$

If the element in the secondary diagonal were to vanish, then the two kinematic tensors relative to the rotated basis would have the form typical of a viscometric flow (cf. equation (2.23)). The associated extra-stresses would therefore be described by the viscometric functions. But if $\partial a_x/\partial y \neq 0$, then a nearly viscometric flow occurs, and by analogy with the relevant statements in Section 6.1 there occur in addition to the viscometric components terms which are linear in the perturbation quantity $\partial a_x/\partial y$:

$$\tau_{x*y*} = F(\dot{\gamma}) - \beta(\dot{\gamma}) \frac{\partial a_x}{\partial y} \tag{6.27}$$

$$\tau_{x*x*} - \tau_{y*y*} = N_1(\dot{\gamma}) - \kappa_1(\dot{\gamma}) \frac{\partial a_x}{\partial y} \tag{6.28}$$

The functions $\beta(\dot{\gamma})$ and $\kappa_1(\dot{\gamma})$ occurring as coefficients in front of $\partial a_x/\partial y$ are the same as for a one-dimensional unsteady flow (cf. equations (6.11) and (6.15)) because the perturbation term $D\dot{\gamma}/Dt$ which appeared there could just as well be written in the form $\partial a_x/\partial y$.

Now the expression for the stress tensor relative to the local 'natural' basis is of less interest than that relative to the fixed wall Cartesian basis. We therefore once more use the transform formulas (6.19) to (6.21), from which within the scope of the approximation the following inversion formulas result:

$$A_{xy} = A_{x*y*} + (A_{x*x*} - A_{y*y*})\,\psi$$
$$A_{xx} - A_{yy} = A_{x*x*} - A_{y*y*} - 4\,A_{x*y*}\,\psi$$

Hence by use of equations (6.25), (6.27) and (6.28) one obtains the following constitutive equations for plane boundary layer flows of a short memory liquid:

$$\tau_{xy} = \eta(\dot{\gamma})\dot{\gamma} - \nu_1(\dot{\gamma})\dot{\gamma} \frac{\partial u}{\partial x} - \beta(\dot{\gamma}) \frac{\partial a_x}{\partial y} \tag{6.29}$$

$$\tau_{xx} - \tau_{yy} = \nu_1(\dot{\gamma})\dot{\gamma}^2 + 4\eta(\dot{\gamma}) \frac{\partial u}{\partial x} - \kappa_1(\dot{\gamma}) \frac{\partial a_x}{\partial y} \tag{6.30}$$

Hence as a consequence once again terms of the order $(h/\ell)^2$ have been neglected. According to equation (6.29) the shear stress τ_{xy} besides depending on the viscometric flow function (first term) also depends on the first normal stress function and on the second flow function, provided that $\partial u/\partial x \neq 0$ and $\partial a_x/\partial y \neq 0$. The influence of the viscometric normal stress properties (second term in equation (6.29)) which was evident in the derivation follows as a consequence of the fact that the fixed wall basis in general does not coincide with the natural basis. The third term on the right-hand side of equation (6.29) represents the memory of the liquid, which was assumed to be short.

The following equations of motion are valid for a plane flow if volume forces are neglected (cf. equations (1.113) and (1.114)):

$$\rho a_x = -\frac{\partial p}{\partial x} + \frac{\partial \tau_{xx}}{\partial x} + \frac{\partial \tau_{xy}}{\partial y} \tag{6.31}$$

$$\rho a_y = -\frac{\partial p}{\partial y} + \frac{\partial \tau_{yy}}{\partial y} + \frac{\partial \tau_{xy}}{\partial x} \tag{6.32}$$

In equation (6.32) the left-hand side and the last term on the right are small for boundary layer type flows ($h/\ell \ll 1$), i.e. the normal stress $-p + \tau_{yy}$ varies only slightly in a direction normal to the wall. One can therefore replace equation (6.32) by the statement

$$-p + \tau_{yy} = -\tilde{p}(x) \tag{6.33}$$

Hence the equation of motion in the x-direction takes the following simplified form:

$$\rho a_x = -\frac{d\tilde{p}}{dx} + \frac{\partial}{\partial x}(\tau_{xx} - \tau_{yy}) + \frac{\partial \tau_{xy}}{\partial y} \tag{6.34}$$

Because the normal stress difference is differentiated here in the direction parallel to the wall, so that it becomes a quantity of the order h/ℓ, and because terms of higher order in h/ℓ can be consistently neglected, it suffices in the evaluation of the 'boundary layer equation' (6.34) to abbreviate the constitutive law (6.30) to the viscometric component

$$\tau_{xx} - \tau_{yy} = \nu_1(\dot{\gamma})\dot{\gamma}^2 \tag{6.35}$$

Thus all the field equations have been derived which are necessary for the calculation of plane nearly viscometric flows of incompressible short memory liquids. These are the continuity equation

$$\frac{\partial u}{\partial x} + \frac{\partial v}{\partial y} = 0 \tag{1.102}$$

and the equation of motion (6.34) into which the constitutive equations (6.29) and (6.35) would have to be inserted. Hence note that $\dot{\gamma}$ in equation (6.24) is to be identified with $\partial u/\partial y$ and a_x according to equation (1.13) with $u\partial u/\partial x + v\partial u/\partial y$. Some boundary conditions naturally are added in the actual case to the field equations. In the following examples considered they are explained at a suitable place, and are included in the discussions.

6.2.1 Stagnation point boundary layer

The relationships derived above are first of all applied to the boundary layer in the vicinity of a stagnation point (Fig. 6.4). Assuming that the extra-stresses only act within a thin boundary layer near the wall and that the flow beyond that proceeds without friction, the function $\tilde{p}(x)$ introduced by equation (6.33) can be identified with the pressure distribution at the wall, as obtained from a frictionless theory. It is connected via the Bernoulli equation with the velocity distribution $\tilde{u}(x)$ which would be present for frictionless flow at the wall. As is well known the latter increases linearly in the case of the stagnation point flow with the distance x from the stagnation point, $\tilde{u} = Ux/\ell$ (cf. Section 1.1, where the properties of the frictionless flow towards a stagnation point have been discussed in detail). Therefore the following applies

$$\tilde{p} + \frac{1}{2}\rho U^2 x^2/\ell^2 = \text{constant or} \quad d\tilde{p}/dx = -\rho U^2 x/\ell^2$$

U and ℓ are constants which have the dimensions of velocity and length respectively.

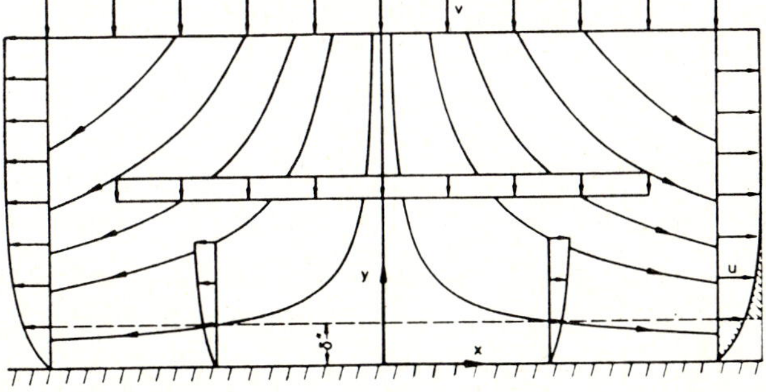

Fig. 6.4 Plane stagnation point flow with boundary layer near wall

The object of the following treatment will be to determine the velocity field in the boundary layer and the quantities resulting therefrom, like the shear stress at the wall and the boundary layer thickness. Considering the fact that the shear rate $\partial u/\partial y$ vanishes at the stagnation stream line and in consequence is small within a certain volume surrounding the stagnation point, we put zero values for the functions η, ν_1 and β. Besides the Newtonian flow properties we therefore consider the relevant normal stress difference as well as a short memory, but describe these physical properties in the simplest non-trivial form. Thus the analysis is comprehensive and we obtain a glimpse of the influence of the normal stress difference and the memory of the liquid.

If one therefore replaces the functions $\eta(\dot{\gamma})$, $\nu_1(\dot{\gamma})$ and $\beta(\dot{\gamma})$ by the constant values η_0, ν_{10} and β_0 and notes that β_0 is equal to $\nu_{10}/2$ (equation (6.14)), then the boundary layer equation (6.34) can be converted by use of the equations (6.29) and (6.35) and of the previously established expression for the pressure change $d\tilde{p}/dx$ into

$$uu_x + vu_y = \frac{U^2}{\ell^2}x + \frac{\eta_0}{\rho}u_{yy} + \frac{\nu_{10}}{2\rho}[u_y u_{xy} - u_x u_{yy} - uu_{xyy} - vu_{yyy}] \tag{6.36}$$

Hence partial derivatives with respect to x and y are indicated by way of exception by subscripts.

The trial function

$$u = xf'(y), \qquad v = -f(y) \tag{6.37}$$

satisfies the continuity equation ($u_x + v_y = 0$), and reduces equation (6.36) to the ordinary differential equation

$$f'^2 - ff'' = \frac{U^2}{\ell^2} + \frac{\eta_0}{\rho}f''' + \frac{\nu_{10}}{2\rho}[f''^2 - 2f'f''' + ff^{IV}] \tag{6.38}$$

The assumption (6.37) is therefore successful because it also allows the boundary conditions, to which the velocity field in the boundary layer is to be subjected, to be satisfied. The adhesion condition at the wall obviously requires that $f(0) = f'(0) = 0$. The smooth contact of the velocity profile parallel to the wall with the frictionless flow outside the boundary layer leads because $\tilde{u} = Ux/\ell$ to the conditions $f' \to U/\ell$ and $f'' \to 0$ at the boundary layer edge.

It is expedient to introduce a dimensionless coordinate

$$\zeta := \sqrt{\frac{\rho U}{\eta_0 \ell}}\, y \tag{6.39}$$

normal to the wall, a dimensionless velocity function

$$\varphi(\zeta) := \sqrt{\frac{\rho \ell}{\eta_0 U}}\, f(y) \tag{6.40}$$

and a dimensionless quantity for the non-Newtonian properties, the *Weissenberg number*

$$We := \frac{\nu_{10} U}{\eta_0 \ell} \tag{6.41}$$

The equation (6.38) and the associated boundary conditions take on the following normalised form

$$\varphi'^2 - \varphi\varphi'' = 1 + \varphi''' + \frac{1}{2} We(\varphi''^2 - 2\varphi'\varphi''' + \varphi\varphi^{IV}) \tag{6.42}$$

$$\varphi(0) = \varphi'(0) = \varphi''(\infty) = 0, \qquad \varphi'(\infty) = 1 \tag{6.43}$$

It is clear that for We = 0 the boundary layer problem for a Newtonian fluid is included as a special case.

Without going into the details of the solution method for this non-linear boundary layer problem, the results of the numerical integration will be recorded here. Fig. 6.5 shows φ' as a function of ζ and hence the velocity profile parallel to the wall in a suitably normalised form for various values of the Weissenberg number. Equation (6.37) expesses a *similarity solution*, i.e. the velocity profile has for a fixed numerical value We at each place x the same form, only the velocity amplitude varies along the wall linearly with the distance x from the stagnation point. Hence it follows that the boundary layer thickness has the same value overall. The displacement thickness δ^* which is defined by the equality of the hatched areas in Fig. 6.4, corresponds to a quite definite value of the dimensionless ζ-coordinate in the direction normal to the wall, which is denoted by Δ:

$$\sqrt{\frac{\rho U \ell}{\eta_0}}\, \frac{\delta^*}{\ell} = \int_0^\infty (1 - \varphi'(\zeta))d\zeta =: \Delta \tag{6.44}$$

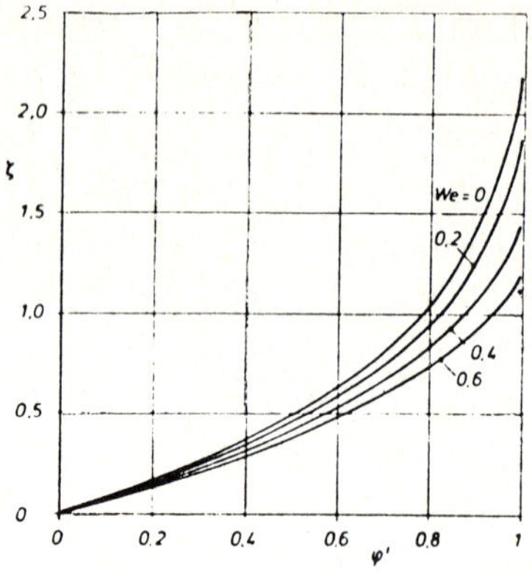

Fig. 6.5 Normalised velocity profiles of stagnation point flow

Fig. 6.5 shows that this value decreases with increasing Weissenberg number. Increasing elasticity of the liquid (for constant viscosity) thereby expresses itself by the stagnation point boundary layer becoming thinner. Hence the velocity gradient at the wall increases, represented by the coefficient $\varphi''(0)$, so that the local wall shear stress

$$\tau_w = \eta_0 \left.\frac{\partial u}{\partial y}\right|_{y=0} = \varphi''(0)\rho U^2 \sqrt{\frac{\eta_0}{\rho U \ell}}\,\frac{x}{\ell} \tag{6.45}$$

increases with Weissenberg number, if all other conditions remain constant. One can without voluminous calculations obtain a general account of the dependence of the quantities Δ and $\varphi''(0)$ on the parameter We as follows. The true velocity function $\varphi(\zeta)$ is replaced by the simplest possible analytical expression, which satisfies all the boundary conditions, gives a correct qualitative description of the velocity profile, but leaves the matter of the boundary layer thickness unsettled.

Such a trial function for the velocity in the direction parallel to the wall would for example be

$$\varphi'(\zeta) = 1 - e^{-\zeta/\Delta} \tag{6.46}$$

The 'form parameter' Δ occurring in this is the previously introduced dimensionless quantity for the displacement thickness (cf. equation (6.44)). By noting the boundary condition $\varphi(0) = 0$ there follows

$$\varphi(\zeta) = \zeta + \Delta\, e^{-\zeta/\Delta} - \Delta \tag{6.47}$$

Hence the differential equation (6.42) is not integrated, but with the introduction of this function one is establishing a deficiency.

Thus the true solution indicates that the deficiency vanishes overall and therefore also its weighted integral. One can now determine the parameter Δ of the approximate solution to the example by stipulating that the integral of the square of the deficiency, which of course can only be positive, lies as near as possible to the optimum zero, hence is as small as possible. If one therefore inserts the expression (6.47) into equation (6.42), brings all the terms to one side, squares the deficiency and finally integrates with respect to ζ from $\zeta = 0$ to $\zeta = \infty$, then there results an algebraic expression in Δ and We. If one looks for its minimum (with respect to Δ for constant We) one will obtain the following relationship between Δ and We

$$\Delta = \sqrt{\frac{8 - 12\,\mathrm{We} + 5\,\mathrm{We}^2}{12 - 10\,\mathrm{We}}} = \frac{1}{\varphi''(0)} \qquad (6.48)$$

Therefore the displacement distance decreases with the Weissenberg number, i.e. with increasing elasticity of the liquid, as long as $\mathrm{We} < 0.8$, and increases without limit when $\mathrm{We} \to 1.2$. The wall shear stress behaves reciprocally, because the associated dimensionless coefficient $\varphi''(0)$ (equation (6.45)) is equal to Δ^{-1} for the trial function according to equation (6.46). For $\mathrm{We} = 1.2$ the expression in equation (6.48) becomes imaginary. This indicates that the boundary value problem (6.42), (6.43) for $\mathrm{We} \gtrsim 1.2$ has no (real) solution, but probably the approximation process described does not yield the exact limit for We.

The assumed flow form (steady similarity solution) therefore has no existence at all under these conditions. One may possibly oppose this statement that for $\mathrm{We} = O(1)$ the supposition of the flow to be slowly varying and hence the assumption of the constitutive equations (6.29) to be applicable are infringed, and that it is therefore not valid to insert values of the order of unity for We. In fact with real fluids under such conditions in general more complicated properties make their appearance. One can however imagine an ideal material which has the properties referred to, also when $\mathrm{We} = O(1)$. At the least this material for a certain value of Weissenberg number would have to flow in the vicinity of a stagnation point in another way from that assumed. One learns from this that laminar flows which one observes in Newtonian liquids can become unstable for viscoelastic liquids, if the elasticity number exceeds a critical value.

In order to assess the goodness of the described integral approximation methods, their results are compared with those of a numerical integration of the boundary value problem (6.42), (6.43) (Fig. 6.6). In the case of a Newtonian fluid (We = 0) the analytical approximation method for the dimensionless displacement thickness Δ yields the value $\sqrt{2/3} = 0.817$, and for the wall friction coefficient $\varphi''(0) = \sqrt{1.5} = 1.225$, whereas 0.648 and 1.233 respectively are the numerically exact values.

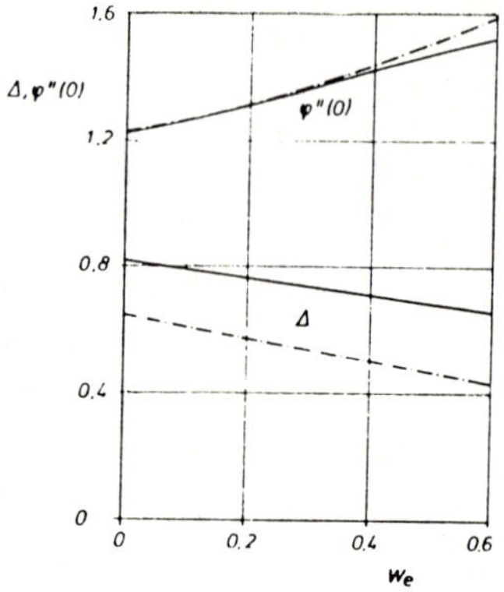

Fig. 6.6 Normalised displacement thickness Δ and normalised wall shear stress $\varphi''(0)$ as functions of the Weissenberg number We

—— : analytical integral method

-·- : numerical integration

Otherwise the dependence of both quantities on We described by the approximation formula (6.48) is qualitatively correct and in the case of the wall friction coefficient even quantitatively surprisingly good. The relatively large differences for the displacement thickness arise from the fact that the trial function (6.46) implies the simple relationship $\Delta \cdot \varphi''(0) = 1$, but for the true solution the product $\Delta \cdot \varphi''(0)$ depends on We and clearly works out to be less than unity.

6.2.2 Modified lubricating film theory

We return to the problem of the journal bearing flow which was considered in Section 3.2.4 under the single heading of viscometric flow behaviour. There is naturally some doubt about this procedure because the flow of the lubricating oil is in fact not viscometric at all, as seen in the shape of the lubricated gap in Fig. 3.20. For a narrow gap we shall look on the process much more as a plane nearly viscometric flow, and if the liquid has a short memory the general constitutive equations derived in Section 6.2 can be established. The purpose of the following statements is therefore to improve the lubricating film theory, taking into account the first normal stress difference and the second flow function. In particular the question is of interest, whether and in what way the load capacity of the bearing is also affected by these properties.

In order to arrange the analysis as clearly as possible, we limit ourselves to the case of small eccentricities ($\varepsilon \ll 1$ in Fig. 3.20). Because in thin lubricating films the inertia of the liquid can in general be neglected, one considers the gap to be unrolled so that the surface of the rotating shaft becomes a plane wall. From the start we must not thereby limit ourselves to the case in which the local height of the gap, as would correspond to that in Fig. 3.20, varies as a cosine form in the direction of motion, but we can formulate the problem at first in more general terms without more effort. The viscoelastic liquid is located in the gap between a plane wall moving at a velocity U and a fixed, slightly tilted or wavy wall, where h(x) describes the local height of the gap (Fig. 6.7). The gap either has a finite length ℓ (one may think about a Michell bearing say) or a 'wavelength' ℓ , at which it periodically repeats. We introduce the mean gap dimension h_0

$$h_0 := \frac{1}{\ell} \int_0^\ell h(x)dx \tag{6.49}$$

and put, because the 'waviness' shall be small, by use of a value $\varepsilon \ll 1$:

$$h(x) = h_0(1 + \epsilon H(x)) \tag{6.50}$$

It is also assumed that for $\varepsilon = 0$ (parallel gap) a pure drag flow exists. Then for $\varepsilon \neq 0$ a weak pressure gradient builds up, which alters the pure drag flow a little. It is expedient first of all to remember once more how under such conditions a non-Newtonian liquid without any memory and without normal stress properties will behave.

According to the statements which have been made in Sections 2.1 and 3.2.1 it behaves in the quasi-Newtonian approximation aimed for here, linear in ε, whereby the differential viscosity $\hat{\eta}(U/h_0)$ depending on the base shear takes over the role of the Newtonian viscosity. In particular therefore there holds for the discharge per unit width (cf. equation (3.38))

$$\frac{\dot{V}}{b} = \frac{Uh(x)}{2} - \frac{h^3(x)}{12\hat{\eta}}\frac{dp}{dx} = \frac{U\bar{h}}{2} \tag{6.51}$$

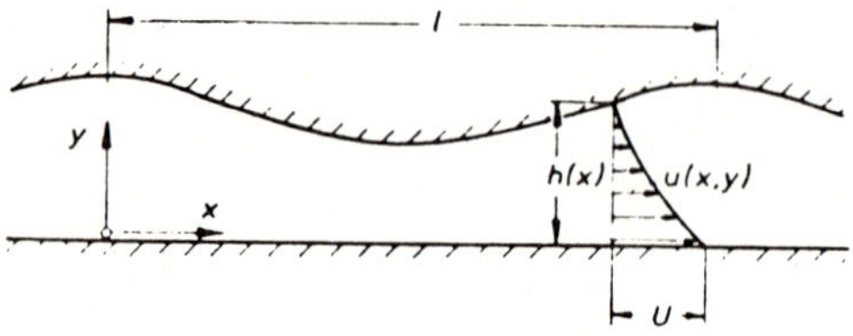

Fig. 6.7 Lubricating film notation

Because this expression must be independent of x for reasons of continuity, we express it by the drag velocity U and a not yet determined length $\bar{h}$. By equation (6.51) $\bar{h}$ is the height of the gap at that place where the pressure gradient vanishes. Because dp/dx vanishes with ε, and therefore represents a perturbation quantity, h_0 can be substituted for h(x) in the leading factor. Hence equation (6.51) gives

$$\frac{dp}{dx} = \frac{6\hat{\eta}U}{h_0^3}(h(x) - \bar{h}) \tag{6.52}$$

Because the pressures at x = 0 and x = ℓ are equal, an integration of this expression leads to the statement that $\bar{h}$ coincides with the mean gap dimension h_0 defined in equation (6.49). Now the velocity profile of the combined pressure-drag flow parallel to the wall in the case of a Newtonian fluid can be expressed in the following way (cf. equation (3.39)):

$$u = U\left(1 - \frac{y}{h}\right) - \frac{1}{2\hat{\eta}}\frac{dp}{dx}(hy - y^2) \tag{6.53}$$

If one eliminates dp/dx here by using equation (6.52), and then replaces $h(x) - h_0$ by $h_0\varepsilon H(x)$ and 1/h by $(1 - \varepsilon H)/h_0$ according to the equation (6.50), then one obtains the explicit representation

$$\frac{u}{U} = 1 - \frac{y}{h_0} + \epsilon H(x)\left(-2\frac{y}{h_0} + 3\frac{y^2}{h_0^2}\right) \tag{6.54}$$

By using the equation of continuity ($\partial v/\partial y = -\partial u/\partial x$) and noting the boundary condition $v = 0$ for $y = 0$, the velocity field normal to the wall can also be calculated:

$$\frac{v}{U} = \epsilon\, h_0 \frac{dH(x)}{dx}\left(\frac{y^2}{h_0^2} - \frac{y^3}{h_0^3}\right) \tag{6.55}$$

For the acceleration component $a_x := u\partial u/\partial x + v\partial u/\partial y$ there results as an approximation of the order ε^1:

$$\frac{a_x}{U^2} = \epsilon \frac{dH}{dx}\left(-2\frac{y}{h_0} + 4\frac{y^2}{h_0^2} - 2\frac{y^3}{h_0^3}\right) \tag{6.56}$$

After this preparatory work we can now deal with the main object and consider in detail a liquid with a short memory and with normal stress properties. We consider first the expression

$$\frac{\partial}{\partial x}(\tau_{xx} - \tau_{yy}) = \frac{d(\nu_1\dot\gamma^2)}{d\dot\gamma} \cdot \frac{\partial^2 u}{\partial x \partial y} \tag{6.57}$$

in which the constitutive equation (6.35) was referred to. Because now more generalised laws apply, here and in what follows $u(x,y)$ is no longer that expression given in equation (6.54), but is a still undetermined velocity field. The second factor on the right-hand side however in any case represents a disturbance function $O(\varepsilon)$ because it vanishes in the limiting case of the parallel gap (for a viscometric flow $\varepsilon = 0$). For this reason the first factor can again be made constant. We therefore consider henceforth the coefficient $d(\nu_1\dot\gamma^2)/d\dot\gamma$ formed with the base shear rate $\dot\gamma_0 := -U/h_0$, and shall suppress for the sake of brevity in the formulas an index which would draw attention to that.

We now turn to the equation of motion (6.34). As was mentioned above in the consideration of lubricating films, the inertia of the liquid can in general be neglected ('creeping flow'), so that the shortened form of the boundary layer equation

$$0 = -\frac{d\tilde{p}}{dx} + \frac{\partial}{\partial x}(\tau_{xx} - \tau_{yy}) + \frac{\partial \tau_{xy}}{\partial y} \tag{6.58}$$

may be used. If one inserts the expression (6.57) here, then one finds that the equation of motion can be integrated once. Thus one obtains the expression for the shear stress

$$\tau_{xy} = \tau_w + \frac{d\tilde{p}}{dx} y - \frac{d(\nu_1 \dot{\gamma}^2)}{d\dot{\gamma}} \cdot \frac{\partial u}{\partial x} \tag{6.59}$$

τ_w is a constant of integration dependent on x and signifying the shear stress at the moving wall for y = 0. On the other hand the constitutive equation (6.29) applies. If one again uses there temporarily the abbreviation $F(\dot{\gamma})$ for $\eta(\dot{\gamma})\dot{\gamma}$ and notes that within the scope of the approximation $F(\dot{\gamma})$ can be replaced by $F(\dot{\gamma}_0) + \hat{\eta}(\dot{\gamma}_0)[\dot{\gamma} - \dot{\gamma}_0]$ and the other material coefficients can be regarded as constants, then one obtains

$$\tau_{xy} = F(\dot{\gamma}_0) + \hat{\eta}(\dot{\gamma}_0)\left[\frac{\partial u}{\partial y} - \dot{\gamma}_0\right] - \nu_1(\dot{\gamma}_0)\dot{\gamma}_0 \frac{\partial u}{\partial x} - \beta(\dot{\gamma}_0)\frac{\partial a_x}{\partial y} \tag{6.60}$$

If one eliminates τ_{xy} from both expressions and substitutes $-\partial v/\partial y$ (continuity!) for $\partial u/\partial x$, then once again one can integrate with respect to y. Considering the boundary conditions $v = a_x = 0$ and $u = U$ for $y = 0$, one thus obtains the relationship

$$\tau_w y + \frac{d\tilde{p}}{dx}\frac{y^2}{2} + \frac{d(\nu_1 \dot{\gamma}^2)}{d\dot{\gamma}} v = F(\dot{\gamma}_0) y + \hat{\eta}(u - U - \dot{\gamma}_0 y) + \nu_1 \dot{\gamma}_0 v - \beta a_x \tag{6.61}$$

This expression applies particularly to the fixed wall (for y = h) where the velocity components u,v and the acceleration a_x vanish:

$$\tau_w h + \frac{d\tilde{p}}{dx}\frac{h^2}{2} = F(\dot{\gamma}_0) h - \hat{\eta}(U + \dot{\gamma}_0 h) \tag{6.62}$$

Now the quantity τ_w can be eliminated from equations (6.61) and (6.62) and the result solved for u:

$$u = U\left(1 - \frac{y}{h}\right) - \frac{1}{2\hat{\eta}}\frac{d\tilde{p}}{dx}(hy - y^2) + \frac{1}{\hat{\eta}}\left(\frac{d(\nu_1\dot{\gamma}^2)}{d\dot{\gamma}} - \nu_1\dot{\gamma}_0\right) v + \frac{\beta}{\hat{\eta}} a_x \tag{6.63}$$

This formula generalises equation (6.53) which was valid in the case of a purely viscous fluid ($\nu_1 = \beta = 0$). For the calculation of the last two terms which have now been included, and which describe the effect of the elasticity of the liquid, one can use the previously established explicit results (6.55) and (6.56). By integrating over the channel cross-section one then obtains the following expression for the discharge per unit width:

$$\frac{\dot V}{b} = \frac{Uh}{2} - \frac{h^3}{12\hat\eta}\frac{d\tilde p}{dx} + \frac{\dot\gamma_0}{12\hat\eta}\frac{d(\nu_1\dot\gamma)}{d\dot\gamma}\epsilon U h_0^2 \frac{dH}{dx} - \frac{\beta}{6\hat\eta}\epsilon U^2 h_0 \frac{dH}{dx} \tag{6.64}$$

If one again puts $U\bar h/2$ for $\dot V/b$ and notes that for the shear rate $\dot\gamma_0 = -U/h_0$ holds, then one finally obtains the following expression for the pressure $\tilde p(x)$ (more accurately: the negative normal stress perpendicular to the wall, cf. equation (6.33)) in the ε^1-approximation:

$$\frac{d\tilde p}{dx} = 6\,\hat\eta\, U \frac{h-\bar h}{h_0^3} - \left(\frac{d(\nu_1\dot\gamma)}{d\dot\gamma} + 2\beta\right)\frac{\epsilon U^2}{h_0^2}\frac{dH}{dx} \tag{6.65}$$

If we now only consider the situation of the journal bearing shown in Fig. 3.20, then x is to be identified with $r\varphi$ and H(x) with $\cos\varphi$ (cf. equations (3.83) and (6.50)). Because the pressure again reaches its original value if one moves round the shaft once in the lubricating gap, the average value of the expression (6.65) vanishes. Because of the periodicity condition it emerges as before that $\bar h$ is equal to h_0, i.e. the volume flux is not altered by the elasticity of the liquid. Accordingly for the pressure distribution in a weakly loaded journal bearing

$$\frac{1}{r}\frac{d\tilde p}{d\varphi} = 6\hat\eta\,\frac{\epsilon U\cos\varphi}{h_0^2} + \left(\frac{d(\nu_1\dot\gamma)}{d\dot\gamma} + 2\beta\right)\frac{\epsilon U^2 \sin\varphi}{h_0^2 r} \tag{6.66}$$

One should remember here that the coefficients ($\hat\eta$ in the first and the bracketed expression in the second term) represent factors which are independent of position, whose magnitude depends on the shear rate of the undisturbed flow $\dot\gamma_0 = U/h_0$. The force resulting from the pressure distribution on the shaft is appropriately resolved into a component parallel to and a component perpendicular to that direction in which the centre point of the shaft moves (cf. equation 3.20). Hence:

$$F_{\parallel} := -\,br\int_0^{2\pi}\tilde p(\varphi)\cos\varphi\, d\varphi = br\int_0^{2\pi}\frac{d\tilde p}{d\varphi}\sin\varphi\, d\varphi \tag{6.67}$$

$$F_{\perp} := br\int_0^{2\pi}\tilde p(\varphi)\sin\varphi\, d\varphi = br\int_0^{2\pi}\frac{d\tilde p}{d\varphi}\cos\varphi\, d\varphi \tag{6.68}$$

Using the expression (6.66) one obtains the force components

$$F_{\parallel} = \pi\left(\frac{d(\nu_1\dot{\gamma})}{d\dot{\gamma}} + 2\beta\right)\frac{\epsilon U^2 br}{h_0^2} \tag{6.69}$$

$$F_{\perp} = 6\pi\,\hat{\eta}\,\frac{\epsilon U b r^2}{h_0^2} \tag{6.70}$$

Accordingly in the journal bearing for liquids with elastic properties in contrast to purely viscous fluids, as have been considered in Section 3.2.4, in addition a force component $F_{\parallel}$ parallel to the direction of the displacement of the shaft appears, and this component in the limiting case of small eccentricity is affected by the viscometric normal stress function $\nu_1(\dot{\gamma})$ and the second flow function $\beta(\dot{\gamma})$. On the other hand the force component $F_{\perp}$ perpendicular to the direction of the displacement is determined only by the viscometric flow property and in fact the same result holds as for a Newtonian fluid, whereby in place of the viscosity of the Newtonian reference fluid, the differential viscosity $\hat{\eta}(\dot{\gamma}_0)$ belonging to the shear rate $\dot{\gamma}_0 = U/h_0$ occurs. Therefore there is an angle ψ differing from 90° between the direction of the external load which equilises the rheodynamic force and the displacement direction of the centre point of the shaft.

With this information as a basis a weakly loaded cylindrical journal bearing of the type illustrated in Fig. 3.20 can be used as a rheometer for the determination of the differential viscosity $\hat{\eta}(\dot{\gamma})$ and the non-viscometric function $[2\beta + d(\nu_1\dot{\gamma})/d\dot{\gamma}]$. A measurement of the two force components would of course permit, by use of equations (6.69) and (6.70), conclusions about the physical quantities quoted. But it is also possible to obtain the wanted information by pressure difference measurements, because it follows from equation (6.66) that the differences of the wall pressure values for $\varphi = \pi/2$ and $\varphi = 3\pi/2$, and for $\varphi = \pi$ and $\varphi = 0$ respectively are a measure of the two material functions:

$$\tilde{p}\left(\frac{\pi}{2}\right) - \tilde{p}\left(\frac{3\pi}{2}\right) = 12\hat{\eta}\,\frac{\epsilon U r}{h_0^2} \tag{6.71}$$

$$\tilde{p}(\pi) - \tilde{p}(0) = 2\left(\frac{d(\nu_1\dot{\gamma})}{d\dot{\gamma}} + 2\beta\right)\frac{\epsilon U^2}{h_0^2} \tag{6.72}$$

In order to obtain their dependence on the shear rate $\dot{\gamma} = U/h_0$, the rotation speed of the rotating shaft must of course be varied. If the first normal stress function of the liquid has already been found in another way, then the method described here represents an access to the second flow function $\beta(\dot{\gamma})$.

6.3 Stability of plane shear flows

Viscometric flows of Newtonian liquids become, as is well known, unstable when the Reynolds number exceeds a critical value; for a pipe flow, e.g. the transition from the laminar to an unsteady turbulent flow occurs say at $Re_{cr} = 2300$. In viscoelastic liquids instabilities can set in at very much smaller Reynolds numbers. It can happen for example that at low velocities the extruded melt emerging from a die has a variable diameter, that the surface is no longer smooth, and that the flow rate varies with time. Such *'viscoelastic turbulence'* phenomena are generally undesired in the plastics industry, and it poses the question about what conditions cause the wanted flow field to become unstable, and what physical properties are the cause thereof. The following considerations are devoted to this matter.

For a theoretical stability study one naturally superposes a small amplitude unsteady perturbation on the undisturbed base flow and then checks whether random perturbations in the form of sinusoidal oscillations grow with time or die away. For the sake of simplicity we restrict the discussion to a shear flow in a straight channel (height h, drag velocity U) with the linear velocity profile Uy/h as the initial state. In addition only plane perturbations are considered, which can be derived on grounds of continuity from a stream function $\Psi(x,y,t)$ (cf. equation (1.103)). As a whole then, the velocity field is defined by

$$u = U\frac{y}{h} + \frac{\partial \Psi}{\partial y}, \qquad v = -\frac{\partial \Psi}{\partial x}, \qquad w = 0 \tag{6.73}$$

In what follows it is critically important to use 'correct' constitutive equations. We must first of all turn to the question, which properties occur in the case of a motion governed by equation (6.73). Because we start by assuming small perturbations, we are concerned with a nearly viscometric flow. We cannot however simply introduce the expressions (6.29) and (6.35) which were derived and used in Section 6.2, because these are applicable to boundary layer type flows. Their use would then only be valid if $|\partial\Psi/\partial x| << |\partial\Psi/\partial y|$ applied. But this property of the perturbation cannot be assumed a-priori, and the perturbations critical for the stability criterion do not in fact have it either. The considerations in the beginning of Section 6.2 concerning the first Rivlin-Ericksen tensor $\mathbf{A}_1$ did not however contain such boundary layer assumptions, so that the formulas (6.16) to (6.22) are now once more applicable. The assumption of small perturbations permits a linearisation in the derivatives of the stream function Ψ, whereby equation (6.18) is simplified to

$$\dot{\gamma} = \frac{\partial u}{\partial y} + \frac{\partial v}{\partial x} \tag{6.74}$$

and equation (6.22) is once more converted to (6.25). Therefore the first Rivlin-Ericksen tensor, which describes the local deformation rates, makes the same contributions to the extra-stresses as in the case of plane boundary layer flows (cf. equations (6.29), (6.30)):

$$\tau_{xy} = \eta(\dot{\gamma})\dot{\gamma} - \nu_1(\dot{\gamma})\dot{\gamma}\frac{\partial u}{\partial x} + \ldots \tag{6.75}$$

$$\tau_{xx} - \tau_{yy} = \nu_1(\dot{\gamma})\dot{\gamma}^2 + 4\eta(\dot{\gamma})\frac{\partial u}{\partial x} + \ldots \tag{6.76}$$

The only difference is that the expression (6.74) is now to be inserted for the shear rate, where for boundary layer flows $\partial u/\partial y$ could be entered. The points represent memory effects, whereby for the consideration of $\mathbf{A}_2$ alone several terms would appear in addition. The following statements become clear if such memory effects are henceforth suppressed. It will in fact appear that a theory which uses the constitutive equations (6.75) and (6.76), which contain the flow function $F(\dot{\gamma}) = \eta\dot{\gamma}$ and the first normal stress function $N_1 = \nu_1\dot{\gamma}^2$, can explain the occurrence of unstable perturbations.

By linearisation relative to Ψ the constitutive equations can be simplified taking into account (6.73) and (6.74)

$$\tau_{xy} = F(\dot{\gamma}_0) + \hat{\eta}(\dot{\gamma}_0)\left(\frac{\partial^2\Psi}{\partial y^2} - \frac{\partial^2\Psi}{\partial x^2}\right) - (\nu_1\dot{\gamma})_0 \frac{\partial^2\Psi}{\partial x\partial y} \tag{6.77}$$

$$\tau_{xx} - \tau_{yy} = N_1(\dot{\gamma}_0) + \left[\frac{d(\nu_1\dot{\gamma}^2)}{d\dot{\gamma}}\right]_0 \left(\frac{\partial^2\Psi}{\partial y^2} - \frac{\partial^2\Psi}{\partial x^2}\right) + 4\eta(\dot{\gamma}_0)\frac{\partial^2\Psi}{\partial x\partial y} \tag{6.78}$$

The subscript 0 denotes the undisturbed state with the shear $\dot{\gamma}_0 = U/h$; $\hat{\eta} = dF/d\dot{\gamma}$ is the differential viscosity already used many times.

Now imagine that the pressure p is eliminated from the equations of motion (6.31) and (6.32) for plane flow, that the previously established expressions (6.77) and (6.78) are inserted for the extra-stresses, and the expressions (1.13) for the acceleration components a_x, a_y which are likewise linearised with respect to Ψ. Because of the linearity of the resulting equation for $\Psi(x,y,t)$ one can decompose any perturbation by the Fourier method into a series of elementary parts which have the form of sinusoidally propagating waves. Therefore one can put

$$\Psi(x, y, t) = \varphi\left(\frac{y}{h}\right) e^{i\frac{\alpha}{h}(x-ct)} \tag{6.79}$$

In this α is a dimensionless real wave number ($2\pi h/\alpha$ is the wavelength) and c is the complex phase velocity, the real part of which describes the wave propagation speed, whilst the imaginary part determines the amplification or the damping of the perturbation. In the limit between increasing and decaying waves (with respect to time) c must be real. Using (6.79) the linear perturbation equation reduces to

$$\begin{aligned} &i\alpha\, Re\left(\frac{y}{h} - \frac{c}{U}\right)(\varphi'' - \alpha^2\varphi) \\ &= \varphi^{IV} - 2\alpha^2(2\eta_v - 1)\varphi'' + \alpha^4\varphi + i\alpha\eta_v We(\varphi''' + \alpha^2\varphi') \end{aligned} \tag{6.80}$$

The left-hand side represents the inertia term; the right-hand side the friction term of the equations of motion. The symbols occurring here for the first time are the Reynolds number $Re := \rho Uh/\hat{\eta}$, formed from the differential viscosity, the viscosity ratio $\eta_v := \eta/\hat{\eta}$ and another dimensionless coefficient,

$$We := \frac{\dot{\gamma}}{\eta}\frac{d(\nu_1\dot{\gamma})}{d\dot{\gamma}} \tag{6.81}$$

It describes the normal stress effect and can again be denoted as the Weissenberg number, because in the special case of a liquid that has shear rate independent viscometric functions ($\eta\to\eta_0, \nu_1\to\nu_{10}$) it corresponds to the earlier used expression $\nu_{10}\dot{\gamma}/\eta_0$. It goes without saying that the coefficients η_v and We have to be taken at the shear rate of the base flow. Because the perturbation velocities vanish at the channel walls, the boundary conditions corresponding to equation (6.80) are

$$\varphi(0) = \varphi'(0) = \varphi(1) = \varphi'(1) = 0 \tag{6.82}$$

If now rather viscous liquids, say polymer melts, are considered, then equation (6.80) can be simplified. Because unstable perturbations with wavelengths of the order of magnitude of the distance between the plates ($\alpha = O(1)$) occur here even for very small Reynolds numbers ($Re \ll 1$), the terms multiplied by $\alpha\, Re\, y/h$ on the left-hand side can be neglected compared with corresponding terms on the right-hand side. For the differential equation thus abbreviated $\alpha Re\, c/U$ plays the part of a complex eigenvalue. One can show that it can only be purely imaginary. On the other hand at the *limit of stability* the wave velocity c is purely real, so that the eigenvalue $\alpha Re\, c/U$ vanishes there. Neutral perturbations, which are neither amplified nor damped and hence limit the stability range against the instability range, therefore satisfy the abbreviated perturbation equation

$$\varphi^{IV} - 2\alpha^2(2\eta_v - 1)\varphi'' + \alpha^4\varphi + i\alpha\eta_v We(\varphi''' + \alpha^2\varphi') = 0 \tag{6.83}$$

Note that the inertia effects are therefore completely removed. However it should be pointed out that for more general shear flows, which have non-linear velocity profiles in the undisturbed flow, say for a combined pressure-drag flow in a straight channel, the simplification just described is no longer justified, because for neutral perturbations $c \neq 0$ applies. For such base flows the stability limit is therefore not only affected by the viscometric properties but also by the inertia of the liquid.

A non-trivial solution for the eigenvalue problem (6.82), (6.83) for any chosen numerical values of the parameters α and η_v exists only when the parameter We takes on certain values dependent on α and η_v. As regards the solution of the problem the amplitude function $\varphi(y/h)$ itself is ultimately of less interest than the much more interesting relationship between α, η_v and We, more accurately, the smallest possible value for We. The dependence $We_{min}(\alpha;\eta_v)$ gives in a We-α diagram a set of *indifference curves*, which separate the range of amplified (unstable) and damped (stable) perturbations one from another. One can show that We_{min} independent of α cannot be smaller than $4/\sqrt{\eta_v}$. For We > $4/\sqrt{\eta_v}$ the fundamental system of the differential equation (6.83) consists of four trigonometric functions of the form $\exp(i\alpha\nu_k y/h)$, in which the numbers ν_k (k=1,2,3,4) as real zero points of a polynomial of the fourth degree depend on the parameters We and η_v, but not on α. The fitting of the coefficients of the general integral constructed from the fundamental system to the boundary conditions (6.82) leads to a determinant which must vanish. The evaluation of this condition provides the above mentioned relationship between the Weissenberg number, the dimensionless wave number and the viscosity ratio.

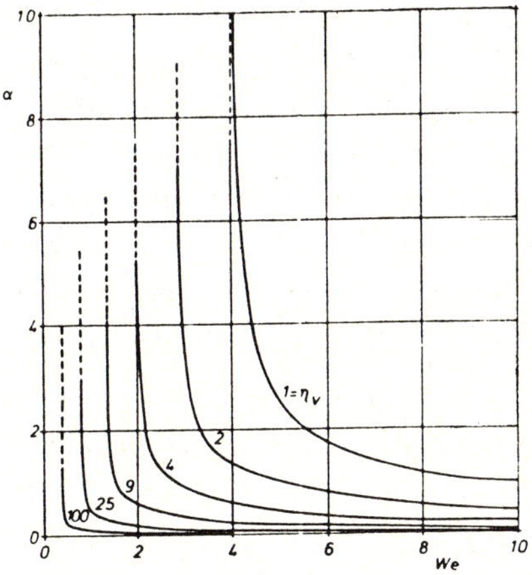

Fig. 6.8 Indifference curves for simple shear flow; Re<<1

Fig. 6.8 shows the results of corresponding numerical calculations. Each indifference curve has a vertical asymptote, namely for We → $4/\sqrt{\eta_v}$. Below this critical Weissenberg number all partial oscillations are damped, above it however some are amplified. In agreement with the observation a shear flow is therefore stable until the shear rate and

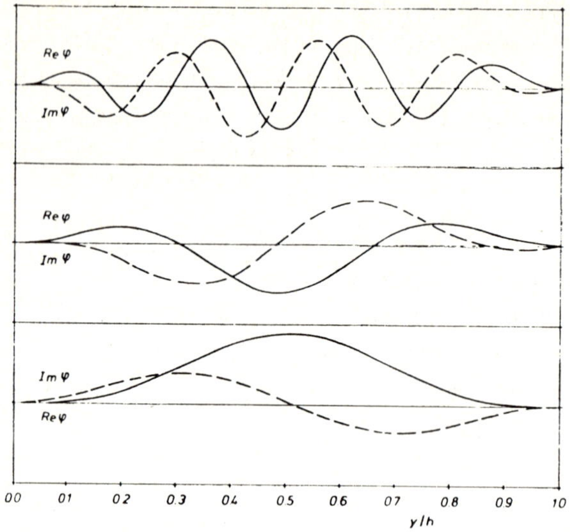

Fig. 6.9 Real and imaginary parts of the eigenfunction $\varphi(y/h)$ for $\eta_v = 1$; below: $\alpha = 1$, centre: $\alpha = 4$; above: $\alpha = 10$

hence the Weissenberg number exceeds a certain limiting value. The appearance of unstable perturbations, which lead to the phenomena of viscoelastic turbulence can therefore be explained by a theory which correctly takes into account the flow and normal stress properties, but which omits memory effects. An unstable region already arises when both viscometric constitutive functions are constant, $\eta = \eta_0$, $\nu_1 = \nu_{10}$ ('second order liquid'). From Fig. 6.8 in this case ($\eta_v = 1$) the critical Weissenberg number and therefore the ratio of the first normal stress difference to the shear stress has the value 4. The plane shear flow is unstable above the related critical shear rate of $4\eta_0/\nu_{10}$. Viscoelastic turbulence would in this special case be a purely normal stress effect. For actual pseudoplastic liquids the viscosity ratio η_v at first increases with the shear rate, but it may decrease again later. As was mentioned above the critical We value thereby falls to $4/\sqrt{\eta_v}$. Pseudoplasticity ($\eta_v > 1$) accordingly has a further destabilising influence. In illustration of the neutral perturbations Fig. 6.9 shows the real and imaginary parts of the function $\varphi(y/h)$ for three different conditions on the indifference curve with the parameter value $\eta_v = 1$. The theory stated here also accounts for the fact that the flow in a rotational viscometer becomes unstable and therefore the measurement results are unreliable if the rotational speed exceeds an upper limit dependent on the viscometric properties.

Problems

6.1 A liquid is in a channel of constant width, but of variable depth h(x) between two steadily rotating rollers (Fig. 3.16). Let the volume flux per unit width passing between the rollers be $r\omega h_3$. So long as the gap is narrow the resolution of the peripheral velocity of the rollers for $y = \pm h/2$ leads to a constant drag velocity $r\omega$ in the x-direction and a transverse variable velocity of magnitude $\pm\frac{1}{2}r\omega\, dh/dx$ in the y-direction, which occur as boundary conditions for the velocity field.

By assuming that the constitutive equation of the second order holds with two physical constants η_0 and $\nu_{10} = 2\beta_0$, the flow field can be evaluated analytically. The velocity field then has the same properties as in the case of a Newtonian fluid, i.e. the component u can be assumed as a quadratic function in y. The component v then follows from the equation of continuity (1.102), the extra-stresses τ_{xy} and $\tau_{xx} - \tau_{yy}$ follow from the constitutive equations (6.29) and (6.35). The equation of motion (6.58) finally gives the pressure rise $d\tilde{p}/dx$. Express all these quantities as functions of η_0, ν_{10}, $r\omega$, h_3, h(x) and y.

6.2 For a pure drag flow in a channel of constant height the shear stress in the gap cross-section is constant. Under isothermal conditions therefore the critical viscosity value and the shear rate are constant with respect to position. If the temperature varies in a direction normal to the wall, then a curved velocity profile arises, because greater shear rates are necessary for the transfer of the shear stress in the regions of lower viscosity than in regions of higher viscosity. If the temperature of the liquid also varies in the flow direction, a pure drag flow can in general no longer satisfy the condition that the discharge along the channel axis stays constant. Therefore there arises in addition an appropriate pressure distribution.

This phenomenon of thermally induced pressure gradients, which also occurs in Newtonian fluids, leads to a nearly viscometric flow, which can be calculated analytically after making some simplifying assumptions. We presume that the normal stresses in the direction perpendicular to the wall are equal for $x = 0$ and $x = \ell$; $\tilde{p}(0) = \tilde{p}(\ell) = 0$. In addition the temperature field is given in the form $\Theta = \Theta_w(1 + \varepsilon\vartheta(x,y))$. The expression $\vartheta = \frac{y}{h}(2 - \frac{y}{h})\frac{x}{\ell}$ simulates the circumstances between a wall heated to a constant temperature ($\Theta = \Theta_w$ for $y = 0$) and a thermally insulated wall ($\partial\Theta/\partial y = 0$ for $y = h$), in which the temperature values increase in the flow direction.

Assuming that ε is sufficiently small, calculate the pressure field and the velocity field in the ε^1-approximation, i.e. the expressions $\tilde{p}'$, u', v' defined by the equations $\tilde{p} = \varepsilon\tilde{p}'(x,y)$, $u = Uy/h + \varepsilon u'(x,y)$, $v = \varepsilon v'(x,y)$ are to be determined. Hence examine first a purely viscous fluid described by the constitutive equation (2.7) and then generalise the results under consideration of the memory and the normal stress properties (constitutive equations (6.29) and (6.35)). The inertia of the fluid can be neglected.

If one remembers that the thermally induced pressure field vanishes on the one hand for $\varepsilon = 0$, and on the other hand also when the zero-shear viscosity of the fluid does not in any way depend on the temperature, then it will be clear that in the results the coefficient $d\eta_0/d\Theta$ occurs as a factor in addition to ε.

6.3. In the stability analysis of a plane pressure-drag flow with a non-linear velocity profile $u_0(y)$ the equation (6.80) has to be generalised. Derive by using the constitutive equations (6.75) and (6.76) this more general perturbation equation. Note that the shear rate of the base flow $\dot{\gamma}_0 = du_0/dy$ and hence the coefficients $\hat{\eta}(\dot{\gamma}_0)$, $\eta(\dot{\gamma}_0)$, etc. in equations (6.77) and (6.78) depend on position, hence for the formation of stress derivatives must therefore also be differentiated.

6.4 Assuming that the influence function $K_\parallel(s,\dot{\gamma}_0)$ occurring in the equation (6.1) can be decomposed into the product of a function of s and a function of $\dot{\gamma}_0$, $K_\parallel$ is determined with the linear viscoelastic properties and the viscometric flow properties. Show that in this case $K_\parallel(s,\dot{\gamma}_0) = G(s)\cdot\hat{\eta}(\dot{\gamma}_0)/\eta_0$ and correspondingly $\eta^*_\parallel(\omega,\dot{\gamma}_0) = \eta^*(\omega)\cdot\hat{\eta}(\dot{\gamma}_0)/\eta_0$ applies. What can be stated about the second flow function $\beta(\dot{\gamma})$?

7 EXTENSIONAL FLOWS

7.1 Theoretical principles

A simple *extensional flow* is in a certain sense the opposite of a simple shear flow. For a shear flow the deformation of a fluid particle of a suitably chosen form as in Fig. 1.4 consists only of an angular change. Correspondingly the tensor of the deformation rates only contains a secondary diagonal element, the shear rate (cf. equation (1.77)). Moreover the fluid particles rotate, so that for a shear flow curl $\mathbf{v} \neq 0$ applies, and in consequence the tensor of the angular velocities does not vanish. A simple extensional flow on the contrary is distinguished by the fact that the particles do not rotate (curl $\mathbf{v} = 0$, so there is in fact a potential flow), and are elongated or compressed in three fixed directions mutually perpendicular (Fig. 1.1). Relative to a Cartesian base coordinated with these directions the deformation rate tensor therefore has a diagonal form:

$$\mathbf{D} = \begin{bmatrix} \dot{\epsilon}_1 & 0 & 0 \\ 0 & \dot{\epsilon}_2 & 0 \\ 0 & 0 & -\dot{\epsilon}_1 - \dot{\epsilon}_2 \end{bmatrix} \tag{7.1}$$

The principal diagonal elements are the *elongation rates* in the different directions. If we express here the third elongation rate in terms of the first two, so that the sum of the three quantities vanishes (tr $\mathbf{D}$ = div $\mathbf{v}$ = 0), then we limit the discussion to constant volume extensional flows and therefore we are considering in particular constant density fluids, which can move only in such a way that div $\mathbf{v}$ =0.

The above stated properties of an isochoric simple extensional flow imply that the two independent elements $\dot{\varepsilon}_1$ and $\dot{\varepsilon}_2$ are constant for any position, but are possibly dependent on time. Therefore all fluid particles at any time are extended or compressed in the same way, which is called *homogeneous deformation.* The term 'simple' extensional flow will indicate this special peculiarity. With a suitable choice of coordinates the velocity field under such conditions is simply expressed as $\mathbf{v} = \mathbf{Dr}$. Some special cases are of particular interest. For a *uniaxial* extensional flow an element of volume lengthens in one direction at the elongation rate $\dot{\varepsilon} > 0$, and shortens in the two other directions at half the elongation rate, i.e. $\dot{\varepsilon}_1 = \dot{\varepsilon}$ and $\dot{\varepsilon}_2 = -\dot{\varepsilon}/2$. Consider a circular section cylindrical bar which elongates in the axial direction, so that the cross-section shrinks, but remains circular. The velocity field of such a uniaxial extensional flow is expressed as follows:

$$u = \dot{\epsilon}\,x, \qquad v = -\frac{\dot{\epsilon}}{2}\,y, \qquad w = -\frac{\dot{\epsilon}}{2}\,z \tag{7.2}$$

The reverse of this process, the shortening of a specimen in a certain direction under elongation in the direction perpendicular thereto with equal elongational velocities $\dot{\varepsilon} > 0$ is known as a *biaxial* extensional flow. A square membrane will serve to illustrate this, which is stretched so that its surface always remains square. For an incompressible material the thickness of the membrane will decrease. This case is contained for $\dot{\varepsilon}_1 = \dot{\varepsilon}$ and $\dot{\varepsilon}_2 = -2\dot{\varepsilon}$ in the class of isochoric simple extensional flows considered, and is expressed by the velocity field

$$u = \dot{\epsilon}\,x, \qquad v = -\,2\dot{\epsilon}\,y, \qquad w = \dot{\epsilon}\,z \tag{7.3}$$

Finally the *plane* extensional flow with the velocity field

$$u = \dot{\epsilon}\,x, \qquad v = -\,\dot{\epsilon}\,y, \qquad w = 0 \tag{7.4}$$

discussed in Section 1.1 represents a special case.

Concerning the stresses which occur in an extensional flow, these are exclusively normal stresses in the directions considered, in which the material is extended or compressed. An accurate proof of this statement can be omitted, because it is surely quite clear to the reader that with isotropic materials, and fluids are isotropic by definition, the deformation shown in Fig. 1.1 can only be brought about by forces normal to the surface of the element. Shear stresses would in fact bring about a non-existent shearing of the element with corresponding changes of angle. We therefore state that not only the deformation rate tensor **D** but also the stress tensor **S** take on the diagonal form relative to the fixed Cartesian base:

$$\mathbf{S} = \begin{bmatrix} \sigma_1 & 0 & 0 \\ 0 & \sigma_2 & 0 \\ 0 & 0 & \sigma_3 \end{bmatrix} \tag{7.5}$$

On grounds of symmetry it is clear that for a uniaxial extensional flow $\sigma_2 = \sigma_3$ and in the case of a biaxial extensional flow as expressed by equation (7.3) $\sigma_1 = \sigma_3$ applies. The differences between two normal stresses, particularly $\sigma_1 - \sigma_2$, are extra-stresses. Their magnitudes at time t for simple fluids depend on the deformation history, in the present case on the time development of the elongational rate in the past, therefore at time t - s ($s \geq 0$). By analogy to the relationship (2.3) concerning unsteady shear flow we have to consider therefore say for a uniaxial extensional flow a constitutive equation of the form

$$\sigma_1(t) - \sigma_2(t) = \mathop{\mathbf{F}}_{s=0}^{\infty} [\dot{\epsilon}(t - s)] \qquad (7.6)$$

Because it is impossible to determine by a finite number of experiments the functional completely, one is limited to making certain standard tests in the testing of the material. Of particular interest then is the process in which the test specimen remains unstressed up to time t = 0, and is then deformed at a constant rate of elongation $\dot{\varepsilon}_0$. The stress behaviour thus recorded as a function of time obviously provides some information about the right-hand side in equation (7.6), that is the expression $\mathop{\mathbf{F}}_{s=0}^{t} [\dot{\varepsilon}_0]$. This is not the place to consider rheometers which provide the wanted information. We limit ourselves to an account of some important difficulties which one meets in bringing about extensional flows: the effect of gravity must be excluded; at the ends of the test volume a tension force must be produced; an initially cylindrical test specimen is to remain cylindrical during the extension, i.e. the cross-section shall reduce in size in the same way overall; finally surface tension, inertia and compressibility of the test fluid can be involved.

As regards applications, the physical properties which occur under steady conditions are of special significance. With homogeneous deformations steady signifies at the same time 'steady in material coordinates' because each individual fluid particle looks back in this case to a constant deformation history which in the course of time does not vary, and one therefore speaks of a 'motion with constant stretch history'. In a simple fluid with fading memory the extra-stresses under these circumstances depend only on the present time rates of deformation. If therefore the elongation rate $\dot{\varepsilon}$ were constant for a long period compared with the range of the memory, then the stress difference $\sigma_1 - \sigma_2$ will take on a time independent value, and equation (7.6) reduces to a generally non-linear relationship between $\sigma_1 - \sigma_2$ and $\dot{\varepsilon}$. It is usual to define by analogy to equation (2.5) an *'elongational viscosity'* η_E as the ratio of the extra-stress to the rate of elongation.

$$\sigma_1 - \sigma_2 = \eta_E(\dot{\epsilon}) \cdot \dot{\epsilon} \tag{7.7}$$

According to the above statements η_E depends on $\dot{\varepsilon}$, and describes the physical properties arising in a materially steady extensional flow. We use the subscript E only in the case of uniaxial elongation. For a biaxial or a plane extensional flow we write for the purposes of clear distinction in equation (7.7) η_{EB} and η_{EP} instead of η_E. Fig. 7.1 shows the elongational viscosity, and as a comparison also the shear viscosity of a polymer melt as a function of the relevant deformation rate. It so happens that both viscosity curves have qualitatively different properties. The pseudoplastic (shear thinning) material becomes obviously more viscous with increasing elongation rate, as long as the elongation rate does not exceed a certain value. This elongation thickening is moreover the basis of the spinning capability of certain polymers. For $\dot{\varepsilon} \to 0$ the melt behaves in a Newtonian manner where the zero value of the elongational viscosity is just three times as great as the zero value of the shear viscosity.

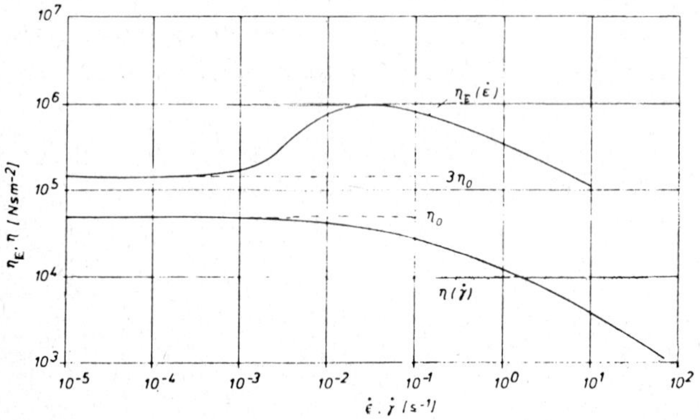

Fig. 7.1 Steady elongational and shear viscosity of a polymer melt (low density polyethylene) as functions of the deformation rate; Θ=150°C (after Laun and Münstedt)

The three elongational viscosities $\eta_E(\dot{\varepsilon})$, $\eta_{EB}(\dot{\varepsilon})$ and $\eta_{EP}(\dot{\varepsilon})$ contain different rheological information and are therefore functions which are independent of each other. Note that all three functions are only defined for $\dot{\varepsilon} > 0$. The biaxial elongational viscosity $\eta_{EB}(\dot{\varepsilon})$ is in a certain sense an extension of the uniaxial elongational viscosity $\eta_E(\dot{\varepsilon})$ for negative elongation rates and vice versa. If one realises in addition that a plane flow with the velocity field (7.4) also remains a plane extensional flow for $\dot{\varepsilon} < 0$, then one learns that the function $\eta_{EP}(\dot{\varepsilon})$ reproduces when $-\dot{\varepsilon}$ is substituted for $\dot{\varepsilon}$. The plane elongational viscosity $\eta_{EP}(\dot{\varepsilon})$ is in consequence an even function of $\dot{\varepsilon}$, but this does not occur for the other two elongational viscosities.

The three previously mentioned constitutive functions are not enough to describe the physical properties of a general steady extensional flow with a deformation rate tensor given by equation (7.1). In order to explain how the law set over is built up one should remember that for such a motion the higher Rivlin-Ericksen tensors are determined by the first according to $\mathbf{A}_n = \mathbf{A}_1^n$ (cf. Problem 1.5). In consequence the deformation history of a fluid particle can be expressed by using equation (1.67) in the form

$$\mathbf{C}_t(s) = e^{-s\mathbf{A}_1} \tag{7.8}$$

For the three critical components this means for the incompressible case (cf. equation (7.1) and note that $\mathbf{A}_1 = 2\mathbf{D}$): $C_{xx} = e^{-2\dot{\varepsilon}_1 s}$, $C_{yy} = e^{-2\dot{\varepsilon}_2 s}$, $C_{zz} = e^{2(\dot{\varepsilon}_1 + \dot{\varepsilon}_2)s}$. These expressions show that line elements which are oriented in one of the three preferred directions lengthen or shorten exponentially with time.

Now it is significant in the above connection that the deformation history $\mathbf{C}_t(s)$ of a steady extensional flow depends only on $\mathbf{A}_1$. Because for a simple fluid the tensor of the extra-stresses $\mathbf{T}$ is determined by $\mathbf{C}_t(s)$, it is obviously only affected by $\mathbf{A}_1$, and therefore by the rate of strain tensor $\mathbf{D}$: $\mathbf{T} = \mathbf{f}(\mathbf{D})$. By using arguments which are stated in Section 8.1 this relationship between $\mathbf{T}$ and $\mathbf{D}$ is considerably simplified. In the case of an incompressible fluid there results as the most general *law for steady extensional flows*

$$\mathbf{S} = -p\mathbf{E} + \varphi_1(I_2, I_3)\mathbf{D} + \varphi_2(I_2, I_3)\mathbf{D}^2 \tag{7.9}$$

in which φ_1 and φ_2 are the two scalar functions of the two non-vanishing principal invariants I_2 and I_3 of the tensor $\mathbf{D}$. Equations (1.44) and (1.45) define I_2 and I_3. Liquids that satisfy the constitutive equation (7.9) are called Reiner-Rivlin fluids. We therefore state that every simple fluid in a steady homogeneous irrotational flow (i.e. a steady simple extensional flow) behaves like a Reiner-Rivlin fluid. The friction properties thus occurring can be completely described by two individual constitutive functions $\varphi_1(I_2, I_3)$ and $\varphi_2(I_2, I_3)$. These naturally include the above mentioned elongational viscosities as special cases. Thus for example the invariants $I_2 = -3\dot{\varepsilon}^2/4$ and $I_3 = \dot{\varepsilon}^3/4$ belong to a uniaxial extensional flow which has the velocity field (7.2), and a comparison between (7.7) and (7.9) yields the relationship

$$\eta_E(\dot{\epsilon}) = \frac{3}{2}\varphi_1\left(-\frac{3}{4}\dot{\epsilon}^2, \frac{1}{4}\dot{\epsilon}^3\right) + \frac{3}{4}\dot{\epsilon}\varphi_2\left(-\frac{3}{4}\dot{\epsilon}^2, \frac{1}{4}\dot{\epsilon}^3\right) \tag{7.10}$$

Hence it is clear that the experimental determination of the uniaxial elongational viscosity $\eta_E(\dot{\varepsilon})$ gives only limited information about the rheological behaviour for steady simple extensional flows. The same holds for the constitutive functions $\eta_{EB}(\dot{\varepsilon})$ and $\eta_{EP}(\dot{\varepsilon})$. Considerable practical difficulties prevent the complete determination of the functions $\varphi_1(I_2, I_3)$ and $\varphi_2(I_2, I_3)$, which would be required for the description of a steady simple extensional flow of the type given by equation (7.1).

7.2 Applications

Simple (homogeneous) extensional flows obviously do not occur in technology. There is however a series of technical processes in which extensional flows which have somewhat more general properties occur. We shall first consider the fibre extrusion process shown in Fig. 7.2. In this the melt extruded from a nozzle is stretched by tension and simultaneously cooled, so that the fibre is thinned and solidifies after travelling a certain distance. The solid fibre is drawn off at the above mentioned tension. The important parameters of this process are in addition to the length of the melt zone ℓ up to solidification, and the tension force F, the reduction ratio of the fibre cross-section $A(\ell)/A(0)$, the fibre drawing velocity and the discharge $\dot{V}$.

Whereas the fluid in the nozzle is chiefly sheared and over the distance ℓ is mainly elongated uniaxially, there is between the two zones a more or less extensive transition zone in which the one type of deformation changes into the other. The flow of the melt in the zone of the elongation is considered in more detail below: its start can be taken to be the place of the greatest widening of the extrudate ($x = 0$). Because the fibre is generally thin and narrow, all fluid particles move in the zone of elongation in the same way as those on the axis of symmetry. They are displaced chiefly in the axial direction by an axial velocity $u(x)$. Therefore an approximately irrotational uniaxial extensional flow occurs. It can be steady, but represents no homogeneous deformation, because the elongation rate $\dot{\varepsilon} = du/dx$ in fact falls to zero at the point of solidification, and thus obviously depends on position. Therefore the process is not at all steady in material coordinates. For a liquid that has a memory the extra-stresses at the point x are therefore not only affected by the local elongation rate, but also by the elongation rates at $x' < x$, possibly even by the shear deformation in the nozzle. The previous deformation history of a fluid particle will however have less effect on the present stresses the shorter is the memory of the fluid, and the more slowly does the deformation state of the particle vary. Assuming a correspondingly short memory it is therefore sufficient to consider the local elongation rate in the law, i.e. one can refer back to equation (7.7) (inelastic approximation). The situation is similar to that in Sections 3.2.3 to 3.3, where in fact non-viscometric flows were described by the consideration of the viscometric properties alone. Accordingly we consider here for the description of a non-homogeneous extensional flow only the characteristic physical properties of homogeneous extensional flows, and therefore omit the effect of the memory of the liquid.

It is useful to invert the relationship (7.7) and by analogy to equation (2.16) to express it in the dimensionless form

$$\frac{\eta_*}{\tau_*}\dot{\epsilon} = f_E\left(\frac{\sigma_1 - \sigma_2}{\tau_*}\right) \tag{7.11}$$

In the situation illustrated in Fig. 7.2 $\dot{\varepsilon}(x)$ would be identified with $du(x)/dx$, as already mentioned; $\sigma_1(x)$ corresponds to the axial,

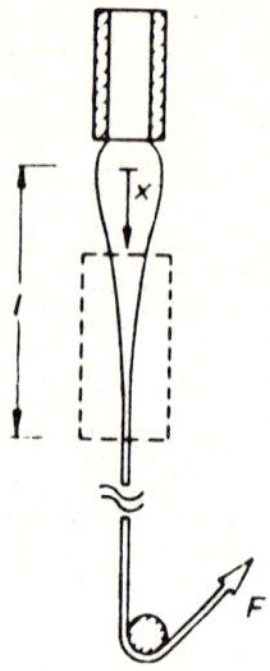

Fig. 7.2 Model of a fibre drawing process

σ_2 to the radial normal stress. By neglecting surface tension at the free surfaces of the slender fibre $-\sigma_2$ is equal to the constant external pressure, which without limiting the generality can be made equal to zero ($\sigma_2 = 0$).

In order to find the velocity distribution u(x) and therefore the variation of the fibre cross-section A(x) associated with it, we formulate a momentum equation for the control volume shown in Fig. 7.2. This states that the difference between the exiting and the entering momentum fluxes is equal to the sum of all the forces acting on the liquid in the control volume. For the downwards flowing fibre this means:

$$\rho\dot{V}[u(\ell) - u(x)] = F - \sigma_1(x)A(x) + \int_x^{\ell} \rho g A(\xi)d\xi \tag{7.12}$$

The local values of the velocity and the fibre cross-section are related to each other by the continuity equation

$$u(x)\,A(x) = \dot{V} \tag{7.13}$$

Under certain circumstances the weight of the fibre in the control space (the integral on the right-hand side of equation (7.12)) and the entering and exiting momentum fluxes can be neglected compared with the force that acts downwards. Hence the influence of the inertia disappears ('creeping flow'). Under these conditions the tension force F acts in every cross-section of the fibre, and according to equations (7.12) and (7.13) the associated stress is simply connected with the local velocity by

$$\sigma_1(x) = \frac{F}{\dot{V}} u(x) \tag{7.14}$$

Note that the relationships (7.12) to (7.14) apply, independently of the constitutive law, hence also for materials with wide-ranging memory. If we now combine equation (7.14) with (7.11), then we restrict ourselves to the above described approximation for fluids that do not have a memory. In this case the following differential equation results for the velocity field u(x):

$$\frac{\eta_*}{\tau_*} \frac{du}{dx} = f_E\left(\frac{F u}{\dot{V}\tau_*}\right) \tag{7.15}$$

It must be once more remembered here that the actual process is connected with the cooling of the melt, which brings about the solidification of the fibre. For this reason the reference viscosity η_* represents a critical temperature dependent factor which varies with x. As long as details of the heat transfer are not considered one can hold the view that $\eta_*(x)$ is known and it would be significant to assume that $\eta_* \to \infty$ as $x \to \ell$. The integration of equation (7.15) yields the following implicit result for the velocity distribution u(x):

$$\int_{\frac{Fu(0)}{\dot{V}\tau_*}}^{\frac{Fu(x)}{\dot{V}\tau_*}} \frac{d\sigma}{f_E(\sigma)} = \frac{F}{\dot{V}} \int_0^x \frac{dx'}{\eta_*(x')} \tag{7.16}$$

In order to obtain a concrete result from this, one must obviously specify the elongation function $f_E(\sigma)$ of the material and the temperature and the viscosity distribution $\eta_*(x)$ in the axial direction. Thus follows for example in the case of a Newtonian fluid ($f_E(\sigma)=\sigma/3$) with the assumption that the viscosity increases in the flow direction, $\eta_*(x) = \eta_*(0)/(1 - x/\ell)$, for the reduction ratio of the fibre cross-section

$$\frac{A(\ell)}{A(0)} = \frac{u(0)}{u(\ell)} = \exp\left[-\frac{F\ell}{6\dot{V}\eta_*(0)}\right] \tag{7.17}$$

If in fact a smaller cross-section contraction is specified, then this depends most of all on the property of the elongation thickening of the polymer melt (Fig. 7.1). The reader may solve equation (7.16) for an actual model of the elongation function $f_E(\sigma)$ of a polymer melt and possibly with modified assumptions about $\eta_*(x)$

The *film blowing* process is related to the fibre extrusion. In this the melt comes out of an annular nozzle and forms a closed rotational symmetric tube, which solidifies on cooling at some distance from the nozzle (Fig. 7.3). A low supporting pressure inside the tube contributes to stability. In order to give the extrudate the desired thickness the solidifed material is drawn with a suitably set tension. Thus the melt is stretched in the tube-forming zone not only in the longitudinal direction but also round its circumference. Therefore there is an extensional flow with two elongation rates independent of each other, so that for the description of the process in the context of an inelastic material approximation the general constitutive equation (7.9) would be used.

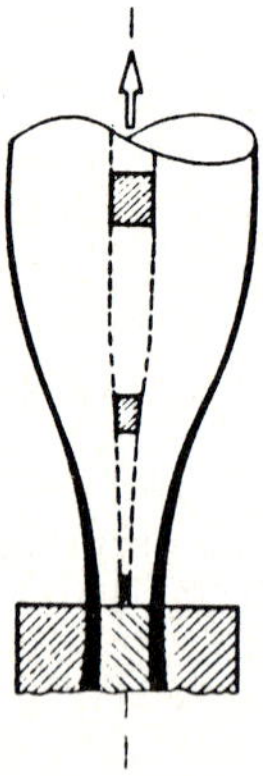

Fig. 7.3 Longitudinal and transverse elongations in film blowing

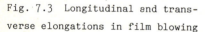

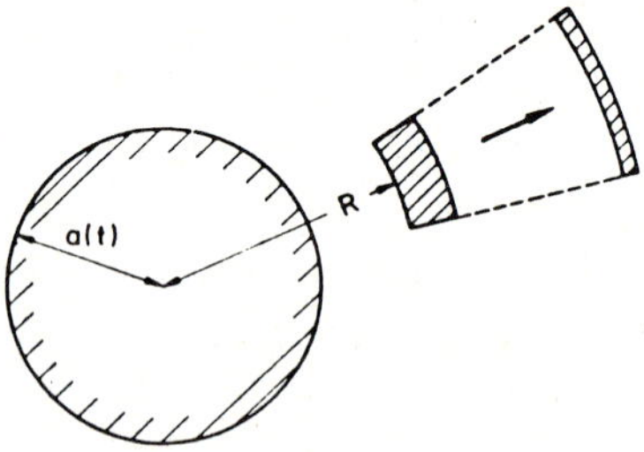

Fig. 7.4 Expanding sphere

We now turn to a process in which phases of uniaxial elongation and phases of biaxial elongation follow one another in turn. The radially symmetric motion of liquid outside a 'breathing' sphere is meant by this, e.g. a gas bubble which has a size that varies with time (Fig. 7.4). Assuming that the material points move purely radially, only the velocity component $v_R(R,t)$ occurs. For an incompressible liquid for reasons of continuity v_R decreases with increasing distance R from the centre of the sphere as $1/R^2$. By using the radius a(t) of the breathing sphere and the velocity $\dot{a}(t) := da/dt$ at which its surface expands, the velocity field of the liquid can therefore be expressed in the form

$$v_R(R, t) = \frac{a^2(t)\dot{a}(t)}{R^2} \tag{7.18}$$

It is clear that with this motion a material volume element is compressed without rotating in the radial direction and is similarly elongated in the two directions perpendicular thereto when the sphere expands (Fig. 7.4). Correspondingly the deformation rate tensor has with reference to a base connected with the spherical coordinates R, ϑ, φ, the diagonal form where two of the elements are identical (cf. Formula Appendix):

$$\mathbf{D} = \mathbf{L} = \begin{bmatrix} \frac{\partial v_R}{\partial R} & 0 & 0 \\ 0 & \frac{v_R}{R} & 0 \\ 0 & 0 & \frac{v_R}{R} \end{bmatrix} \tag{7.19}$$

Accordingly only the two principal diagonal elements σ_{RR}, $\sigma_{\vartheta\vartheta}$ and $\sigma_{\varphi\varphi}$ occur in the stress tensor, where $\sigma_{\vartheta\vartheta}$ and $\sigma_{\varphi\varphi}$ always have the same values. Without limiting the generality we can identify the normal stress in the radial direction with the pressure, $p = -\sigma_{RR}$.

Because this concerns a one-dimensional motion, one has to evaluate a single relevant equation of motion. This is:

$$\rho\left(\frac{\partial v_R}{\partial t} + v_R\frac{\partial v_R}{\partial R}\right) = -\frac{\partial p}{\partial R} + \frac{1}{R}(2\sigma_{RR} - \sigma_{\vartheta\vartheta} - \sigma_{\varphi\varphi}) \tag{7.20}$$

(cf. Formula Appendix; volume forces are neglected).

The acceleration term on the left-hand side leads by use of equation (7.18) to a simple expression in R, whereby in addition to a(t), $\dot{a}$ and $\ddot{a}$ also occur. For the normal stress differences on the right-hand side we require the constitutive law for a biaxial extensional flow (provided that $\dot{a} > 0$). For a liquid without memory the biaxial elongational viscosity $\eta_{EB}(\dot{\varepsilon})$ operates, and in fact (analogous to equation (7.77)) $\sigma_{\vartheta\vartheta} - \sigma_{RR} = \sigma_{\varphi\varphi} - \sigma_{RR} = \eta_{EB}(\dot{\varepsilon})\cdot\dot{\varepsilon}$ with $\dot{\varepsilon} = v_R/R = a^2\dot{a}/R^3$ in accordance with equations (7.19) and (7.18). If one inserts these expressions in equation (7.20) and notes that $3\dot{\varepsilon}/R = -\partial\dot{\varepsilon}/\partial R$, then one can integrate once with respect to R:

$$\rho\left[-\frac{(a^2\dot{a})^{\cdot}}{R}+\frac{(a^2\dot{a})^2}{2R^4}\right] = -p+p_\infty+\frac{2}{3}\int_0^{a^2\dot{a}/R^3}\eta_{EB}(\dot{\epsilon})d\dot{\epsilon} \tag{7.21}$$

This relationship combines the pressure $p = -\sigma_{RR}$ in the liquid at a distance R from the centre with the time law a(t) according to which the bubble expands. p_∞ denotes the pressure far from the bubble.

For the derivation of equation (7.21) it was assumed that the bubble expanded. However if the radius contracts ($\dot{a}<0$) then a uniaxial extensional flow of elongation rate $\dot{\varepsilon} = \partial v_R/\partial R$ (cf. equation (7.19)) is involved, hence $\dot{\varepsilon} = -2a^2\dot{a}/R^3$ according to equation (7.18). In this case in equation (7.21) naturally the uniaxial elongational viscosity $\eta_E(\dot{\varepsilon})$ occurs (instead of $\eta_{EB}(\dot{\varepsilon})$). One can easily show that equation (7.21) is also valid for $\dot{a}<0$, if one replaces the friction term on the right-hand side by the expression $H(a^2\dot{a}/R^3)$, whereby the function $H(\alpha)$ is defined thus:

$$H(\alpha) := \begin{cases} \dfrac{2}{3}\displaystyle\int_0^{\alpha}\eta_{EB}(\dot{\epsilon})d\dot{\epsilon} & \text{for } \alpha>0 \\ -\dfrac{2}{3}\displaystyle\int_0^{-2\alpha}\eta_{E}(\dot{\epsilon})d\dot{\epsilon} & \text{for } \alpha<0 \end{cases} \tag{7.22}$$

The value of the pressure at the surface of the expanding sphere (for R = a(t)) can be of particular interest. It is denoted by p_a. With the above described generalisation it follows from equation (7.21) that

$$p_a - p_\infty = \rho\left(a\ddot{a}+\frac{3}{2}\dot{a}^2\right)+H\left(\frac{\dot{a}}{a}\right) \tag{7.23}$$

If we now assume that the expanding or contracting sphere is a gas bubble, then the liquid pressure p_a can be replaced by the gas pressure p_G. By considering capillary forces, which are particularly noticeable in small bubbles, $p_a = p_G - 2\sigma/a$, in which σ represents the surface tension characteristic for the surface which separates gas from liquid. If one measures the size of the bubble a(t) for given values of gas pressure $p_G(t)$, then on the basis of equation (7.23) the functions

$\eta_{EB}(\dot{\varepsilon})$ and $\eta_E(\dot{\varepsilon})$ respectively can be found. On the other hand if the elongation properties of the liquid are known, then equation (7.23) allows the bubble radius in the course of time to be calculated. If in particular the bubble consists of a constant amount of an ideal gas and the gas state varies isentropically (adiabatic reversibly), then as is known the product of the gas pressure p_G and the power of the gas volume V_G formed with the isentropic exponent κ is constant. If one considers that $V_G \sim a^3$, then it follows that the gas pressure is related in a simple way to the radius of the bubble:

$$p_G = p_{G0}\left(\frac{a_0}{a}\right)^{3\kappa} \tag{7.24}$$

The constants a_0 and p_{G0} are values of the bubble radius and gas pressure which at any time, say at the start of the motion, occur simultaneously. Therefore equation (7.23) can be written as

$$p_{G0}\left(\frac{a_0}{a}\right)^{3\kappa} - \frac{2\sigma}{a} - p_\infty = \rho\left(a\ddot{a} + \frac{3}{2}\dot{a}^2\right) + H\left(\frac{\dot{a}}{a}\right) \tag{7.25}$$

This is a non-linear differential equation for the time variation of the radius $a(t)$ of the bubble. The ambient pressure $p_\infty(t)$ thus represents an external excitation which is to be regarded as given. Together with suitable initial conditions (e.g. $a = a_0$, $\dot{a} = 0$ for $t = 0$) equation (7.25) describes the forced radial oscillation of a gas bubble in an incompressible liquid, the pressure p_∞ of which changes with time in a known way - one is thinking say of a sudden pressure change or a sinusoidally oscillatory ambient pressure. The rheological properties of the liquid are reflected in the function $H(\dot{a}/a)$. Note that in this way phases of the compression of the gas ($\dot{a}<0$) are affected by the uniaxial elongational viscosity; phases of the expansion ($\dot{a}>0$) on the other hand by the biaxial elongational viscosity. In the special case of a Newtonian liquid (for which $\eta_E = 3\eta_0$ and $\eta_{EB} = 6\eta_0$) the friction term $H(\dot{a}/a)$ moreover reduces independently of the sign of the argument to $H(\dot{a}/a) = 4\eta_0\dot{a}/a$.

Problems

7.1 Equation (7.9) is the most general law of an incompressible simple fluid for steady homogeneous extensional flow. How are the three elongational viscosities $\eta_E(\dot{\varepsilon})$, $\eta_{EB}(\dot{\varepsilon})$ and $\eta_{EP}(\dot{\varepsilon})$ related to the two functions $\varphi_1(I_2, I_3)$ and $\varphi_2(I_2, I_3)$ which occur in it? Use the results to demonstrate that the lower Newtonian limiting values (for $\dot{\varepsilon} \to 0$) of the three elongational viscosities are in the ratios 3:6:4.

7.2 It is conceivable that in the film blowing process shown in Fig. 7.3 the tube is stretched only in the axial direction, but not in the tangential direction, i.e. it represents a thin circular section cylinder of variable wall thickness.

Derive for this special case, neglecting the effect of gravity, the inertia and the memory of the liquid, a relationship analogous to equation (7.16), which combines the drawing ratio $u(\ell)/u(0)$ and hence the reduction ratio $s(\ell)/s(0)$ of the film thickness with the drawing force F, the flow rate $\dot{V}$, the length ℓ of the sleeve-forming zone, the radius r, and the original wall thickness s(0) of the cylinder, as well as the critical physical properties.

7.3 There has been a proposal to create simple (homogeneous) uniaxial extensional flows in circular section nozzles of suitable form. For this a thin lubricating medium allows the test liquid to slip at the wall.

Determine, neglecting the thickness of the lubricating film, the necessary cross-section distribution A(x) of the nozzle as a function of the axial coordinate x, the desired elongation rate $\dot{\varepsilon}$ and the flow rate $\dot{V}$. Then deduce, neglecting gravity and inertia, a relationship between the local wall pressure $p_W(x)$, the elongational viscosity η_E and the above mentioned quantities x, $\dot{\varepsilon}$ and $\dot{V}$. The result shows that a difference pressure measurement between two different places in the nozzle leads to a conclusion about the elongational viscosity.

Hint. First express p_W in terms of the two principal stresses σ_{xx}, σ_{rr} and the slope of the nozzle contour, and conclude from this that under the given assumptions σ_{xx} and σ_{rr} are spatially constant for any position.

8 SPECIAL RHEOLOGICAL LAWS

So far we have primarily been dealing with flows that have simple kinematic properties, particularly viscometric flows, unsteady shear flows of small amplitude and steady extensional flows. We have reached a point where the general constitutive equations for simple fluids are to be attributed to a few functions which can be determined in suitable rheometers. These laws applied to the whole class of simple fluids.

The individual rheological properties of a certain liquid express themselves in the particular form of the functions, say the three viscometric functions, the linear viscoelastic relaxation function or the elongational viscosities.

Now the motions so far considered for the applications are certainly of particular importance, but of course there are also complex flow forms which are also of interest. In this respect one meets a major difficulty because the constitutive equation controlling the flow is not known. Even if one knows the previously mentioned physical properties of kinematically simple flows there is still uncertainty about how such a real liquid behaves in more general situations. As long as it is a matter of processes that depart only a little from one of the basic flow forms, one will be able with some justification to draw on the appropriate law for the 'adjacent' reference flow. For example if the fluid particles are predominantly sheared and the shear rate of the particles measured over the range of the memory varies only slightly with time, then one will most conveniently use equation (2.36), valid for viscometric flows. If on the other hand we are considering a motion in which the particles are predominantly being stretched, and the elongation rates are changing only relatively slowly with time, then equation (7.9), valid for steady extensional flows, represents the simplest significant approximation of the effective law. We have already considered this point of view in Sections 3.2 and 7.2.

However if the actual flow cannot be accurately classified in the vicinity of a kinematically simple reference flow, then the uncertainty consists of what the appropriate law looks like. In such cases there is no alternative to using special rheological laws which on the one hand describe important properties of real liquids, but on the other hand are not so complex that the mathematical treatment of actual flow problems would in principle be impossible. Such models, which of course only apply to a more or less highly constricted part of the simple fluids, are described in this section.

8.1 Fluids without memory

In order to describe processes for which the memory effect probably plays a subordinate role, one will possibly wish to bring in the model of a liquid without a memory. We therefore first consider fluids that have the property that the extra-stresses within any arbitrarily chosen fluid particle at the present time only depend on the instantaneous deformation rates of the particle. We therefore postulate a constitutive equation of the form

$$\mathbf{T} = \mathbf{F}(\mathbf{D}) \tag{8.1}$$

This general relationship between the extra-stress tensor and the deformation rate tensor can be considerably simplified by noting that a fluid is isotropic, i.e. has no physical preferred directions. As a result the local principal axis systems of both tensors coincide (cf. the concluding remarks about equation (7.4)). If one uses this general principal axis system, then equation (8.1) reduces to three scalar relationships between the eigenvalues of **T** and **D** which are denoted by τ_i and $\dot{\varepsilon}_i$ (i = 1,2,3). These relationships run: $\tau_i = f_i(\dot{\varepsilon}_1, \dot{\varepsilon}_2, \dot{\varepsilon}_3)$. However they can just as well be expressed in the following form:

$$\tau_i = \varphi_0 + \varphi_1 \dot{\varepsilon}_i + \varphi_2 \dot{\varepsilon}_i^2 \qquad (i = 1, 2, 3) \tag{8.2}$$

In this φ_0, φ_1, φ_2 are three functions which depend on $\dot{\varepsilon}_1$, $\dot{\varepsilon}_2$, $\dot{\varepsilon}_3$ and hence on the three principal invariants of the deformation rate tensor $I_1(\mathbf{D})$, $I_2(\mathbf{D})$ and $I_3(\mathbf{D})$. We can bring these relationships together into

$$\mathbf{T} = \varphi_0(I_1, I_2, I_3)\mathbf{E} + \varphi_1(I_1, I_2, I_3)\mathbf{D} + \varphi_2(I_1, I_2, I_3)\mathbf{D}^2 \tag{8.3}$$

In the transformation to any rotated coordinate system with respect to the principal axis system the components of the extra-stress tensor and of the deformation rate tensor vary, but the invariants and hence the values of the scalar factors φ_0, φ_1, φ_2 remain unchanged. Equation (8.3) then generally represents six scalar relationships, which one obtains when one inserts the expression (1.24) for **D**, and evaluates the components of the relationship. The materials characterised by equation (8.3) are called *Reiner-Rivlin fluids*.

The above statements also embrace compressible fluids. If we now restrict the discussion to constant density liquids, then only motions with I_1 = div **v** = 0 are realisable, so that I_1 vanishes as a parameter, and the scalar quantity φ_0 can be combined with the pressure. Therefore in an incompressible Reiner-Rivlin fluid the following applies to the total stresses:

$$\mathbf{S} = -p\,\mathbf{E} + \varphi_1(I_2, I_3)\mathbf{D} + \varphi_2(I_2, I_3)\mathbf{D}^2 \tag{8.4}$$

In order to understand how such a fluid behaves under viscometric conditions we compare (8.4) with the general constitutive equation (2.36) for viscometric flows. On the basis of the expression (1.77) for $\mathbf{A}_1 = 2\mathbf{D}$ one finds that for a viscometric flow $I_2 = -\dot{\gamma}^2/4$ and $I_3 = 0$. Thus the expressions $\frac{1}{2}\varphi_1(-\dot{\gamma}^2/4,0)$ and $\frac{1}{4}\varphi_2(-\dot{\gamma}^2/4,0)$ represent the two viscometric functions $\eta(\dot{\gamma})$ and $\nu_2(\dot{\gamma})$. The first normal stress coefficient however vanishes, $\nu_1(\dot{\gamma}) = 0$. This unrealistic property shows that the model is not suitable for the description of normal stress effects, apart from those which are determined by $\nu_2(\dot{\gamma})$ alone.

Hence there is no significant loss of the generality when we concern ourselves henceforth only with the model that has no normal stress properties ($\varphi_2 = 0$). The remaining coefficient φ_1 will in general depend on both invariants $I_2(\mathbf{D})$ and $I_3(\mathbf{D})$. Whereas the dependence on I_2 can be determined in a viscometer, it is much more difficult to demonstrate a dependence on I_3, and there are reasons for the supposition that in the equation $\mathbf{T} = \varphi_1\mathbf{D}$ the invariant I_3 can hardly occur at all.

In the case of isothermal processes the entropy production rate up to a constant factor Θ^{-1}, which we can formally put at unity, is equal to the dissipation function Φ defined in equation (1.124). In the phraseology of irreversible thermodynamics Φ is a sum of products of 'forces' (the extra-stresses) and 'fluxes' (the deformation rates). By use of the rheological law the entropy production rate can be represented by the fluxes alone: in the present case $\Phi = \mathrm{tr}(\mathbf{T}\,\mathbf{D}) = \varphi_1\,\mathrm{tr}(\mathbf{D}^2) = -2\varphi_1 I_2$. A hypersurface is defined by Φ = constant in the configuration space of the fluxes. Consider the tangential plane to an arbitrary point on such a surface. For linear dissipative processes (in the above case that means φ_1 = constant) the vector of the forces belonging to the chosen point is always perpendicular to this tangential plane. It is conjectured that this orthogonality also exists in non-linear processes. In the present case this means $\tau_{ik} = \lambda(I_2, I_3)\partial[\varphi_1(I_2, I_3)I_2]/\partial D_{ik}$, in which λ signifies a scalar factor of proportionality. This relationship between **T** and **D** is however only consistent with the constitutive equation $\mathbf{T} = \varphi_1\mathbf{D}$ when there is no dependence on I_3 for the coefficient φ_1 (and therefore also for λ). The described orthogonality principle corresponds moreover for stiff viscoplastic materials (with singular function φ_1 for $I_2 \to 0$) which conform to the generally accepted hypothesis that the plastic flux is perpendicular to the flow surface.

We therefore start from the fact that for a Reiner-Rivlin liquid without normal stress properties φ_1 depends only on $I_2(\mathbf{D})$, but not on $I_3(\mathbf{D})$. As we have already seen, $\varphi_1(I_2) = 2\eta(\dot\gamma)$, where $\dot\gamma^2 = -4I_2$. The constitutive equation therefore runs:

$$\mathbf{S} = -p\mathbf{E} + 2\eta(\dot\gamma)\mathbf{D}; \qquad \dot\gamma^2 = -4I_2(\mathbf{D}) \tag{8.5}$$

It contains as a special case (if η is constant with respect to $\dot\gamma$) the law of incompressible Newtonian fluids. One therefore speaks of (incompressible) *generalised Newtonian fluids* in connection with equation (8.5). The generalisation consists of the consideration of non-linear flow properties in the form of the viscometric function $\eta(\dot\gamma)$ instead of the velocity independent viscosity in the case of a Newtonian fluid.

8.1.1 Minimum principle for generalised Newtonian fluids

The extra-stresses of a generalised Newtonian fluid can be derived from a scalar *potential* Ω, which depends on the second invariant of the rate of strain tensor:

$$\Omega := \frac{1}{2}\int_0^{-4I_2} \eta(\dot\gamma)\,d\dot\gamma^2 \tag{8.6}$$

One can show that the term $2\eta\mathbf{D}$ in the equation (8.5) by use of Ω can be written in the form $\partial\Omega/\partial\mathbf{D}$, so that $\tau_{ik} = \partial\Omega/\partial D_{ik}$. The potential is always positive and in addition increases monotonically with the (positive) variable $-4I_2$. For a Newtonian fluid 2Ω is the same as the dissipation function. For generalised Newtonian fluids however this is no longer so. Much more does an analogously formed integral based on the differential viscosity enter into the relationship between the dissipation function Φ and the potential Ω :

$$\Phi = \frac{1}{2} \int_0^{-4I_2} (\eta + \hat{\eta}) d\dot{\gamma}^2 \tag{8.7}$$

This relationship shows once more that the differential viscosity $\hat{\eta}$ is the natural partner of the usual viscosity η.

The potential Ω is in the centre of the following remarks. They refer to processes in which the inertia of the liquid plays no part, because either the material points remain totally unaccelerated, or the Reynolds number is sufficiently small so that the inertia term can be neglected. Such flows are described by the equation of motion $\mathrm{div}\, \mathbf{S} + \mathbf{f} = 0$. It is assumed that the volume force field $\mathbf{f}$ is independent of the velocity field. In addition we start from the fact that the boundary limiting the liquid resolves into two components A_v and A_t in which the flow velocity $\mathbf{v}$ and the stress vector $\mathbf{t}$ respectively are given quantities (Fig. 8.1). Hence A_t for example can be totally absent. The following statement applies to the motions characterised in this way.

For all incompressible flow fields which satisfy the boundary conditions on A_v and A_t, the true flow field (which satisfies the equation of motion) makes the following expression a minimum:

$$\int_V (\Omega - \mathbf{f} \cdot \mathbf{v}) dV - \int_{A_t} \mathbf{t} \cdot \mathbf{v}\, dA \tag{8.8}$$

Note that there is no integral extended over A_v.

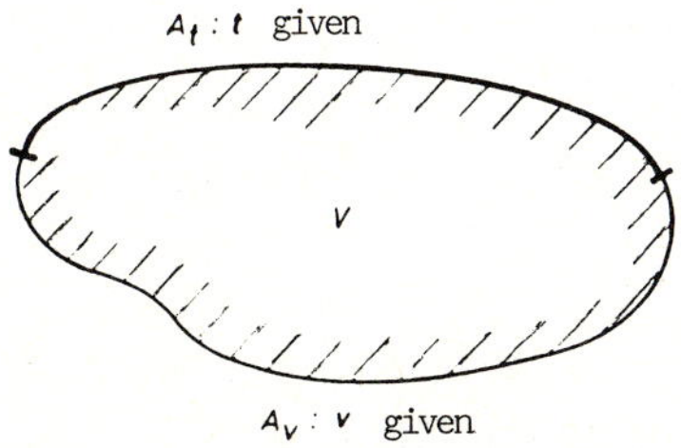

Fig. 8.1 Illustration of the minimum principle

In the derivation of the statement one considers the velocity fields that differ from the true field by $\delta \mathbf{v}$ which are kinematically possible ($\text{div}\,\delta \mathbf{v} = 0$), and which satisfy all boundary conditions. If one multiplies the equation of motion scalar by $\delta \mathbf{v}$, integrates the resulting expression over the whole volume and transforms the term with the stresses by partial integration, then one finds that the first variation of the expression (8.8) for the true flow vanishes. For the proof that there is in fact a minimum one shows that the second variation is positive. Hence one needs to assume that both viscosities, η and $\hat{\eta}$ are positive.

In the case where no stresses occur on the boundary (imagine for instance a closed container), the surface integral in (8.8) vanishes. If in addition volume forces play no part (conservative forces can be combined with the pressure and therefore also disappear), there remains

$$\int_V \Omega \, dV = \text{Min!} \tag{8.9}$$

In the case of a Newtonian fluid this is the principle of minimum total energy dissipation. For generalised Newtonian fluids Ω however has, as was stated at the beginning, another significance, i.e. the solution of the boundary value problem does not in fact make the dissipated energy a minimum, but it leads to a minimum for the potential of the extra-stresses. One can therefore in the concrete case probably denote isochoric flow fields with lower energy dissipation, which certainly do not satisfy the equations of motion.

If one limits the discussion to plane flows, then it is obvious that the left-hand side of equation (8.9) is an integral over the flow plane. For the invariant of the rate of strain tensor we have $-4I_2 = 4u_x^2 + (u_y + v_x)^2$ (cf. equation (6.16) and note that $\mathbf{A}_1 = 2\mathbf{D}$). In addition if one brings in a stream function Ψ based on (1.103), then

$$-4I_2 = 4\left(\frac{\partial^2\Psi}{\partial x \partial y}\right)^2 + \left(\frac{\partial^2\Psi}{\partial y^2} - \frac{\partial^2\Psi}{\partial x^2}\right)^2 \tag{8.10}$$

The class of steady axial shear flows through straight pipes with the velocity field $u(y,z)$ forms another special case. Because these are unaccelerated motions the neglect of the inertia term does not involve any limitation, i.e. the minimum principle holds for any value of Reynolds number, provided that the flow is laminar. If the volume force remains constant in the axial direction, and if the pressure drop is denoted by k as in Section 4.4, then the statement reduces to

$$\int_{A_0} (\Omega - ku)dA = \text{Min!} \tag{8.11}$$

in which the integral is to be extended over the cross-section of the pipe A_0. The argument of Ω is now (cf. equation (4.30))

$$-4I_2 = \left(\frac{\partial u}{\partial y}\right)^2 + \left(\frac{\partial u}{\partial z}\right)^2 \tag{8.12}$$

The significance of the minimum principle is that it permits one to obtain approximate solutions to certain flow problems without solving the non-linear equations of motion. For this purpose one choses a suitable trial function for the velocity field, which satisfies the kinematic boundary conditions and contains a number of independent parameters. One calculates the integral and determines the independent parameters from the condition that the integral value shall be as small as possible. Therefore the determination of the flow field is reduced essentially to an algebraic optimisation problem. One will in general not obtain the exact solution to the problem, but if the trial function was 'reasonable' one would expect in this way to reach a useful approximation.

8.2 Integral models

In Section 2 simple fluids were defined by the statement that the extra-stresses acting at the present time on any arbitrary fluid particle depend on the history of the relative deformation gradient. For compressible fluids the density $\rho(t)$ would also be considered. However we restrict the discussion from now on to constant density fluids, so that their density does not appear as a parameter. In the continuum mechanics books it is demonstrated that the actual stress condition cannot depend in any way on the relative deformation gradient $\mathbf{F}_t(\mathbf{r},s)$, but only on the symmetrical product $\mathbf{C}_t := \mathbf{F}_t^T\mathbf{F}_t$, which is the relative right-Cauchy-Green tensor. This reduction is not established here in detail. Apart from isotropy the above mentioned principle of material objectivity plays a part. We confine ourselves to a reference to Section 1.5, in which it was demonstrated that the deformation gradient itself represents no suitable measure of the strain, but rather the right-Cauchy-Green tensor. The general law of a rheologically simple liquid is therefore

$$\mathbf{T}(\mathbf{r},t) = \underset{s=0}{\overset{\infty}{\mathbf{F}}} [\mathbf{C}_t(\mathbf{r},s)] \tag{8.13}$$

For motions that give rise to only small strains (the displacement of the fluid particles and their rotation may be arbitrarily large) the right-hand side to a first approximation can be replaced by a linear functional in $\mathbf{C}_t - \mathbf{E}$, which in general can be expressed as an integral:

$$\mathbf{T}(\mathbf{r},t) = \int_0^\infty \frac{dG(s)}{ds}(\mathbf{C}_t(\mathbf{r},s) - \mathbf{E})ds \tag{8.14}$$

The term $\mathbf{E}$ has to be subtracted because the extra-stresses in the state of rest, i.e. when $\mathbf{C}_t = \mathbf{E}$, vanish. Because of the isotropy the kernel of the integral dG/ds is a scalar function, hence for all six tensor components it is the same. In order to identify the influence function G(s) we consider an unsteady simple shear flow which has the velocity field $\mathbf{v}(\mathbf{r},t) = \dot{\gamma}(t)y\mathbf{e}_x$. Thus we find the material point which at time t is at the position x,y,z, at time t - s from

$$x^* = x - \int_0^s \dot{\gamma}(t-\sigma)y\, d\sigma, \qquad y^* = y, \qquad z^* = z \tag{8.15}$$

For this motion the relative deformation gradient is

$$\mathbf{F}_t = \begin{bmatrix} 1 & -\gamma & 0 \\ 0 & 1 & 0 \\ 0 & 0 & 1 \end{bmatrix} \tag{8.16}$$

and the relative right-Cauchy-Green tensor is

$$\mathbf{C}_t = \begin{bmatrix} 1 & -\gamma & 0 \\ -\gamma & 1+\gamma^2 & 0 \\ 0 & 0 & 1 \end{bmatrix} \tag{8.17}$$

Thus the abbreviation is made:

$$\gamma(t,s) := \int_0^s \dot{\gamma}(t-\sigma)d\sigma \tag{8.18}$$

If one now compares the constitutive equation for the shear stress contained in (8.14) with equation (5.6), then one will find that the quantity G(s) used here is the relaxation function of the linear viscoelasticity introduced in Section 5.1. Equation (8.14) therefore expresses a generalisation of the linear constitutive equation (5.6) valid for shear flows.

For motions with small strains the generalisation can be formulated just as well by use of the inverse measure of strain $\mathbf{C}_t^{-1}$, which describes the deformation of a fluid element at the present time t related to the configuration at the earlier time t - s:

$$\mathbf{T}(\mathbf{r}, t) = -\int_0^\infty \frac{dG(s)}{ds}(\mathbf{C}_t^{-1}(\mathbf{r}, s) - \mathbf{E})ds \tag{8.19}$$

In the scope of a consideration linear with respect to the elements of the tensor $\mathbf{F}_t$ - $\mathbf{E}$ both expressions are equal. Note that the constitutive equation of each simple fluid can be approximated in the form (8.14) or (8.19) if the strains occurring in the fluid particle during the motion are sufficiently small. On the other hand equations (8.14) and (8.19) contain finite measures of strain, which are meaningful also for large deformations and can therefore serve as models to describe arbitrary motions. Because the extra-stresses in both cases are linked to a linear relationship with such a finite measure of strain, namely $\mathbf{C}_t$ - $\mathbf{E}$ and $\mathbf{C}_t^{-1}$ - $\mathbf{E}$, one speaks of constitutive equations of *finite linear viscoelasticity* in connection with (8.14) and (8.19).

The linear viscoelastic properties are of course correctly described by the function G(s). Furthermore the two models have special properties because the stresses do in fact depend in a special way on the deformation history, and all coefficients are determined by the relaxation function G(s) alone. If for example one considers a viscometric flow (with $\dot{\gamma}$ = constant in equation (8.18)), then one finds that both models possess Newtonian flow properties and quadratic normal stress properties. The three viscometric coefficients η, ν_1 and ν_2 are therefore constant (cf. Problem 8.3).

For the model (8.14) in addition $\nu_2 = -\nu_1$, which is quite unrealistic. For the model (8.19) on the other hand $\nu_2 = 0$. In this respect equation (8.19) is preferable. Because both models in the case of a shear flow provide a linear expression for the shear stress, they cannot describe real non-linear flow properties like for example the overshoot of the viscosity in the stress growth experiment beyond the equilibrium value, and a dependence on the shear rate (cf. Fig. 5.5).

The inadequacies of the constitutive equations of the finite linear viscoelasticity can be considerably overcome if one combines both expressions and in addition concedes that the memory functions depend on the first two invariants I_1, I_2 of the inverse relative right-Cauchy-Green tensor $\mathbf{C}_t^{-1}$ (the third invariant always has the value 1 for a constant density fluid):

$$\mathbf{T} = \int_0^\infty [m_1(s, I_1, I_2)(\mathbf{C}_t^{-1} - \mathbf{E}) + m_2(s, I_1, I_2)(\mathbf{C}_t - \mathbf{E})]\,ds \qquad (8.20)$$

This constitutive equation and a form equivalent to it have been developed by Kaye, also by Bernstein, Kearsley and Zapas independently and are therefore known as the K-BKZ model. With a suitable choice of the two memory functions $m_1(s,I_1,I_2)$ and $m_2(s, I_1,I_2)$ quite different rheological properties can be correctly described. Provided that one disregards the second normal stress difference, one can put $m_2 = 0$. There are also indications that the remaining memory function m_1 (in any case for certain polymer melts) can be factorised, i.e. can be represented as the product of a pure time dependent component and a component which depends only on the strain invariants. The time dependent factor must be the same as that in equation (8.19). The strain dependent factor represents the non-linear physical behaviour:

$$\mathbf{T} = -\int_0^\infty \frac{dG(s)}{ds} h(I_1, I_2)(\mathbf{C}_t^{-1} - \mathbf{E})\,ds \qquad (8.21)$$

In the special case of a shear flow the invariants I_1 and I_2 are of the same magnitude and are determined by the effective shear γ defined in equation (8.18): $I_1 = I_2 = 3 + \gamma^2$. Therefore the factor $h(I_1,I_2)$ reduces to a (even) function of γ. For motions with the deformation history given by (8.17) therefore

$$\mathbf{T} = -\int_0^\infty \frac{dG(s)}{ds} h(\gamma)(\mathbf{C}_t^{-1} - \mathbf{E})\,ds \qquad (8.22)$$

applies.

For a pseudoplastic fluid h decreases with γ monotonically. The model (8.22) can for example qualitatively correctly describe real physical properties such as occur during sudden changes in shear rate.

In a stress growth experiment, say with a history of motion given by equation (5.19), there holds for the effective shear defined by equation (8.18)

$$\gamma(t, s) = \begin{cases} \dot{\gamma}_0 s & \text{for } 0 < s < t \\ \dot{\gamma}_0 t & \text{for } 0 < t < s \end{cases}$$

A material that is described by the constitutive equation (8.22) thus reacts with the following time process of shear stress (according to equation (8.17) $+\gamma$ is the relevant secondary diagonal element of the tensor $C_t^{-1} - E$):

$$\tau(t) = -\int_0^t \frac{dG(s)}{ds} h(\dot{\gamma}_0 s)\dot{\gamma}_0 s \, ds - \int_t^\infty \frac{dG(s)}{ds} ds \cdot h(\dot{\gamma}_0 t)\dot{\gamma}_0 t$$

Because $G(\infty) = 0$ the second integral on the right-hand side gives the expression $-G(t)$. If one differentiates the expression with respect to time, then two terms cancel each other out and there results the remarkable formula

$$\frac{d\eta^+(t; \dot{\gamma}_0)}{dt} = G(t) \left. \frac{d(\gamma h(\gamma))}{d\gamma} \right|_{\gamma = \dot{\gamma}_0 t}$$

Thus as described in Section 5.1.1 the stress growth viscosity η^+ has been introduced for the quotient $\tau/\dot{\gamma}_0$. According to the relationship the occurrence of a stress maximum in the stress growth experiment is bound up with the existence of a maximum of the constitutive function $\gamma h(\gamma)$. The appropriate value of the shear γ represents a material constant. The formula shows that the product of the shear rate $\dot{\gamma}_0$ and the time t_m at which the stress maximum occurs agrees with this value of the shear, hence $t_m \cdot \dot{\gamma}_0$ = constant, independent of $\dot{\gamma}_0$. The experimental results for a polymer solution reproduced in Fig. 5.5 confirm this relationship qualitatively because t_m decreases when $\dot{\gamma}_0$ increases. For the quantitative comparison however, one realises that the product $t_m \cdot \dot{\gamma}_0$ does not remain constant at all, but takes up values between 3 and 13 when $\dot{\gamma}_0$ varies between $1.67 s^{-1}$ and $52.8 s^{-1}$. This discrepancy between theory and experiment shows that the model (8.22) describes only incompletely the polymer solution considered. For more viscous liquids, particularly polymer melts, the comparison may turn out more favourably. Note that the result $t_m \cdot \dot{\gamma}_0$ = constant has been derived without special assumptions about the two constitutive functions G(s) and h(γ) (except that $\gamma h(\gamma)$ has a maximum) from equation (8.22). An analogous relationship with corresponding consequences holds moreover for the first normal stress difference, whereby in the formula on the right-hand side instead of $\gamma h(\gamma)$ the expression $\gamma^2 h(\gamma)$ occurs.

Integral constitutive equations which have been discussed here are only of limited utility in fluid mechanics because in the application to actual boundary value problems considerable mathematical difficulties arise. They arise from the fact that in the calculation of the extra-stresses the fluid particles have to be retraced in time, their motion however in general is not known initially, and not until the equations of motion have been determined. Even for unsteady shear flows, which represent a special case because the path lines are known from the first, the use of an integral law leads to a partial integro - differential equation for the velocity field. One example is the periodic pipe flow with the equation of motion (5.100) considered in Section 5.2, where the expression

$$\tau = -\int_0^\infty \frac{dG(s)}{ds} h(\gamma)\gamma \, ds \tag{8.23}$$

had to be inserted for the shear stress. The effective shear $\gamma(r,t,s)$ is given by equation (8.18) from the history of the shear rate $\dot{\gamma}(r,t-\sigma)$ which for a pipe flow agrees with the velocity derivative $\partial u(r,t-\sigma)/\partial r$. For more complex motions the matter is more complex in that the path lines are a-priori unknown, and therefore their differential equations (1.2) must be integrated together with the equations of motion. The appropriate formulation of an initial value problem requires not only data about the velocity field at the starting time, but its whole past history must be known. Thus without considerable effort only kinematically relatively simple flows can be analysed, and in fact preferably within the scope of linear viscoelasticity theory.

8.2.1 Flow between eccentric rotating discs

In the measurement of the linear viscoelastic properties it is convenient to use the following method. The fluid is located between two parallel circular discs with slightly displaced central axes relative to each other. The gap between them is small compared with the radius of the discs. The discs rotate at the same angular velocity ω round their respective axes. As regards the kinematics of the flow thus created, one can assume that the flow field consists of a set of parallel layers z = constant, which are all rotating at the same angular velocity without internal strain. The centres of rotation of the various layers lie on the straight line joining the mid-points of the two discs (Fig. 8.2). The angle between this straight line and the z-axis is assumed to be small, $|\Psi| \ll 1$.

Because the arbitrary layer z = constant rotates round the point $x = 0$, $y = \psi z$, all path lines are circles and the position of x^*, y^*, z^* occupied by a material point at time t - s is therefore related to the position x,y,z at time t as follows:

$$\begin{aligned} x^* &= x\cos\omega s + (y - \psi z)\sin\omega s \\ y^* &= \psi z - x\sin\omega s + (y - \psi z)\cos\omega s \\ z^* &= z \end{aligned} \qquad (8.24)$$

To this motion, which in fact has already formed the basis of Problem 1.7, belongs

$$\mathbf{C}_t = \begin{bmatrix} 1 & 0 & -\psi\sin\omega s \\ 0 & 1 & -\psi(1-\cos\omega s) \\ -\psi\sin\omega s & -\psi(1-\cos\omega s) & 1 + 2\psi^2(1-\cos\omega s) \end{bmatrix} \qquad (8.25)$$

as the history of the relative right-Cauchy-Green tensor.

This relationship refers to the Cartesian x-y-z system shown in Fig. 8.2. Because the flow is steady the past time s, but not the present time t occurs here. If one incorporates these expressions into the constitutive equation of finite linear viscoelasticity (8.14), then one finds that three extra-stresses vanish, and the others are constant with respect to position, i.e. are independent of x, y, z. Therefore div **T** = 0, and the equations of motion, provided that the effect of inertia can be neglected, lead to the statement that the

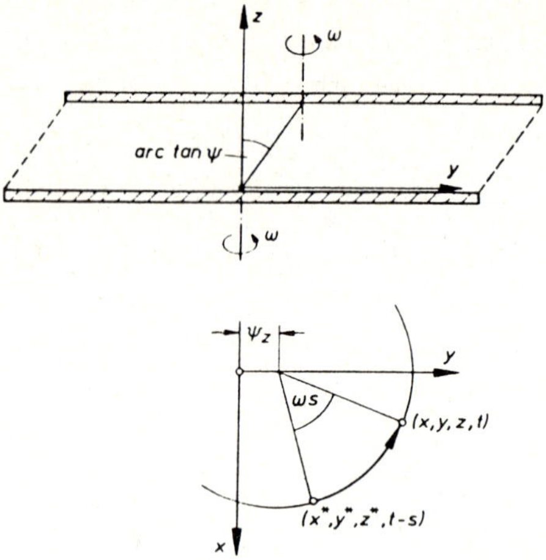

Fig. 8.2 Flow kinematics between eccentric rotating discs

pressure too is constant over the field. For the two non-vanishing shear stresses then

$$\tau_{xz} = \psi\omega \int_0^\infty G(s) \cos \omega s \, ds \tag{8.26}$$

$$\tau_{yz} = \psi\omega \int_0^\infty G(s) \sin \omega s \, ds \tag{8.27}$$

We have already met the two integrals in Section 5.1.3, where we learned that they are the real component and the imaginary component of the complex viscosity (cf. equation (5.38)). For the shear stresses acting on the discs therefore $\tau_{xz} = \psi\omega\eta'(\omega)$ and $\tau_{yz} = \psi\omega\eta''(\omega)$. From that it follows that the dynamic viscosity $\eta'(\omega)$ and the dynamic shear modulus $G'(\omega) = \omega\eta''(\omega)$ of a viscoelastic liquid can be determined with the help of the device described, although the discs do not actually oscillate but rotate at constant velocity, and hence create a steady flow (*Maxwell-Chartoff rheometer*). The deformation of an individual fluid particle depends of course on the time and corresponds, apart from a rotation, to an oscillatory sinusoidal shear of angular velocity ω.

The use of the law of the linear viscoelasticity is only valid when the shear amplitude ψ is small compared with unity. Under this condition however it suffices to drive one of the two circular discs at the angular velocity ω. The other disc, free to rotate about its centre axis, is then dragged round by the fluid, but does not rotate at exactly the same angular velocity, but rather more slowly. The slip $\delta\omega$ can only be an even function of ψ, because the reading of one and the same situation from opposite directions only changes the sign of ψ; $\delta\omega$ stays. Hence for $|\psi|<<1$ the slip is quadratically small, $\delta\omega = O(\psi^2)$ and can therefore be neglected in the context of a theory linear with respect to the strain amplitude ψ.

The quantities measured in the Maxwell-Chartoff rheometer are the two tangential components F_x and F_y of the force on the circular disc as it is dragged round at a certain angular velocity ω. Because the associated shear stresses τ_{xz} and τ_{yz} (apart from boundary effects) are spatially constant, simple relationships exist between them, the force components and the surface A of a disc: $\tau_{xz} = F_x/A$, $\tau_{yz} = F_y/A$. Hence one can obtain the two linear viscoelastic properties $\eta'(\omega)$ and $G'(\omega)$ from the expressions

$$\eta'(\omega) = \frac{F_x(\omega)}{A\psi\omega}, \quad G'(\omega) = \frac{F_y(\omega)}{A\psi} \tag{8.28}$$

If the angular velocity is varied, then the thus evaluated force measurement leads to expressions of the two constitutive functions as shown in Fig. 5.9.

8.2.2 Boundary layer at a plane wall with homogeneous suction

The equations of motion for Newtonian fluids have an exact solution which describes the asymptotic condition near a plane plate adjacent to a longitudinal flow with homogeneous suction remote from the front edge of the plate. The flow field is isochoric, steady and plane. The velocity component normal to the wall is spatially constant (equal to the suction velocity -V), and that parallel to the wall varies with the distance from the wall (Fig. 8.3). The flow has a boundary layer character, the displacement distance depends on the suction velocity, the viscosity and the density of the liquid.

The purpose here is to investigate whether, and under what conditions, a flow of the form described, hence with the velocity field $u = u(y)$, $v = -V$, $w = 0$ is also possible in linear viscoelastic liquids, how the 'asymptotic suction profile' $u(y)$ looks, and to find which physical quantities affect the boundary layer thickness. The history of the relative right-Cauchy-Green tensor has already been determined for this motion in Section 1.5. If one inserts the result (1.74) in the constitutive equation (8.14) of the finite linear viscoelasticity, then one obtains for the shear stress $\tau_{(xy)}$ at a distance y from the wall

$$\tau(y) = -\frac{1}{V}\int_0^\infty \frac{dG(s)}{ds}[u(y+Vs) - u(y)]\, ds \qquad (8.29)$$

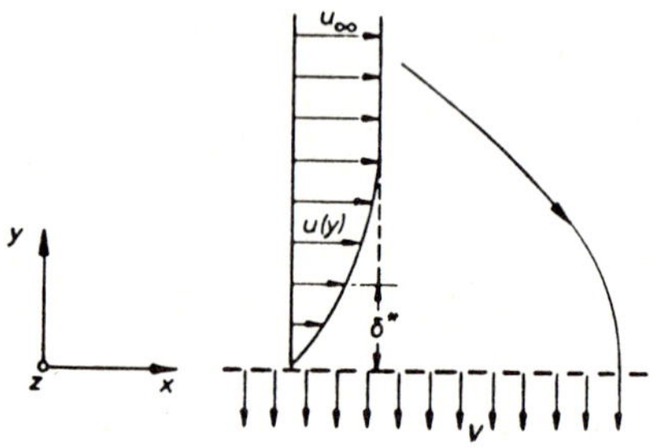

Fig. 8.3 Flow near a plane wall for nomogeneous suction

The equation of motion in the direction parallel to the wall combines the stress derivative $d\tau/dy$ with the acceleration $a_x = Du/Dt$, which only contains the convective part $-V du/dy$:

$$-\rho V\frac{du}{dy} = \frac{d\tau}{dy} \qquad (8.30)$$

Here volume forces are neglected and it is assumed that a pressure gradient in the direction parallel to the wall is absent. The normal stress differences contained in the constitutive equation (8.14) depend, like τ only on y and therefore vanish from the equation of motion. We furthermore assume that the liquid is sucked only in a direction normal to the wall, so that $u(0) = 0$ applies, and interpret the velocity u_∞ beyond the boundary layer as a given quantity. Under these conditions the path lines of the material points outside the boundary layer are straight lines of slope $-V/u_\infty$, inside they are curved, and on the wall they are perpendicular to it (Fig. 8.3).

Equation (8.30) can be integrated once to give

$$\tau(y) = \rho V[u_\infty - u(y)] \tag{8.31}$$

From this one learns that the wall shear stress $\tau(0) = \rho V u_\infty$ does not at all depend on the extra-stress behaviour of the liquid. If one eliminates the shear stress from equations (8.29) and (8.31) there results an integral equation of the second kind for the velocity profile:

$$(\rho V^2 - G(0))\,[u_\infty - u(y)] = \int_0^\infty \frac{dG(s)}{ds}\,[u_\infty - u(y + Vs)]\,ds \tag{8.32}$$

It is known, and one can easily demonstrate on the basis of equation (8.31) that for a Newtonian fluid of viscosity η_0 (with $\tau = \eta_0 du/dy$) the velocity field parallel to the wall is described by the analytical expression $u(y)/u_\infty = 1 - \exp(-\rho V y/\eta_0)$. It is therefore obvious to try the following function as a solution of equation (8.32)

$$u(y) = u_\infty\left[1 - e^{-\frac{\rho V y}{\eta_0 \Delta}}\right] \tag{8.33}$$

In this η_0 denotes, as previously, the zero-shear viscosity of the liquid which is connected with the relaxation function G(s) by the relation (5.7). The constant factor Δ is a dimensionless quantity for the displacement thickness δ^* of the boundary layer because

$$\delta^* := \int_0^\infty \left(1 - \frac{u(y)}{u_\infty}\right) dy = \frac{\eta_0}{\rho V}\cdot\Delta \tag{8.34}$$

belongs to the velocity profile (8.33).

In the special case of the Newtonian fluid $\Delta = 1$. With (8.33) equation (8.32) reduces to a conditional equation for Δ:

$$\rho V^2 - G(0) = \int_0^\infty \frac{dG(s)}{ds}\, e^{-\frac{\rho V^2 s}{\eta_0 \Delta}}\, ds \tag{8.35}$$

If for the sake of example we restrict the discussion to a Maxwell fluid with an exponentially decreasing relaxation function (cf. equation (5.8)), then the integral can be evaluated and the following expression for Δ derived:

$$\Delta = 1 - \frac{\rho V^2}{G(0)} \tag{8.36}$$

One learns from this that the boundary layer for a viscoelastic liquid is thinner than for a Newtonian liquid of the same zero-shear viscosity. One can easily demonstrate this phenomenon, starting from the already mentioned fact that different liquids under the same conditions give rise to the same stress value at the wall, namely $\rho V u_\infty$. For a Newtonian fluid a quite definite value of the velocity gradient du/dy at the wall corresponds to this stress. In a liquid with a memory however the wall shear stress orients itself to previous values of the velocity gradient which a material point has experienced before it reached the wall. Under the idealised qualitatively correct concept that the memory expresses itself in a certain dead time λ, the velocity gradient du/dy at a distance $V\lambda$ from the wall would be just as great as in the Newtonian reference field at the wall itself. The result of this consideration is that the boundary layer is compressed by the memory capability.

According to equation (8.36) the boundary layer thickness vanishes when the suction rate reaches the value $\sqrt{G(0)/\rho}$. It moreover corresponds exactly to the propagation rate of high frequency transverse sinusoidal waves (cf. Problem 5.6). For greater suction rates the expression (8.33) is obviously no longer meaningful, because Δ is negative. A steady flow with an unvarying velocity field in a direction parallel to the wall can therefore no longer exist for $V > \sqrt{G(0)/\rho}$. Other more complex flows which might arise instead of this under homogeneous suction cannot be investigated in more detail here. One must in any case expect that the solution of the equations of motion branches out and the assumed flow pattern becomes unstable when the suction parameter $\rho V^2/G(0)$ controlled by the elasticity of the liquid reaches a certain critical value (<1). On the basis of this example one learns once more that laminar flows, as one knows them in hydrodynamics, can become unstable in viscoelastic fluids, as soon as the elasticity coefficient becomes too large.

8.3 Differential models

In the study of unsteady shear flows in Section 5.1 we learned that the constitutive equation of linear viscoelastic fluids can possibly be formulated differentially. In the case of a Maxwell fluid for example the relationship between the shear stress and the shear rate runs thus: $\tau(t) + \lambda\dot{\tau}(t) = \eta_0\dot{\gamma}(t)$ (cf. equation (5.12)). Whereas in the appropriate integral form (5.6) the stress at the present time is linked to the motion of the earlier time, here only quantities at the present time enter, besides the stress itself of course also the time derivative of it. In order to be able to describe more general motions, such a constitutive equation must be generalised. It should in fact be valid not only for one-dimensional but also for any three-dimensional process. That means the scalar relationship between τ, $\dot{\tau}$ and $\dot{\gamma}$ must be replaced by an expression which combines the tensor of the extra-stresses **T** and its time derivative with the rate of strain tensor **D**. In order that the resulting constitutive equation should satisfy the 'principle of material objectivity', the partial time derivative of **T** alone may not be used, but an 'objective', for example the Jaumann time derivative defined by equation (1.56) must be used. A possible generalisation of equation (5.12) therefore reads:

$$\mathbf{T} + \lambda \overset{\circ}{\mathbf{T}} = 2\eta_0 \mathbf{D} \tag{8.37}$$

Although the two coefficients, the zero-shear viscosity η_0 and the relaxation time λ are constant, this is in general no linear relationship between the extra-stresses and the velocity components, because the Jaumann derivative $\overset{\circ}{\mathbf{T}}$ contains the terms $-\mathbf{WT} + \mathbf{TW}$, hence bilinear expressions of stresses and velocity derivatives (**W** was the vorticity tensor; cf. equation (1.25)). Thus this 2-constant model simulates certain non-linear physical properties, particularly non-linear flow behaviour and normal stress differences under viscometric conditions.

In order to determine the viscometric functions belonging to the model (8.37) one best considers the simplest case of a steady plane shear flow with the velocity gradient tensor (1.76) and the symmetric and skew-symmetric components derived therefrom

$$\mathbf{D} = \begin{bmatrix} 0 & \dot{\gamma}/2 & 0 \\ \dot{\gamma}/2 & 0 & 0 \\ 0 & 0 & 0 \end{bmatrix}, \quad \mathbf{W} = \begin{bmatrix} 0 & \dot{\gamma}/2 & 0 \\ -\dot{\gamma}/2 & 0 & 0 \\ 0 & 0 & 0 \end{bmatrix} \tag{8.38}$$

As is known only one shear stress component $\tau_{(xy)}$ occurs, so that the extra-stress tensor **T** has four independent elements:

$$\mathbf{T} = \begin{bmatrix} \tau_{xx} & \tau & 0 \\ \tau & \tau_{yy} & 0 \\ 0 & 0 & \tau_{zz} \end{bmatrix} \tag{8.39}$$

Because the motion considered here is steady in material coordinates ($D\mathbf{T}/Dt = 0$), $\overset{\circ}{\mathbf{T}} = -\mathbf{WT} + \mathbf{TW}$ holds for the Jaumann stress derivative, and by matrix multiplication one obtains

$$\overset{\circ}{\mathbf{T}} = -\mathbf{WT} + \mathbf{TW} = \begin{bmatrix} -\tau\dot{\gamma} & (\tau_{xx} - \tau_{yy})\dot{\gamma}/2 & 0 \\ (\tau_{xx} - \tau_{yy})\dot{\gamma}/2 & \tau\dot{\gamma} & 0 \\ 0 & 0 & 0 \end{bmatrix} \tag{8.40}$$

If one enters all this into equation (8.37) and collects together corresponding components of the three matrices, then one obtains a linear system of equations for the four stress components as functions of $\dot{\gamma}$. Its solution leads to the viscometric functions

$$\tau = \frac{\eta_0\dot{\gamma}}{1 + \lambda^2\dot{\gamma}^2} \equiv \eta(\dot{\gamma}) \cdot \dot{\gamma} \tag{8.41}$$

$$\tau_{xx} - \tau_{yy} = \frac{2\lambda\eta_0\dot{\gamma}^2}{1 + \lambda^2\dot{\gamma}^2} \equiv N_1(\dot{\gamma}) \tag{8.42}$$

$$\tau_{yy} - \tau_{zz} = -\frac{\lambda\eta_0\dot{\gamma}^2}{1 + \lambda^2\dot{\gamma}^2} \equiv N_2(\dot{\gamma}) \tag{8.43}$$

Certain deficiencies in the model of course appear here. The flow law (8.41) can in fact simulate shear thinning, but it is only meaningful for relatively low shear rates, $|\dot{\gamma}| < 1/\lambda$. Above that the differential viscosity would be negative, which for real liquids cannot occur for reasons of stability. The normal stress functions are qualitatively realistic for pseudoplastic substances. The value 1/2 for the ratio $-N_2/N_1$ does not however correspond to the experimental results, according to which the absolute value of the second normal stress difference works out in general considerably smaller than the first normal stress difference.

In order to assess the goodness of a model it is appropriate to determine also its elongational properties. Because steady homogeneous extensional flows are free of rotation ($\mathbf{W} = 0$) and at the same time are materially steady ($D\mathbf{T}/Dt = 0$), the Jaumann derivative $\overset{\circ}{\mathbf{T}}$ vanishes under these conditions. The model (8.37) has accordingly the same elongational properties as a Newtonian fluid. Elongational thickening as can be seen in Fig. 7.1 cannot therefore be simulated.

Equation (8.37) can be arbitrarily generalised by the addition of suitable objective terms. Thus for example it has been suggested that $\overset{\circ}{\mathbf{D}}$ should be considered besides $\overset{\circ}{\mathbf{T}}$, and all permissible bilinear expressions that can be formed from the extra-stresses and the velocity gradient should be taken into account. This leads to an *8-constant model* which is named after *Oldroyd* (seven time scales occur in addition to the zero-shear viscosity η_0):

$$\begin{aligned}&\mathbf{T} + \lambda_1 \overset{\circ}{\mathbf{T}} + \mu_0 (\mathrm{tr}\,\mathbf{T})\mathbf{D} - \mu_1 (\mathbf{DT} + \mathbf{TD}) + \kappa_1 (\mathrm{tr}\,\mathbf{TD})\mathbf{E} \\ &= 2\eta_0 [\mathbf{D} + \lambda_2 \overset{\circ}{\mathbf{D}} - 2\mu_2 \mathbf{D}^2 + \kappa_2 (\mathrm{tr}\,\mathbf{D}^2)\mathbf{E}] \qquad (8.44)\end{aligned}$$

It is obvious that because of the greater number of constant parameters rheological data can be more realistically approximated. However one should not overlook the fact that each empirical constitutive equation will naturally describe some phenomena more or less well, but others only unsatisfactorily. If one adapts the constant parameters for example to the viscometric data, then one ought naturally not expect that the whole model thus specified also correctly describes the elongational viscosities of the material or its unsteady shear properties quantitatively. Such an adaptation would therefore be inappropriate if the elongational properties for the complex flow under study are particularly important. For the determination of the constants therefore the question of which measurable rheological properties probably affect the flow critically, and which are unimportant, plays a part.

If in equation (8.44) one puts $\mu_1 = \lambda_1 = \lambda$ and further $\mu_0 = \kappa_1 = \lambda_2 = \mu_2 = \kappa_2 = 0$, then one obtains as a special case the *Maxwell-Oldroyd model*.

$$\mathbf{T} + \lambda(\overset{\circ}{\mathbf{T}} - \mathbf{DT} - \mathbf{TD}) = 2\eta_0 \mathbf{D} \qquad (8.45)$$

This is formed in the same way as the model (8.37) mentioned at the beginning, whereby of course instead of the Jaumann time derivative $\overset{\circ}{\mathbf{T}} := \dot{\mathbf{T}} - \mathbf{WT} + \mathbf{TW}$ the expression in brackets occurs, which one can represent by using the velocity gradient tensor $\mathbf{L}$ also in the form $\dot{\mathbf{T}} - \mathbf{LT} - \mathbf{TL}^T$, and which is conveniently called an Oldroyd time derivative. Equation (8.45) again represents a permissible generalisation of equation (5.12). The significance of the two constants has not changed; η_0 denotes the zero-shear viscosity and λ the relaxation time of the liquid. Under viscometric conditions the model has Newtonian flow behaviour and a non-zero first normal stress function with constant coefficient:

$$\eta(\dot{\gamma}) = \eta_0, \qquad N_1(\dot{\gamma}) = 2\lambda\eta_0\dot{\gamma}^2, \qquad N_2(\dot{\gamma}) = 0 \tag{8.46}$$

It is therefore suited to the qualitative study of motions that are critically affected by the memory and by the first normal stress difference, so long as shear thinning or shear thickening does not matter.

The advantage of differential laws results from the disadvantage of integral models described in Section 8.2. The local formulation makes it unnecessary to trace backwards in time the material points along their path lines. The memory of the fluid expresses itself instead in an integral over the past motion in time derivatives at the present time. Therefore the stress components are in fact no longer explicitly expressed by the velocity field and its gradient. The constitutive equations form much more a system of differential equations for the stresses which in general cannot be solved, so that the elimination of the stresses from the equations of motion is not immediately possible. One has therefore to integrate simultaneously the equations of motion and the constitutive equations as a system. This presents no major difficulty in the numerical treatment of actual flow problems. Where the elimination of the stresses can be carried out it shows that the integration of the resulting material specific field equations of higher order requires a comparable expenditure of effort. Initial value problems have the usual form, whence the quantities occurring in the field equations (velocities, stresses), and if necessary their time derivatives at the initial time are to be given. The past history of the motion, which with the use of an integral constitutive equation must be known, therefore plays no part in it. Differential models compared with integral ones therefore have considerable advantages in respect of the numerical treatment of more complex flow processes.

8.4 Approximation for slow and slowly varying processes

In Section 5.2.1 we were able, as an example of an unsteady shear flow, to recognise that the general constitutive equation of a simple fluid could be simplified when the flow proceeded slowly, and, measured at the range of the memory, only slightly varied with time. Under these assumptions the memory of the fluid could be considered by the fact that besides the shear rate at the present time, the time derivatives of it were taken into account in the constitutive equation (cf. equation (5.118)). As was explained in more detail there occurred $\dot{\gamma}(t)$ as a quantity of the first order, $\dot{\gamma}^2$ and $\partial\dot{\gamma}/\partial t$ as quantities of the second order, $\dot{\gamma}^3$, $\dot{\gamma}\partial\dot{\gamma}/\partial t$ and $\partial^2\dot{\gamma}/\partial t^2$ as quantities of the third order, etc. It is not difficult to transfer the conclusions made on the basis of a one-dimensional flow to more general motions, so that we can be really brief here.

What is involved here is the approximation of the functional (8.13) for slow motions which vary only slightly within the characteristic memory time (cf. once again Fig. 5.20). In such cases the history of the relative right-Cauchy-Green tensor can be expanded in accordance with equation (1.67), in which the Rivlin-Ericksen tensors for the present time occur as expansion coefficients. Therefore the functional (8.13) reduces to a function of these kinematic tensors. As previously explained however $\mathbf{A}_1$, $\mathbf{A}_2$,....$\mathbf{A}_n$ represent quantities of the first, second order, etc. In the first approximation one therefore only needs to consider $\mathbf{A}_1$, in the second order $\mathbf{A}_2$ besides $\mathbf{A}_1$, etc. Under certain regularity conditions, which in general can be considered as satisfied, the constitutive function $\mathbf{F}(\mathbf{A}_1)$ can be approximated for $\mathbf{A}_1 \to 0$ by linear expressions in $\mathbf{A}_1$. Because fluids are isotropic, only $\mathbf{A}_1$ itself and (tr $\mathbf{A}_1$)$\mathbf{E}$ are involved. For constant density fluids, which permit only isochoric flows (with tr $\mathbf{A}_1 = 0$), one thus obtains as an approximation of the first order

$$\mathbf{S} + p\,\mathbf{E} \equiv \mathbf{T} = \eta_0 \mathbf{A}_1 \tag{8.47}$$

This constitutive equation is characteristic of *Newtonian* liquids. One learns here that 'almost every' simple liquid in very slow and very slowly varying flow behaves in a Newtonian manner, in which the zero-shear viscosity η_0 is of course critical. In the context of a second order approximation one linear term in $\mathbf{A}_2$ and one in $\mathbf{A}_1^2$ are added:

$$\mathbf{S} + p\mathbf{E} = \eta_0 \mathbf{A}_1 + \alpha_1 \mathbf{A}_2 + \alpha_2 \mathbf{A}_1^2 \tag{8.48}$$

The other still permissible isotropic terms of the second order, say $(\mathrm{tr}\,\mathbf{A}_2)\mathbf{E}$, are proportional to the unit tensor and can therefore be added to the spherical symmetric pressure stresses $-p\mathbf{E}$, which must be added to the extra-stresses $\mathbf{T}$ in order to obtain the total stresses $\mathbf{S}$. In connection with the constitutive equation (8.48) one conveniently speaks of *'liquids of the second order'*. This term is of course rather unfortunate because it is not the special physical properties of a few liquids which lead to equation (8.48), but the limitation to a certain class of motion. In contrast to the constitutive equations given in Sections 8.2 and 8.3, this is not so much a matter of a model, but much more is almost every simple liquid under the described conditions a second order liquid. It would therefore be more appropriate to denote equation (8.48) as an *approximation of the second order* for slow and slowly varying flows. The applicability of the approximation depends quite critically on the flow process, and in fact the maximum shear rate $\dot{\gamma}_m$ in the flow field must necessarily be so small that its product with the material specific relaxation time $(|\alpha_1| + |\alpha_2|)/\eta_0$, the Weissenberg number $(|\alpha_1| + |\alpha_2|)\dot{\gamma}_m/\eta_0$, is small compared with unity.

The formula (8.48) remarkably resembles the constitutive equation (2.36) which is valid for viscometric flows of arbitrary strength. Note however the different significance of the two expressions. A comparison shows that a second order liquid has Newtonian flow properties for viscometric flow and both the normal stress coefficients are constant. The associated normal stress functions therefore increase with $\dot{\gamma}^2$:

$$\eta(\dot{\gamma}) = \eta_0 \tag{8.49}$$

$$N_1(\dot{\gamma}) = -2\alpha_1 \dot{\gamma}^2 \tag{8.50}$$

$$N_2(\dot{\gamma}) = (2\alpha_1 + \alpha_2)\dot{\gamma}^2 \tag{8.51}$$

Therefore the second order coefficients α_1 and α_2 of a real liquid are related in a simple way to the lower limiting values of the normal stress coefficients of this liquid:

$$\alpha_1 = -\frac{1}{2}\nu_{10}, \qquad \alpha_2 = \nu_{10} + \nu_{20} \tag{8.52}$$

In the context of a *third order approximation* one has to consider some further terms. Besides the spherical symmetric terms, which can again be combined with the 'undetermined' pressure, these are $\mathbf{A}_3$, $\mathbf{A}_1\mathbf{A}_2$, $\mathbf{A}_2\mathbf{A}_1$, $I_1(\mathbf{A}_2)\cdot\mathbf{A}_1$, $I_1(\mathbf{A}_1^2)\cdot\mathbf{A}_1$, $I_2(\mathbf{A}_1)\cdot\mathbf{A}_1$ and $\mathbf{A}_1^3$, in which I_1 and I_2 symbolise the first two principal invariants of the added kinematic tensors in brackets.

For isochoric motions $\mathbf{A}_1^3$ can be replaced by a spherical symmetric component, and the terms $-I_2(\mathbf{A}_1)\cdot\mathbf{A}_1$. In addition $I_2(\mathbf{A}_1)$ and $I_1(\mathbf{A}_2)$ can be reduced to $I_1(\mathbf{A}_1^2) \equiv \mathrm{tr}\,\mathbf{A}_1^2$, the 'trace' of the tensor $\mathbf{A}_1^2$. Because the stress tensor is symmetrical the two expressions $\mathbf{A}_1\mathbf{A}_2$ and $\mathbf{A}_2\mathbf{A}_1$ cannot appear independently of each other, but only as their sum. The equation of a *'third order liquid'* therefore reads:

$$\mathbf{S} + p\mathbf{E} = \eta_0\mathbf{A}_1 + \alpha_1\mathbf{A}_2 + \alpha_2\mathbf{A}_1^2 + \beta_1\mathbf{A}_3 + \beta_2(\mathbf{A}_1\mathbf{A}_2 + \mathbf{A}_2\mathbf{A}_1) + \beta_3(\mathrm{tr}\,\mathbf{A}_1^2)\mathbf{A}_1 \tag{8.53}$$

Three third order coefficients β_1, β_2 and β_3 are added to the previously described constants η_0, α_1 and α_2. The expansion coefficients can be determined independently of the order of the approximation partly from viscometric flows, partly also from steady extensional flows or suitably unsteady shear flows (cf. Problem 8.7). In the discussion about secondary flows in Section 9 we shall make abundant use of second and third order approximations for slow and slowly varying processes described here.

Problems

8.1 The flow rate through a circular pipe is to be approximately determined for a Bingham material, using the minimum principle (8.11), and compared with the accurate result obtained for Problem 3.8. To this end first find the potential Ω which a Bingham material exhibits, then use the expression $u = u_0(1 - r^2/r_0^2)$ for the velocity field.

Determine the parameter u_0 on the basis of the minimum principle and calculate the discharge, which with suitable normalising depends on a dimensionless pressure parameter.

8.2 Show that for incompressible generalised Newtonian fluids the relationship (8.7) mentioned in the text applies.

8.3 Which viscometric functions $\eta(\dot{\gamma})$, $\nu_1(\dot{\gamma})$, $\nu_2(\dot{\gamma})$ and which elongational viscosities $\eta_E(\dot{\varepsilon})$, $\eta_{EB}(\dot{\varepsilon})$, $\eta_{EP}(\dot{\varepsilon})$ belong to the rheological equations (8.14) and (8.19) of finite linear viscoelasticity? Only consider the special case $G(s) = G(0)\exp(-s/\lambda)$.

8.4 Determine for a special K-BKZ model of the type in equation (8.22) with the relaxation function (5.16) and the 'damping function' $h(\gamma) = e^{-N|\gamma|}$ the shear stresses as a function of the time for the two standard tests with suddenly changing rate of shear, hence for processes in accordance with equations (5.19) and (5.21), and compare with Figs. 5.5 and 5.6.

8.5 Describe the free convectional flow of a linear viscoelastic liquid at a heated vertical wall with homogeneous suction, following the statements in Section 8.2.2. The wall has a constant temperature Θ_0, the liquid outside the boundary layer has the temperature Θ_∞. In addition $u_0 = u_\infty = 0$. Neglecting the dissipation, the energy equation gives the temperature profile $\Theta(y) - \Theta_\infty = (\Theta_0 - \Theta_\infty)\exp(-Pr\rho Vy/\eta_0)$. Pr is the Prandtl number for the liquid. The temperature differences create an inhomogeneous buoyancy force of magnitude $f_x = \rho g\beta(\Theta - \Theta_\infty)$. β denotes the thermal expansion coefficient of the liquid. This volume force parallel to the wall is to be added to the right-hand side of equation (8.30). The resulting velocity profile can be expressed as the sum of two exponential functions. Discuss the solution for velocity and shear stress on the basis of a Maxwell model and then show that even for low values of the elasticity number $\rho V^2/G(0)$ marked differences from the Newtonian reference flow occur. Hence note that the Prandtl number can be quite large for liquids (say $Pr = 10^3$).

8.6 Consider a special Oldroyd fluid given by equation (8.44), for which $\mu_1 = \lambda_1$ and $\kappa_1 = \lambda_2 = \mu_2 = \kappa_2 = 0$ (3-constant model), and show that the constitutive equation for an unsteady simple shear flow with the velocity field $\mathbf{v} = y\dot{\gamma}(t)\mathbf{e}_x$ reduces to the two relations

$$\tau_{xx} - \tau_{yy} + \lambda_1(\dot{\tau}_{xx} - \dot{\tau}_{yy}) - 2\lambda_1\dot{\gamma}\tau_{xy} = 0$$

$$\tau_{xy} + \lambda_1\dot{\tau}_{xy} + \frac{1}{2}\mu_0\dot{\gamma}(\tau_{xx} - \tau_{yy}) = \eta_0\dot{\gamma}$$

Then calculate for the two standard tests described in Section 5.1.1

the shear stress τ_{xy} and the first normal stress difference $\tau_{xx} - \tau_{yy}$ as functions of time, and compare the analytical results with experimental data. Which of the processes describes the model qualitatively correctly, and which not?

8.7 The law (8.53) of an incompressible 'third order liquid' contains six independent coefficients. Which of these constants, or which combination of them, can be determined by use of viscometric flows, of steady extensional flows, and of unsteady shear flows of small amplitude?

8.8 Show that the linear viscoelastic constitutive functions G(s) and G'(ω) as follows are connected with the lower limiting value of the first normal stress coefficient ν_{10}:

$$\int_0^\infty s\, G(s)\, ds = \frac{\nu_{10}}{2} = \lim_{\omega \to 0} \frac{G'(\omega)}{\omega^2}$$

These relationships have already been given without proof in Sections 5.1.2 and 5.1.3.

8.9 Prove that for slow and slowly varying motions the Oldroyd 8-constant model (8.44) can also be approximated to a third order accuracy by the explicit constitutive equation (8.53), and determine how the six expansion coefficients in equation (8.53) are related to the eight constants of the model.

9 SECONDARY FLOWS

9.1 General theory

We now consider certain noteworthy flow phenomena, which one observes in viscoelastic liquids, but which however do not occur in Newtonian fluids. Consider for example the laminar steady flow through a cylindrical pipe. For a Newtonian fluid the velocity vector is independently of the overall shape of the pipe cross-section always parallel to the pipe axis, and the amount of the velocity is constant for each stream line. For non-Newtonian fluids this simple flow form is however in general no longer dynamically possible. Besides the one in the axial direction there is another motion perpendicular to it in the plane of the cross-section. The flow field therefore differs not only quantitatively from that of a Newtonian fluid, but the appearance of the phenomenon is changed. In certain cases however the kinematically simple motion of the Newtonian fluid gives a first approximation for the actual process. One will now be able to understand the actual motion as the superposition of the motion of a Newtonian fluid, known as the *'primary flow'*, and an additional *'secondary flow'*, and apply a perturbation process for the calculation of the secondary effects. This concept of primary and secondary flows now makes the analytical treatment of the flow through circular pipes generally possible.

Because in this problem the primary flow represents an unaccelerated motion, the inertia of the liquid plays no part up to that approximation at which a viscoelastic secondary flow in the plane of the pipe cross-section appears for the first time.

In more complex processes the inertia of the liquid naturally affects the flow form. Because the perturbation process to be described assumes that the flow proceeds sufficiently slowly, and the inertia is often of minor significance, we shall understand by the term primary flow always the creeping flow, i.e. that determined by disregarding the inertia terms of the Newtonian fluid subordinated to the real fluid. We have therefore in the treatment of actual flow problems first of all always to solve the shortened Navier-Stokes differential equations by cancelling the inertia terms in order to determine the primary flow. If the external force $\mathbf{f}$ acts on the fluid particles per unit volume, and if the fluid is incompressible, then the equations of motion are as follows:

$$-\eta_0 \operatorname{curl}\operatorname{curl}\mathbf{v} + \mathbf{f} = \operatorname{grad} p \tag{9.1}$$

From the linearity of this differential equation and of the continuity equation (1.99) to be taken into account for incompressible liquids, it follows that the velocity field of the primary flow always increases linearly with the cause of the motion, e.g. with the pressure drop given from the exterior, or with the drag velocity of the moving wall. In the determination of the viscoelastic secondary flow the approximation (8.53) of the general constitutive equation valid for slow flow is used, in which the leading term is considered to be the dominant one and the other terms as perturbations. This leads to a series expansion of the velocity field and of the pressure field in powers of the Weissenberg number, in which the Newtonian fields occur as an absolute term:

$$\mathbf{v} = \mathbf{v}^{(1)} + \mathbf{v}^{(2)} + \mathbf{v}^{(3)} + \ldots \tag{9.2}$$
$$p = p^{(1)} + p^{(2)} + p^{(3)} + \ldots$$

$\mathbf{v}^{(1)}$ and $p^{(1)}$ signify the Newtonian velocity field and pressure field (primary flow). As an external force acting on the primary flow for many applications, only the force of gravity has to be considered, the volume force vector of which can be expressed in the case of an incompressible liquid as the gradient of a scalar potential, which can be combined with the pressure. With the elimination of the pressure from equation (9.1) therefore this volume force disappears at the same time, so that one has to determine the primary flow field from the equation

$$\operatorname{curl}\operatorname{curl}\operatorname{curl}\mathbf{v}^{(1)} = 0 \tag{9.3}$$

The fields $\mathbf{v}^{(n)}$, $p^{(n)}$ $(n>1)$ characterise those parts of the secondary flow which are proportional to the (n-1)th power of the Weissenberg number, and therefore all together increase non-linearly with the cause of the motion. If one puts these series expansions into the constitutive equation (8.53) and inserts this again into the equation of motion (1.116) shortened by the acceleration term and arranges in powers of the Weissenberg number, then one obtains a series of field equations of the type (9.1), in which $\mathbf{v}^{(n)}$ instead of $\mathbf{v}$, and $p^{(n)}$ instead of p occur, and the vector field $\mathbf{f}$ appears as a function of the velocity fields of lower order, $\mathbf{v}^{(1)}$ to $\mathbf{v}^{(n-1)}$. One has therefore to determine successively creeping flows of a Newtonian liquid, which move under the action of a known volume force field in any case.

We shall go on in the perturbation calculation only up to the point where a contribution to the secondary flow occurs, which differs in a characteristic way from the primary flow. Higher order approximations would indeed lead to more accurate results, but they affect the phenomenon of the secondary flow in general only insignificantly and are therefore of less interest for the explanation of the secondary flow phenomena.

In many applications secondary flow brought about by non-Newtonian behaviour can be immediately explained in the context of a *second order theory*, i.e. the phenomena have already been described by the terms $\mathbf{v}^{(2)}$, $p^{(2)}$. In order to calculate them one needs the second order terms $\alpha_1\mathbf{A}_2 + \alpha_2\mathbf{A}_1^2$ in the equation (8.48), which in dimensionless terms contain the first power of the Weissenberg number as a factor. The causes of the secondary flow of the second order are these extra-stresses, or more accurately, the additional second order stresses caused by the primary flow. In the field equation for $\mathbf{v}^{(2)}$ therefore the following expression occurs as the driving volume force

$$\mathbf{f}^{(2)} = \operatorname{div}\left(\alpha_1\mathbf{A}_2^{(1)} + \alpha_2\mathbf{A}_1^{(1)2}\right) \tag{9.4}$$

In order to indicate that the right-hand side is to be formulated with the primary velocity field, the upper indices (1) are accordingly used.

The purpose of the following considerations is to simplify the expression (9.4). This is done with the help of the following statement: for a velocity field $\mathbf{v}$ (here the primary flow), which describes the creeping flow of an incompressible Newtonian fluid acted on by a conservative volume force field (which can be represented as the gradient of a scalar function), which will be proved later, the following expression applies

$$\operatorname{div}\left(\mathbf{A}_2 - \mathbf{A}_1^2\right) = \operatorname{grad}\left[\frac{1}{\eta_0}\frac{Dp^*}{Dt} + \frac{1}{4}\operatorname{tr}\mathbf{A}_1^2\right] \tag{9.5}$$

p* describes the 'piezometric pressure' composed of the static pressure and the potential of the volume force. Because we consider only the force of gravity for which a potential exists, as an external force, the requirements in the statement are fulfilled for the primary field. If one resolves the right-hand side of equation (9.4) into $\alpha_1 \mathrm{div}(\mathbf{A}_2^{(1)} - \mathbf{A}_1^{(1)2}) + (\alpha_1 + \alpha_2)\mathrm{div}\mathbf{A}_1^{(1)2}$, then the first term affects only the pressure field, because it can be represented by equation (9.5) as a gradient of a scalar field, and only the second term contributes to the secondary flow. If one eliminates the pressure field by performing the operation curl, then one obtains accordingly as the conditional equation for the secondary flow field

$$\mathbf{curl\ curl\ curl\ v}^{(2)} = \frac{\alpha_1 + \alpha_2}{\eta_0}\,\mathbf{curl\ div\ A}_1^{(1)2} \tag{9.6}$$

Note that in this equation of motion only the combination $\alpha_1 + \alpha_2$ of the physical quantities of the second order are considered. It is connected simply with the two normal stress coefficients of the second order ν_{10} and ν_{20}, cf. equation (8.52):

$$\alpha_1 + \alpha_2 = \frac{1}{2}\nu_{10} + \nu_{20} \tag{9.7}$$

This relationship shows that the secondary flows of the second order are normal stress induced phenomena. Because ν_{10} is always positive and ν_{20} is generally negative, but in magnitude always significantly smaller than $\nu_{10}/2$, the constant in equation (9.6) can be henceforth regarded as positive.

For the proof of the statement (9.5) we specify more appropriately a Cartesian coordinate system x_1, x_2, x_3, make use of the index notation, therefore denote the three components of the velocity vector by $v_i (i=1,2,3)$, and make use of the sum convention. Derivatives with respect to the Cartesian coordinates are indicated by indices, whereby for the sake of clarity commas have been inserted between the indices of different significance. Thus for example the symbols $v_{i,k}$ indicate the components of the velocity gradient tensor $\partial v_i / \partial x_k$. For the Cartesian components of the tensor $\mathbf{A}_2 - \mathbf{A}_1^2$ we can (by use of (1.52) and (1.55)) write:

$$(\mathbf{A}_2 - \mathbf{A}_1^2)_{ik} = \frac{\partial}{\partial t}(v_{i,k} + v_{k,i}) + v_\ell(v_{i,k\ell} + v_{k,i\ell}) + v_{\ell,i}v_{\ell,k} - v_{i,\ell}v_{k,\ell} \tag{9.8}$$

By differentiating we obtain for the divergence of the tensor

$$(A_2 - A_1^2)_{ik,k} = \frac{\partial}{\partial t}(\underline{v_{i,kk}} + \underline{v_{k,ik}}) + v_\ell \underline{v_{i,kk\ell}} + v_{\ell,k} v_{k,i\ell} + v_\ell \underline{v_{k,ik\ell}}$$
$$+ v_{\ell,ik} v_{\ell,k} + v_{\ell,i} \underline{v_{\ell,kk}} - v_{i,\ell} \underline{v_{k,k\ell}} \tag{9.9}$$

Now for incompressible fluids only those velocity fields have to be considered whose divergence vanishes, for which therefore $v_{k,k} = 0$. For this reason the terms underlined by interrupted lines in (9.9) vanish. Because in addition this is a matter of a creeping flow of a Newtonian fluid and the volume force has a potential, so that it can be combined with the pressure gradient into grad p^*, the velocity field satisfies the Navier-Stokes differential equations shortened by the inertia term, which with the use of Cartesian coordinates can be written in the form

$$\eta_0 v_{i,kk} = p^*_{,i} \tag{9.10}$$

Hence it follows that the terms underlined in equation (9.9) can be replaced by derivatives of the piezometric pressure and combined. Finally the two remaining (unmarked) terms in equation (9.9) can be written jointly as the gradient of a quadratic scalar quantity in the velocity components. Therefore equation (9.9) transforms into

$$(A_2 - A_1^2)_{ik,k} = \frac{1}{\eta_0}\left(\frac{\partial p^*}{\partial t} + v_\ell p^*_{,\ell}\right)_{,i} + \frac{1}{4}\left[(v_{\ell,k} + v_{k,\ell})(v_{\ell,k} + v_{k,\ell})\right]_{,i} \tag{9.11}$$

and therefore can be expressed symbolically in the previously given form (9.5).

It has been mentioned above that in many cases viscoelastic secondary flow phenomena appear as second order effects, so that the perturbation velocities under the condition of slow flow increase as the square of the external cause of the motion and the coefficient $\alpha_1 + \alpha_2$ plays a decisive part in the secondary flow. Plane flows are exceptions, as are those motions in which the primary flow has the properties of an axial shear flow. Under isothermal conditions the secondary flows involved appear as third and fourth order effects respectively. We shall discuss this in greater detail in Sections 9.3 and 9.4.

The actual secondary flow phenomena to be dealt with below are characterised by having velocity vectors of the primary and secondary flows either located perpendicular to each other or located in the same plane. In addition these motions have a rotational symmetric or plane secondary flow. For the flow through a conical nozzle for example, the velocity vectors of primary and secondary flows lie in the meridian plane through the axis of the nozzle, and the flow is rotational symmetric. For the flow through cylindrical pipes on the other hand the velocity vector of the primary flow lies parallel to the pipe axis, whereas the secondary flow proceeds in the cross-section of the pipe normal to the axis.

9.2 Rotational symmetric flows

We first consider steady flows whose velocity fields relative to a fixed axis possess rotational symmetry. The stream surfaces of such motions are surfaces of rotation with a common axis, the stream lines being more or less complicated three-dimensional curves on these surfaces of rotation. It is expedient first of all to use *cylindrical coordinates*, r, φ, z, which are chosen so that r denotes the distance from the symmetry axis and z, φ the longitudinal coordinate in the axial direction and the angular coordinate in the azimuthal direction respectively. Rotational symmetry means that the velocity components v_r, v_φ, v_z associated with the coordinates are dependent on r and z, but not on φ. For actual applications it is recommended in some cases to use other rotational symmetric coordinates (e.g. spherical coordinates), thus exchanging the coordinates in the meridian section r, z for more suitable ones. The angle coordinate φ however stays. The criterion for rotational symmetric flows, that is the independence from φ of the velocity components associated with the coordinates, naturally remains unchanged.

For isochoric rotational symmetric flows the two velocity components in the meridian plane (hence for example v_r and v_z) can be derived from a *scalar stream function* Ψ (r,z), so that the total velocity field can be expressed in the form

$$\mathbf{v} = v_\varphi \mathbf{e}_\varphi \ \operatorname{curl}\left(\frac{\Psi}{r}\mathbf{e}_\varphi\right) \tag{9.12}$$

With this the continuity equation div $\mathbf{v}$ = 0 is satisfied, so that henceforth it need not be taken into account. For the two velocity components v_r and v_z there follows from (9.12)

$$v_r = -\frac{1}{r}\frac{\partial \Psi}{\partial z}, \qquad v_z = \frac{1}{r}\frac{\partial \Psi}{\partial r} \tag{9.13}$$

The factor 1/r in front of the stream function in equation (9.12) is introduced in order to bring it about that the surfaces Ψ = constant are stream surfaces. From equation (9.13) it follows immediately that the velocity vector and the vector grad Ψ under these conditions are perpendicular to each other, $\mathbf{v}\cdot\mathrm{grad}\,\Psi = 0$, i.e. the local velocity vector is tangential to the surface Ψ = constant running through the point considered, so that the surfaces Ψ = constant are built up from nothing but stream lines, the integral curves on the velocity field, and thus are in fact stream surfaces. In other words each curve Ψ = constant perceptible in the r-z plane represents the projection of a stream line, and because of the steadiness of the motion at the same time represents the projection of a path line in the meridian section. Therefore as soon as the function $\Psi(r,z)$ is determined the stream line picture of the motion in the meridian plane can be expressed by drawing the curves $\Psi(r,z)$ = constant. From (9.13) it follows that the stream function Ψ has the dimensions $[\text{length}]^3/\text{time}$. By the decomposition (9.12) the left-hand sides of the field equations (9.3) and (9.6) split up into two components, whereby the one that contains the stream function contains the unit vector $\mathbf{e}_\varphi$ as a factor. The operator curl transforms in consequence of the rotational symmetry a vector field lying in the meridian plane into a vector field perpendicular to the meridian plane and vice versa. In particular therefore the twofold application to an azimuthal vector field again gives rise to a vector in the φ-direction. This permits us to introduce a scalar differential operator L which is defined by the expression

$$\mathbf{curl\,curl}\left(\frac{\Psi}{r}\mathbf{e}_\varphi\right) = -\frac{\mathbf{e}_\varphi}{r}\mathsf{L}\Psi \tag{9.14}$$

By using cylindrical coordinates there follows

$$\mathsf{L} = r\frac{\partial}{\partial r}\left(\frac{1}{r}\frac{\partial}{\partial r}\right) + \frac{\partial^2}{\partial z^2} \tag{9.15}$$

With (9.12) and (9.14) one first obtains from equations (9.3) and (9.6) by considering the terms multiplied by $\mathbf{e}_\varphi$ the following differential equations for the stream functions of the primary and secondary flows:

$$\mathsf{L}\mathsf{L}\,\Psi^{(1)} = 0 \tag{9.16}$$

$$\frac{1}{r}\,\mathsf{L}\mathsf{L}\,\Psi^{(2)} = \frac{\alpha_1 + \alpha_2}{\eta_0}\,\mathrm{curl}_\varphi\ \mathrm{div}\ \mathbf{A}_1^{(1)2} \tag{9.17}$$

In the example of rotational symmetric motions which are dealt with below the secondary flow consists only of the component described by $\Psi^{(2)}$. Therefore $v_\varphi^{(2)} = 0$, so that it is superfluous to give an account of the general differential equation for $v_\varphi^{(2)}$. As regards the appropriate primary flow, the component $v_\varphi^{(1)}$ is present in two cases. By use of the operator L defined in equation (9.14) the equation for $v_\varphi^{(1)}$ following from equation (9.3) can be written in the form

$$\mathrm{curl}\left(\frac{\mathbf{e}_\varphi}{r}\,\mathsf{L}(r\,v_\varphi^{(1)})\right) = 0 \tag{9.18}$$

Hence it can be concluded that

$$\mathsf{L}(r\,v_\varphi^{(1)}) = -\frac{p'}{\eta_0} \tag{9.19}$$

in which the right-hand side is constant and a comparison with (9.1) shows that the constant p' in the numerator describes the pressure drop in the φ-direction impressed from outside, i.e. $p' = -\partial p/\partial\varphi$.

In the treatment of special rotational symmetric flows the main point now is to solve the field equations (9.16) and (9.19) for the primary flow under the special boundary conditions, to construct the right-hand side of equation (9.17) with the result, and from this relationship to determine the secondary flow, taking into account the boundary conditions involved.

9.2.1 Conical nozzle flow

For the first example we consider the steady inlet flow into a conical nozzle, which we represent in a theoretical model by a cone of semi-angle ϑ_0 (Fig. 9.1). There is an orifice in the apex of the cone through which the volume $\dot{V}$ passes per unit time. At the position of the point of the cone we can therefore consider a pointed sink which takes off the volume flux $\dot{V}$. The velocity vector of the resulting flow lies totally in the meridian plane, so that only the second term of equation (9.12) occurs ($v_\varphi \equiv 0$).

We work more conveniently in *spherical polar coordinates* R, ϑ, φ, which are obviously adapted to the problem (Fig. 9.1).

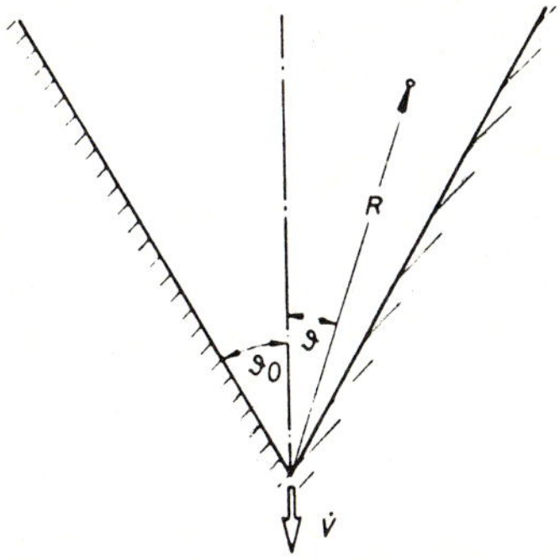

Fig. 9.1 Coordinates of conical nozzle flow

Thus the factor r in equation (9.12) is to be replaced by $R\sin\vartheta$. By use of the formula given in the Appendix for the operator curl in spherical coordinates one obtains for the two velocity components

$$v_R = \frac{1}{R^2 \sin\vartheta}\frac{\partial\Psi}{\partial\vartheta}; \qquad v_\vartheta = -\frac{1}{R\sin\vartheta}\frac{\partial\Psi}{\partial R} \tag{9.20}$$

Thus the critical differential operator defined by equation (9.14) takes on the following form

$$L = \frac{\partial^2}{\partial R^2} + \frac{\sin\vartheta}{R^2}\frac{\partial}{\partial\vartheta}\left(\frac{1}{\sin\vartheta}\frac{\partial}{\partial\vartheta}\right) \tag{9.21}$$

Because the axis of the cone for reasons of symmetry represents a stream line, Ψ must be constant along it. Without limiting the generality we can state that

$$\Psi = 0 \text{ for } \vartheta = 0 \tag{9.22}$$

In the same way Ψ has a constant value at the wall of the cone, whose value is connected with the volume that passes per unit time

$$\begin{aligned}\dot{V} &= -2\pi R^2 \int_0^{\vartheta_0} \sin\vartheta\,(v_R)_{R=\text{const}}\,d\vartheta = -2\pi \int_0^{\vartheta_0}\left(\frac{\partial\Psi}{\partial\vartheta}\right)_{R=\text{const}} d\vartheta \\ &= -2\pi\Psi(R,\vartheta_0)\end{aligned} \tag{9.23}$$

Hence it follows that

$$\Psi = -\frac{\dot{V}}{2\pi} \text{ for } \vartheta = \vartheta_0 \tag{9.24}$$

Therefore v_ϑ vanishes at the wall. In addition the adhesion condition demands that $v_R = 0$ for the conical casing. This leads to the further condition for the stream function

$$\frac{\partial \Psi}{\partial \vartheta} = 0 \text{ for } \vartheta = \vartheta_0 \tag{9.25}$$

Because in the boundary condition only the two constants $\dot{V}$ and ϑ_0 occur, and in the field equation (9.16) no parameters at all are involved, the immediately wanted function $\Psi^{(1)}$ apart from the coordinates R and ϑ depends only on $\dot{V}$ and ϑ_0. Now $\Psi^{(1)}$ has the same dimensions as $\dot{V}$; $\Psi^{(1)}/\dot{V}$ is therefore dimensionless. Furthermore ϑ and ϑ_0 are also dimensionless quantities. From the physical parameters R and $\dot{V}$ one can obviously not form dimensionless ones, so the relationship between $\Psi^{(1)}$ and R, ϑ, $\dot{V}$, ϑ_0 reduces for reasons of dimensional analysis to the form

$$\frac{\Psi^{(1)}}{\dot{V}} = f(\vartheta; \vartheta_0) \tag{9.26}$$

The stream function for the creeping conical nozzle flow is therefore quite independent of the coordinate R. Therefore $v_\vartheta^{(1)} = 0$ (equation (9.20)), i.e. every material point moves purely radially towards the sink. The stream lines of the primary flow are straight lines through the apex of the cone.

Because there is no dependence on R, equation (9.16) reduces to an ordinary fourth order differential equation, the general integral of which can be written down immediately. One of the four fundamental integrals of the differential equation has the property that its derivative with respect to ϑ for $\vartheta \to 0$ decreases only as $\vartheta \cdot \ln \vartheta$, so that by this a component would result for the velocity which is singular on the axis (equation (9.20)). The obvious requirement that the flow velocity on the symmetry axis must take on finite values at a finite distance from the sink excludes the occurrence of this fundamental integral. The other three constants of integration are determined by the boundary conditions (9.22), (9.24) and (9.25). The result is as follows:

$$\psi^{(1)} = -\frac{\dot{V}}{2\pi} \cdot \frac{1 - 3\cos^2\vartheta_0 + \cos\vartheta + \cos^2\vartheta}{(1 + 2\cos\vartheta_0)(1 - \cos\vartheta_0)^2}(1 - \cos\vartheta) \tag{9.27}$$

Fig. 9.2 shows the shape of this function for various values of the orifice angle ϑ_0. From equation (9.20) it follows that for the velocity field of the primary flow $v_\vartheta^{(1)} = 0$, and

$$v_R^{(1)} = -\frac{C\dot{V}}{R^2}(\cos^2\vartheta - \cos^2\vartheta_0) \tag{9.28}$$

whereby

$$C := \frac{3}{2\pi(1 + 2\cos\vartheta_0)(1 - \cos\vartheta_0)^2} \tag{9.29}$$

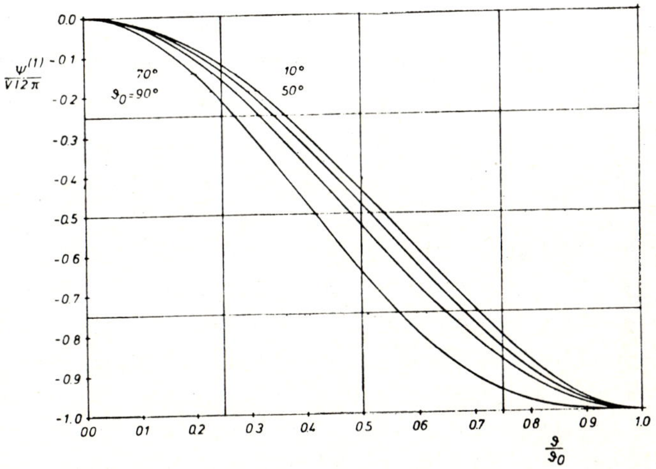

Fig. 9.2 Stream function of the primary flow as a function of the coordinate ϑ

The first Rivlin-Ericksen tensor describing this velocity field has with reference to the specified R-ϑ-φ-coordinate system the matrix representation

$$\mathbf{A}_1^{(1)} = \frac{2C\dot{V}}{R^3}\begin{bmatrix} 2(\cos^2\vartheta - \cos^2\vartheta_0) & \cos\vartheta\sin\vartheta & 0 \\ \cos\vartheta\sin\vartheta & -\cos^2\vartheta + \cos^2\vartheta_0 & 0 \\ 0 & 0 & -\cos^2\vartheta + \cos^2\vartheta_0 \end{bmatrix} \tag{9.30}$$

Therefore the expression $\operatorname{curl}_\varphi \operatorname{div} \mathbf{A}_1^{(1)2}$ can be formed, which occurs on the right-hand side of the equation (9.17) for the secondary flow. If furthermore one replaces the factor r by $R\sin\vartheta$, then one obtains

$$LL\Psi^{(2)} = \frac{16(\alpha_1 + \alpha_2)C^2\dot{V}^2}{\eta_0 R^7} \sin^2\vartheta \cos\vartheta\,(-5\cos^2\vartheta + 6\cos^2\vartheta_0 - 1) \quad (9.31)$$

For the boundary conditions one has to specify once more the expressions (9.22) and (9.25) as well as the regularity condition for $\partial^2\Psi^{(2)}/\partial\vartheta^2$ on the axis; also, instead of (9.24) the condition $\Psi^{(2)} = 0$ for $\vartheta = \vartheta_0$, because the stream function of the primary motion at the wall takes on the value dictated by (9.24). A dimensional analysis shows that the dimensionless field function $\Psi^{(2)}/\dot{V}$ can depend only on the dimensionless parameters $(\alpha_1 + \alpha_2)\dot{V}/\eta_0 R^3$ (the local Weissenberg number), ϑ and ϑ_0. Because the second order theory is linear with respect to the Weissenberg number, there follows

$$\Psi^{(2)} = \frac{(\alpha_1 + \alpha_2)\dot{V}^2}{\eta_0 R^3} \cdot G(\vartheta; \vartheta_0) \quad (9.32)$$

Therefore in contrast to the primary flow there is a dependence on R, so that the velocity vector also has a ϑ-component. The viscoelastic fluid therefore in contrast to the Newtonian fluid does not flow along straight path lines into the sink. This leads to a much more complicated stream line picture. By (9.32) the dependence on the coordinate R is made apparent, so that equation (9.31) reduces to an ordinary fourth order differential equation for $G(\vartheta,\vartheta_0)$. This is:

$$30y + \sin\vartheta \frac{d}{d\vartheta}\left(\frac{1}{\sin\vartheta}\frac{dy}{d\vartheta}\right) = 16C^2 \sin^2\vartheta \cos\vartheta\,[-5\cos^2\vartheta + 6\cos^2\vartheta_0 - 1] \quad (9.33)$$

in which for abbreviation:

$$y := 12G + \sin\vartheta \frac{d}{d\vartheta}\left(\frac{1}{\sin\vartheta}\frac{dG}{d\vartheta}\right) \quad (9.34)$$

The solution as a polynomial of the sixth degree in $\cos\vartheta$ can be written down explicitly, but the coefficients have after fitting to the described boundary conditions a rather complicated form, so that the analytical result will not be reproduced here. Instead of that Fig. 9.3 shows the result in graphical form. The superposition of the primary and secondary flow fields $\Psi = \Psi^{(1)} + \Psi^{(2)}$ leads to the stream lines shown in Fig. 9.4 (graphs of Ψ = constant). At sufficiently great distance from the sink there is hardly any noticeable secondary flow because the stream lines are straight and directed towards the sink. One quickly realises this when one remembers that the local Weissenberg number $(\alpha_1 + \alpha_2)\dot{V}/\eta_0 R^3$, the ratio of characteristic second and first order extra-stresses at a greater distance from the apex is always small compared with unity. The flow there is so slow that the fluid behaves in an almost Newtonian manner and the extra-stresses multiplied by the Weissenberg number cannot have any effect.

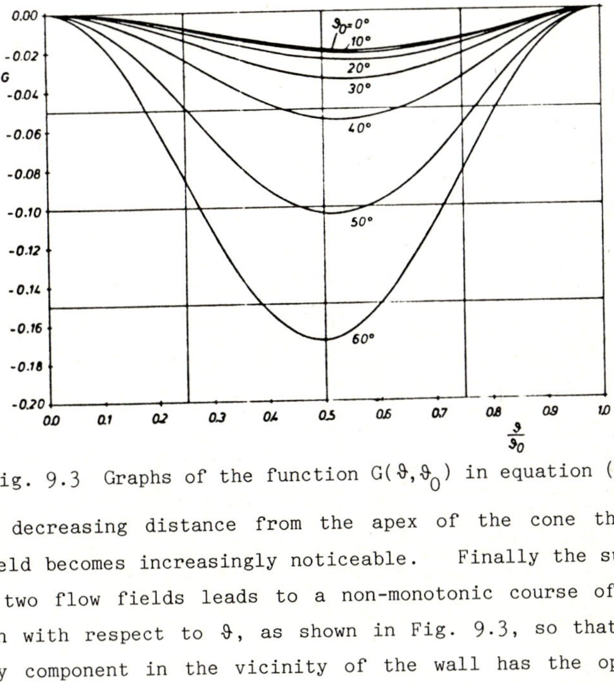

Fig. 9.3 Graphs of the function $G(\vartheta, \vartheta_0)$ in equation (9.32)

With decreasing distance from the apex of the cone the secondary flow field becomes increasingly noticeable. Finally the superposition of the two flow fields leads to a non-monotonic course of the stream function with respect to ϑ, as shown in Fig. 9.3, so that the radial velocity component in the vicinity of the wall has the opposite sign to that on the axis. For fluids with a positive coefficient $(\alpha_1 + \alpha_2)$ there results from that the stream line pattern illustrated with a ring-shaped vortex located at the wall of the cone, and with converging stream lines in the middle.

The stream line pattern shown in Fig. 9.4 has unrealistic properties in one detail. It describes the stream line inside the vortex near the wall as closed, but overall as curves traversing the apex

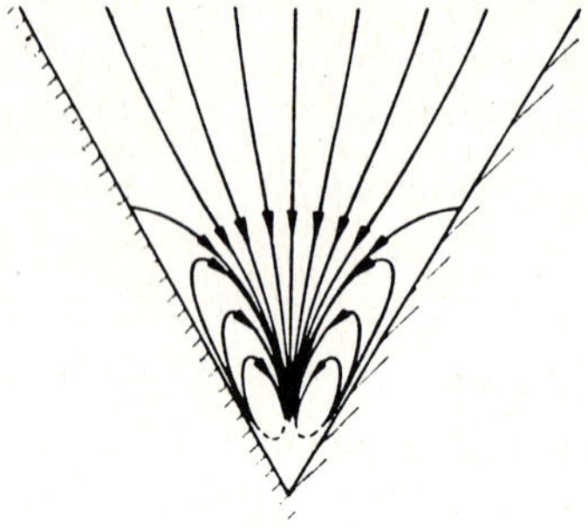

Fig. 9.4 Formation of stream lines in the convergent conical nozzle flow; result of the second order theory; $\vartheta_0 = 30^\circ$

of the cone. This defect in theory would be eliminated if the approximation were taken a step further by considering also the third order contribution to the flow field. The pattern would then change in the vicinity of the sink, so that the stream lines within the vortex would appear as closed loops not touching the apex of the cone, as is clearly indicated in Fig. 9.4 by the interrupted lines. Representation of the calculation of the third order contribution is not however done here, because that would be beyond the scope of this treatment.

An estimate of the radial extension R_* of the vortex is obtained from the consideration that within this region the amplitudes of the primary and secondary flows have the same order of magnitude, so that for the Weissenberg number $(\alpha_1 + \alpha_2)\dot{V}/\eta_0 R_*^3 \sim 1$ holds. The extent of the vortex is therefore related as follows to the relaxation time $(\alpha_1 + \alpha_2)/\eta_0$, and to the volume flux

$$R_* \sim \sqrt[3]{\frac{\alpha_1 + \alpha_2}{\eta_0}\dot{V}} \tag{9.35}$$

The factor which has to be introduced here in order to formulate an equation from this estimate depends slightly on the orifice angle and has the order of magnitude unity for small angles.

One might possibly query the described method of calculating the flow field. Firstly one could query that it is permissible to neglect the inertia of the liquid, because the primary flow is an accelerated motion. In fact this viewpoint leads to limiting conditions. Neglect of the acceleration in comparison with the Newtonian friction terms presumes in fact that the local Reynolds number $\rho \dot{V}/\eta_0 R$ is small compared with unity, i.e.

$$R \gg \frac{\rho}{\eta_0}\dot{V} \tag{9.36}$$

This condition is always violated in the vicinity of the sink, so that the result of the theory can only apply at a sufficiently great distance from the apex of the cone, where (9.36) is satisfied. For this reason the region in the immediate vicinity of the apex was omitted from the representation of the flow in Fig. 9.4. On the other hand the neglect of the acceleration terms in contrast to the second order extra-stresses taken into account is only permissible when the local Reynolds number is small compared with the local Weissenberg number. This leads to the condition

$$R^2 \ll \frac{\alpha_1 + \alpha_2}{\rho} \tag{9.37}$$

which is obviously infringed at a great distance from the apex of the cone. Note that on the right-hand side of this criterion only material properties occur, but not the flow rate. According to these statements the theory provides 'correct' results at the place where the visco-elastic secondary flow makes its presence noticeable, supposing that conditions (9.36) and (9.37) apply, if for R the value (9.35) is inserted which is characteristic of the extent of the vortex region. Hence there follows for the volume flux the limitation

$$\dot{V}^{2/3} \ll \frac{\eta_0}{\rho}\left(\frac{\alpha_1 + \alpha_2}{\eta_0}\right)^{1/3} \tag{9.38}$$

In a fluid which has marked viscoelastic properties, whose kinematic viscosity and relaxation time have the values $\eta_0/\rho = 1m^2/s$ and $(\alpha_1+\alpha_2)/\eta_0 =$ 1s respectively, there would result for the right-hand sides of (9.37) and (9.38) the numerical values $1m^2$ and $1m^2/s^{2/3}$. The condition (9.38) would be met for example when $\dot{V} = 10^{-4} m^3/s$. The characteristic radius for the extent of the vortex would then have the value $R_* \sim 4.6 \cdot 10^{-2} m$; the condition (9.36) would give $R \gg 10^{-4} m$.

A further argument against the method of calculation which must be considered seriously comes from the statement that a series expansion (with respect to the Weissenberg number) has been truncated after the linear term and the resulting two term sum has been used as an approximate solution, although the expansion parameter in the interesting region has the order of magnitude unity. This objection would lead to the demand to calculate higher order approximations, and to consider them in the solution. Hence the result would of course change quantitatively, and in fact all the more so when going towards the apex of the cone. It turns out however that the basic flow form illustrated in Fig 9.4 and based on the first two terms corresponds qualitatively to the observations of the exit flow of viscoelastic liquids from conical nozzles. A quantitative comparison between theory and experiment is difficult, because a point-shaped sink as has been specified for this consideration cannot in fact be realised, but is always replaced by an opening of finite extent. So far we are able here to explain with second order theory the observed phenomenon qualitatively correctly as a normal stress induced phenomenon. However one has to appreciate the fact that in respect of the numerical results one has to reckon with an increase in error when the distance from the apex of the cone decreases.

In addition we shall briefly consider how the flow pattern changes when the flow passes through the cone in the reverse direction, i.e. from the apex outwards. We take as a basis once more the second order result $\Psi = \Psi^{(1)} + \Psi^{(2)}$, and note that now $\dot{V}<0$. Thus we see that the primary flow changes its sign (9.27); the secondary flow remains unchanged because it is a squared term (9.32). The superposition of the two leads to the stream line pattern shown in Fig. 9.5, which has a central ring-shaped vortex and stream lines which are deflected towards the walls. Such zones of circulation however have so far not actually been observed in divergent nozzle flows, but there is doubtless a divergence of the stream lines towards the wall near the source. The last is rather reminiscent of the onion-shaped widening of a jet emerging from a nozzle (cf. Section 4.3), and it surely has similar origins. For a divergent nozzle flow however the results of the second order theory agree far less with the observations than in the case of convergent flows. The cause of this discrepancy is obvious when one considers which motion a particular fluid particle remembers. In the case of convergent flows these are gradually increasing deformation rates, so that the assumption of a slowly varying previous history, which forms the basis of the constitutive equation (8.48) is quite justified within the previously discussed limits. For a divergent conical flow the fluid particle on the contrary has experienced the exit from the source as a singular event, thus one cannot talk of a slowly varying history. Therefore one should not be surprised that theory in this case has failed.

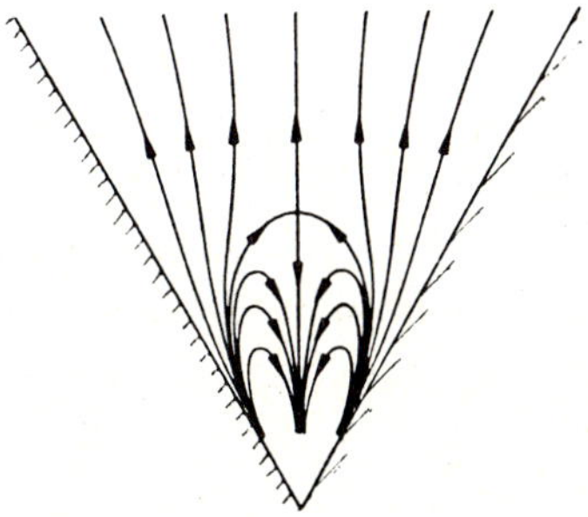

Fig. 9.5 Formation of stream lines in the divergent conical nozzle flow; result of the second order theory; $\vartheta_0 = 30^\circ$

9.2.2 Flow round a rotating body

The previously considered inlet flow in a conical nozzle is characterised by the fact that the velocity vectors of the primary and secondary flows both lie in the same meridian plane. We now turn to motions for which this is no longer so, but the two velocity vectors are much more perpendicular to each other,and are located in such a way that the secondary flow is still in the meridian plane, but the primary flow is along the φ-direction ($\Psi^{(1)} \equiv 0$ and $v_\varphi^{(2)} \equiv 0$ in equation (9.12)).

We first study processes which occur in such a way that certain rotational symmetric boundary surfaces, which confine the liquid, rotate with constant angular velocity round the common axis of symmetry and thus drag the liquid along with them. We shall consider in particular the case of a uniformly rotating sphere in a liquid of infinite extent. There is no pressure drop in the φ-direction, so that the equation (9.19) for the primary field reduces to

$$L(r\, v_\varphi^{(1)}) = 0 \tag{9.39}$$

The primary flow, as a creeping motion of a Newtonian fluid, is a pure drag flow, for which every material point revolves at its own angular velocity $v_\varphi^{(1)}/r$ in a circle whose centre lies on the axis of rotation. All particles that rotate with the same angular velocity lie on one surface of rotation and retain their relative positions. The flow field therefore consists of a set of rotational symmetric layers which rotate at various angular velocities round the common axis, so that adjacent layers slide over each other. Apart from a rotation, there is an overall shear flow at any place, for which the shear rate

$$\dot{\gamma} = r \left| \operatorname{grad} \frac{v_\varphi^{(1)}}{r} \right| \tag{9.40}$$

for any arbitrary particle is always the same, and the primary flow is perceived as viscometric as defined in Section 1.6.

With this knowledge the cause of a viscoelastic secondary flow can be clearly explained and at the same time the right-hand side of equation (9.17) expressed directly as a function of $v_\varphi^{(1)}$.

The extra-stresses for a shear flow depend only on the shear rate, as is well known. In the context of a second order theory the non-Newtonian behaviour expresses itself in two normal stress differences proportional to the square of the shear rate. Because we have already established that for the secondary flow the sum of the two coefficients of the second order is critical, it is sufficient to consider a hypothetical fluid whose second normal stress difference vanishes. The first normal stress difference corresponds to a tensile stress of magnitude $\nu_{10}\dot{\gamma}^2$ in the circular stream lines of the primary motion superposed on the pressure field. Each stream line therefore behaves as a stretched elastic band: it tries to contract. In competition with the other stretched bands those will chiefly prevail whose stress is particularly high. Because the stress increases with the shear rate these are regions of relatively high shear rate. In the flow circulating round the rotating sphere in a liquid of infinite extent for example, the shear rate reaches its maximum on the equator of the sphere and vanishes at the poles. One will therefore expect that the fluid under the action of this normal stress difference flows inwards in the vicinity of the equator, and outwards at the poles. The result is that two ring-shaped vortices form, which are separated from each other by the equatorial plane (Fig. 9.6).

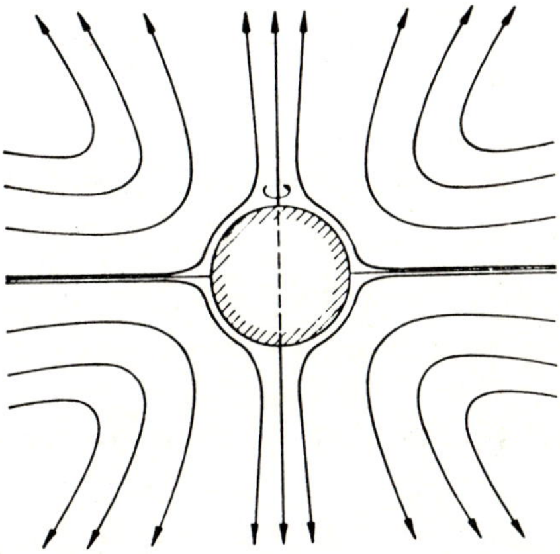

Fig. 9.6 Viscoelastic secondary flow of the second order round a rotating sphere

To calculate the secondary flow one has first of all to form the right-hand side of the critical differential equation (9.17) with the primary velocity field $\mathbf{v}^{(1)} = v_\varphi^{(1)}\mathbf{e}_\varphi$. The computation required for that can be shortened by the following consideration. The non-Newtonian behaviour expresses itself, because of the simplicity of the primary flow, in the previously discussed tensile stress $\nu_{10}\dot{\gamma}^2$ in the φ-direction. It corresponds to a force per volume of magnitude $-(\nu_{10}\dot{\gamma}^2/r)\mathbf{e}_r$ (cf. Fig. 4.3; put there $N_1 = \nu_{10}\dot{\gamma}^2$ and $N_2 = 0$). The curl of this volume force field, which because of the rotational symmetry is a vector in the φ-direction, has the magnitude $-(\nu_{10}/r)\partial\dot{\gamma}^2/\partial z$, and describes up to a factor η_0 the right-hand side of equation (9.17). If one considers equation (9.40) and notes that ν_{10} for more general fluids is to be replaced by $\nu_{10} + 2\nu_{20}$ or by $2(\alpha_1 + \alpha_2)$(equation (9.7)), then one finds as the equation of the viscoelastic secondary flow for rotational symmetric flows with a purely azimuthal primary velocity field

$$LL\Psi^{(2)} = -\frac{2(\alpha_1 + \alpha_2)}{\eta_0} r^2 \frac{\partial}{\partial z} \left| \text{grad} \frac{v_\varphi^{(1)}}{r} \right|^2 \tag{9.41}$$

One has to specify as boundary conditions that $\Psi^{(2)}$ be constant, in general equal to zero at the fixed walls, and that the derivative in the normal direction vanishes.

For certain geometrically simple arrangements the solutions to the boundary value problems for $v_\varphi^{(1)}$ and $\Psi^{(2)}$ can be expressed by elementary functions. We restrict ourselves here to the case of a sphere of radius R_0 rotating at the angular velocity Ω_0, so that the fluid remote from the sphere is at rest. We thus use spherical coordinates as more convenient, in which the differential operator **L** has the form given by equation (9.21). The connection with the cylindrical coordinates is established by the relationship $z = R\cos\vartheta$, $r = R\sin\vartheta$. The primary velocity field as the solution of equation (9.39) depends in a simple way on R and ϑ according to

$$v_\varphi^{(1)} = R_0^3\,\Omega_0\,\frac{\sin\vartheta}{R^2} \tag{9.42}$$

Hence the sliding surfaces which rotate at constant angular velocity are spheres, and the angular velocity of these spheres decreases outwards as R^{-3}. Therefore the right-hand side of equation (9.41) is proportional to $R^{-7}\sin^2\vartheta\cos\vartheta$. The trial function $\Psi^{(2)} = f(R)\sin^2\vartheta\cos\vartheta$ reduces the partial differential equation (9.41) to an ordinary fourth order differential equation. Its general integral contains two parts which increase with increasing distance from the sphere and which because of the requirement that the solution should die away to zero at infinity should not be present. The two remaining constants of integration are determined from the boundary conditions at the surface of the sphere, namely $\Psi^{(2)} = \partial\Psi^{(2)}/\partial R = 0$ for $R = R_0$. The analytical result runs thus

$$\Psi^{(2)} = \frac{(\alpha_1 + \alpha_2)}{2\eta_0} R_0^3 \Omega_0^2 \left(1 + 2\frac{R_0}{R}\right)\left(1 - \frac{R_0}{R}\right)^2 \sin^2\vartheta \cos\vartheta \tag{9.43}$$

Fig. 9.6 shows the corresponding stream line pattern (graphs of $\Psi^{(2)}$ = constant) of the viscoelastic second order secondary flow created by a rotating sphere. It consists of two ring-shaped vortices separated by the equatorial plane and has with regard to the direction of rotation those properties which have been previously explained by qualitative arguments.

On this normal stress effect there is superposed a 'classical' inertia effect which also occurs with Newtonian fluids, and which we shall study shortly. Because in the primary flow each fluid particle follows a circular path, it experiences a centrifugal force per unit volume, of magnitude

$$\mathbf{f}_I^{(2)} = \rho \frac{v_\varphi^{(1)2}}{r} \mathbf{e}_r \tag{9.44}$$

in which ρ designates the density of the liquid. The subscript I indicates that this concerns the force field caused by the inertia of the liquid. The superscript index (2) indicates that the term is proportional to the square of the primary velocity field. This volume force field also gives rise to a perturbation motion whose amplitude compared with that of the primary flow is proportional to the Reynolds number. The qualitative properties of the resultant secondary flow are easily seen by considering the example of a rotating sphere. The centrifugal force thus vanishes on the axis of rotation and at infinity, and has its greatest value at the equator of the sphere. One will therefore expect that under its action the liquid at the equator is thrown outwards, and at the poles flows inwards again. Hence there arises the same stream line pattern as for viscoelastic secondary flow (Fig. 9.6), but with reversed rotation direction.

For the calculation of the inertia induced secondary flow one has to use the differential equation

$$\mathrm{LL}\Psi_I^{(2)} = \frac{\rho}{\eta_0} \frac{\partial v_\varphi^{(1)2}}{\partial z} \tag{9.45}$$

instead of equation (9.41), in which $\Psi_I^{(2)}$ is to be subject to the same boundary conditions as $\Psi^{(2)}$ was previously.

In the case of a rotating sphere the right-hand side is proportional to $R^{-5}\sin^2\vartheta\cos\vartheta$ because of equation (9.42). A trial function of the product form $\Psi_I^{(2)} = f(R)\sin^2\vartheta\cos\vartheta$ leads again to an ordinary, easily solved differential equation. One obtains in this way

$$\Psi_I^{(2)} = -\frac{\rho}{8\eta_0} R_0^5 \Omega_0^2 \left(1 - \frac{R_0}{R}\right)^2 \sin^2\vartheta \cos\vartheta \qquad (9.46)$$

The secondary flow thus described consists, like the normal stress induced secondary flow, of two symmetric ring-shaped vortices, and leads to a stream line pattern similar to that shown in Fig. 9.6. The direction of rotation of the vortices wholly corresponds to the concept that the fluid is thrown outwards by the centrifugal forces in the equatorial plane. The different signs of the expressions in equations (9.43) and (9.46) show in fact that both effects have opposing tendencies in respect of the flow direction.

The effective secondary flow of a viscoelastic fluid for slow rotation of the sphere results from the superposition of both components. One finds the stream lines involved from the expression $\Psi^{(2)} + \Psi_I^{(2)} =$ constant. Therefore the properties of the resultant stream line pattern depend on the relative magnitudes of the two effects, or more accurately, on the ratio of the Reynolds number to the Weissenberg number, i.e. on the numerical parameter $\rho R_0^2/(\alpha_1 + \alpha_2)$. Note that this parameter apart from the physical properties (density ρ, viscoelastic coefficient $\alpha_1 + \alpha_2$), depends only on the radius of the sphere, but not on its rate of rotation, and therefore for a given material can only be affected by the size of the rotating sphere. If the Reynolds number and the Weissenberg number differ from each other appreciably, then one of the two parts is dominant and imprints its character on what happens. Thus the stream line pattern illustrated in Fig. 9.6 reproduces the real secondary flow for $\rho R_0^2/(\alpha_1 + \alpha_2) \ll 1$. Under certain conditions however the superposition leads to a different situation. If the parameter $\rho R_0^2/(\alpha_1 + \alpha_2)$ has a value between 4 and 12 (which can always be obtained by a suitably sized sphere) then a spherically shaped stagnation stream surface occurs in the liquid, where the number of the annular vortices doubles (Fig. 9.7). The normal stress effect is dominant inside, so that the same direction results as shown in Fig. 9.6. The externally prevailing inertia effect causes the outer vortices to rotate in the opposite direction.

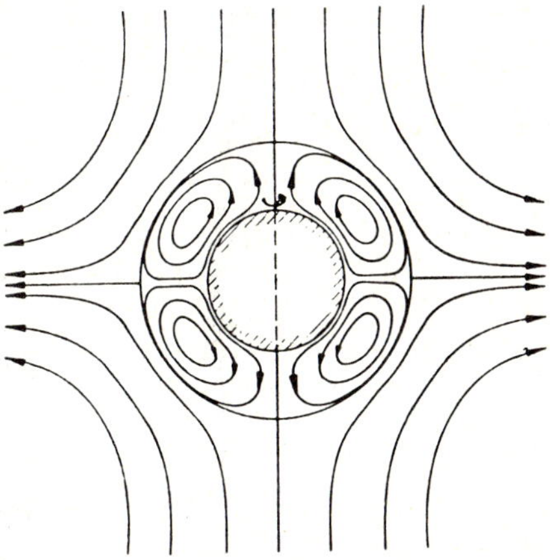

Fig. 9.7 Superposition of the secondary flow caused by normal stresses and inertia for We<<1 and Re<<1; $\rho R_0^2/(\alpha_1 + \alpha_2) = 7$

9.2.3 Flow through curved pipes

Steady flow through a curved pipe is closely related to the previously considered situation. The essential difference is that the flow in the pipe is not brought about by the drag effect of the moving walls, but by an externally impressed pressure drop along the pipe axis in the φ-direction. For a 'fully developed' flow with a velocity field independent of φ the pressure derivative in the φ-direction, $\partial p/\partial\varphi$, is spatially constant.

Now, as has been clearly explained at the start of this chapter, for low pressure drops the creeping flow of a Newtonian fluid of viscosity η_0 (the zero-shear viscosity of the actual fluid) gives a first approximation for the effective flow. This primary flow occurs in such a way that each material point flows perpendicular to the pipe cross-section. For the velocity vector therefore $\mathbf{v}^{(1)} = v_\varphi^{(1)}\mathbf{e}_\varphi$. The differential equation for the field $v_\varphi^{(1)}$ has already been found with

(9.19).

Because of the adhesion one has to add $v_\varphi^{(1)} = 0$ at the pipe wall as a boundary condition. Whatever the solution of this boundary value problem may look like, the primary flow is in every case a steady shear flow because each surface $v_\varphi^{(1)}/r$ = constant during the motion remains undeformed, because all the material points lying on it move round the z-axis at the same angular velocity. The resultant motion is therefore the same as that of infinitesimally thin rotational symmetric layers which rotate round a common axis at different angular velocities, thus sliding over each other. This is a viscometric flow. Therefore the primary flow has the same kinematic properties as in the case of a motion round a rotating body. The normal stress and the inertia forces which occur in such a shear flow produce a secondary flow, which takes place again in the meridian plane, and can consequently be described by one stream function each, $\Psi^{(2)}$ and $\Psi_I^{(2)}$. To find these one has to bring in again the previously stated equations (9.41) and (9.45), for the derivation of which only the above mentioned properties of the primary flow have been used and which therefore apply also to flow through curved pipes. The dissimilarity from flow round rotating bodies consists of the boundary value problem for the primary field (compare equation (9.39) with (9.19).

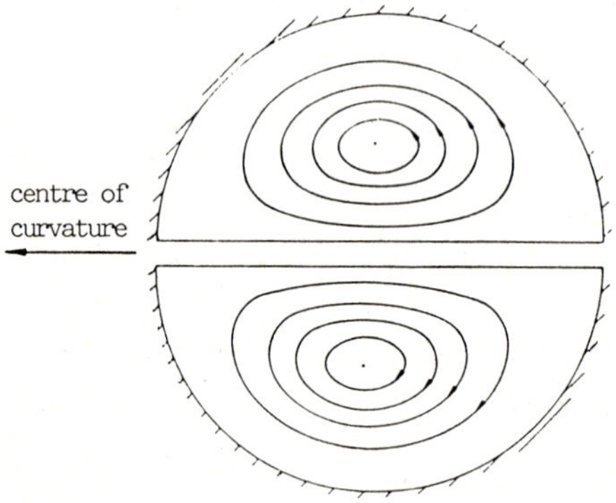

Fig. 9.8 Secondary flow in a curved circular section pipe; above: the inertia effect for Re<<1; below: the normal stress effect for We<<1; the ratio of the pipe radius to the radius of curvature is 0.2

Fig. 9.8 shows the stream line pattern of the secondary flow in a curved pipe of circular cross-section. Not only the inertia forces but also the normal stresses give rise to two annular vortices which are separated from each other by the plane formed by the curved pipe axis and which reproduce mutually by reflection there. It is therefore sufficient to show only one of these annular vortices at any time. In the upper part of the figure one sees the secondary flow in a Newtonian fluid which is caused by the inertia forces alone. For the clear proof of the rotation direction noted therein one considers that the flow velocity and with it the centrifugal acceleration at the pipe wall vanishes, and in the vicinity of the pipe axis reaches its maximum. One will therefore expect that in the centre of the pipe the fluid is accelerated outwards and flows back again in the vicinity of the wall. Fig. 9.8 confirms this concept. The additional component of the secondary flow in viscoelastic fluids caused by normal stresses has similar properties, which are shown in the lower part of the diagram. The stream lines of the primary flow form, as has been previously shown in more detail, are arcs of circles, which in consequence of the first normal stress difference are exposed to a tensile force. They tend to shorten themselves as a result, like stretched elastic bands, i.e. they try to move nearer to the axis of rotation (to the left in Fig. 9.8). For reasons of continuity however all stream lines cannot succeed in doing this simultaneously. It is feasible that in this attempt those regions are chiefly more successful which exhibit a particularly large tension.

Because the shear rate, and with it the tensile force, vanish in the vicinity of the pipe axis and increase outwards to the wall, it is understandable that the liquid in the vicinity of the pipe wall flows inwards; in the centre of the pipe consequently the flow is outwards. Thus arises inclusive of the rotation direction the same pattern as in the centrifugal effect, so that in the superposition both components mutually reinforce. Whereas for a slightly curved circular pipe the centres of the annular vortices practically lie on a diameter of the circular cross-section parallel to the rotation axis, and the stream line pattern is symmetrical with respect to this straight line, asymmetry becomes noticeable with marked pipe curvature, whereby the centres of the vortices are displaced towards the axis of rotation, to the left in Fig. 9.8.

9.3 Plane flows

We now consider plane flows for which the velocity vector is everywhere parallel to the plane z = 0, and only depends on the Cartesian coordinates x and y. For such flows the first Rivlin-Ericksen tensor has the Cartesian matrix

$$A_1 = \begin{bmatrix} 2\frac{\partial u}{\partial x} & \frac{\partial u}{\partial y} + \frac{\partial v}{\partial x} & 0 \\ \frac{\partial u}{\partial y} + \frac{\partial v}{\partial x} & 2\frac{\partial v}{\partial y} & 0 \\ 0 & 0 & 0 \end{bmatrix} \tag{9.47}$$

Because with incompressible fluids for reasons of continuity $\partial v/\partial y = -\partial u/\partial x$ by taking the square there arises a matrix, which differs from zero at only two places on the principal diagonal, and these elements are of equal size

$$A_1^2 = \frac{1}{2}\,\mathrm{tr}\,A_1^2 \begin{bmatrix} 1 & 0 & 0 \\ 0 & 1 & 0 \\ 0 & 0 & 0 \end{bmatrix} \tag{9.48}$$

Therefore

$$\mathrm{div}\,A_1^2 = \mathrm{grad}\left(\frac{1}{2}\,\mathrm{tr}\,A_1^2\right) \tag{9.49}$$

so that in the differential equation (9.6) for the second order secondary flow the right-hand side vanishes. This shows that for plane motions the second order stress field induced by the primary flow only affects the pressure field. A second order secondary flow does not arise under isothermal conditions unless the fluid has free boundaries at which boundary conditions for the pressure are to be satisfied, as is the case say for a stream of liquid emerging from a slit nozzle. For plane motions without free surfaces on the liquid there can occur according to the above statements a secondary flow only within the scope of a *third order theory*, i.e. $\mathbf{v}^{(2)} = 0$, but $\mathbf{v}^{(3)} \neq 0$. Accordingly the second order coefficients α_1 and α_2 in the expansion (8.53) are not really involved. Much more affected are the observed phenomena in the lowest approximation by the third order coefficients β_1, β_2 and β_3. The additional third order stresses induced by the primary flow are therefore the cause of the secondary flow the divergence of which occurs as a driving volume force in the differential equation for the velocity field $\mathbf{v}^{(3)}$

$$\mathbf{f}^{(3)} = \operatorname{div}[\beta_1 \mathbf{A}_3^{(1)} + \beta_2(\mathbf{A}_1^{(1)}\mathbf{A}_2^{(1)} + \mathbf{A}_2^{(1)}\mathbf{A}_1^{(1)}) + \beta_3(\operatorname{tr}\mathbf{A}_1^{(1)2})\mathbf{A}_1^{(1)}] \tag{9.50}$$

Now one can prove that for a plane flow the tensor $\mathbf{A}_1\mathbf{A}_2 + \mathbf{A}_2\mathbf{A}_1 - (\operatorname{tr}\mathbf{A}_1^2)\mathbf{A}_1$, just as $\mathbf{A}_1^2$, is proportional to the two-dimensional unit tensor:

$$\mathbf{A}_1\mathbf{A}_2 + \mathbf{A}_2\mathbf{A}_1 - (\operatorname{tr}\mathbf{A}_1^2)\mathbf{A}_1 = \frac{1}{2}\operatorname{tr}(\mathbf{A}_1\mathbf{A}_2 + \mathbf{A}_2\mathbf{A}_1)\begin{bmatrix} 1 & 0 & 0 \\ 0 & 1 & 0 \\ 0 & 0 & 0 \end{bmatrix} \tag{9.51}$$

This part of the stress tensor therefore affects for a plane flow only the pressure field, but not the velocity field. If one inserts the volume force field (9.50) into equation (9.1), one obtains by considering equation (9.51) as a conditional equation for the secondary flow of the lowest order for plane motions:

$$\eta_0 \operatorname{curl}\operatorname{curl}\operatorname{curl}\mathbf{v}^{(3)} = \beta_1 \operatorname{curl}\operatorname{div}\mathbf{A}_3^{(1)} + (\beta_2 + \beta_3)\operatorname{curl}\operatorname{div}[(\operatorname{tr}\mathbf{A}_1^{(1)2})\mathbf{A}_1^{(1)}] \tag{9.52}$$

The third order coefficients β_1 and $(\beta_2 + \beta_3)$ are therefore critical for the velocity field $\mathbf{v}^{(3)}$. They are related on the one hand in a simple way to the properties of the shear viscosity for a steady shear flow; on the other hand to the properties of the elongational viscosity for a steady plane extensional flow. Deviations from Newtonian behaviour are described for the shear viscosity in the lowest approximation by the coefficient $(\beta_2 + \beta_3)$, for the elongational viscosity on the contrary by the coefficient $(\beta_1 + 2\beta_2 + 2\beta_3)$ (cf. Problem 8.7). By measuring the two viscosity functions the constants needed in equation (9.52) can therefore be determined.

Instead of the extensional flow which is difficult to create experimentally, a simple unsteady relaxation test can also be used because the constant β_1 can be derived from the linear viscoelastic relaxation function which such a test provides.

With plane flows of constant density fluids one can introduce, as is well known, in accordance with

$$\mathbf{v} = \text{curl}\,(\Psi\,\mathbf{e}_z) \tag{9.53}$$

a stream function of the dimensions [length]^2/time.

If we take as a basis, as in the following example to be considered, polar coordinates r, φ within the flow plane, then the two velocity components can be found as follows from Ψ(r,φ):

$$v_r = \frac{1}{r}\frac{\partial \Psi}{\partial \varphi}, \qquad v_\varphi = -\frac{\partial \Psi}{\partial r} \tag{9.54}$$

Because Ψ is independent of z, the following vector identity applies, the validity of which one can easily prove:

$$\text{curl curl}\,(\Psi\,\mathbf{e}_z) = -\mathbf{e}_z\,\text{div grad}\,\Psi =: -\mathbf{e}_z \Delta\Psi \tag{9.55}$$

For the sake of brevity we signify the *Laplace operator* in the plane by the usual symbol Δ. By using polar coordinates r, φ

$$\Delta = \frac{\partial^2}{\partial r^2} + \frac{1}{r}\frac{\partial}{\partial r} + \frac{1}{r^2}\frac{\partial^2}{\partial \varphi^2} \tag{9.56}$$

By considering the representation (9.53) we can thus introduce the Laplace operator twice on the left-hand sides of equations (9.3) and (9.52), and obtain for the stream function of the primary flow the homogeneous bipotential equation

$$\Delta\Delta\Psi^{(1)} = 0 \tag{9.57}$$

and for the stream function of the secondary flow the inhomogeneous bipotential equation

$$\Delta\Delta\Psi^{(3)} = \frac{\beta_1}{\eta_0}\,\text{curl}_z\,\text{div}\,\mathbf{A}_3^{(1)} + \frac{\beta_2+\beta_3}{\eta_0}\,\text{curl}_z\,\text{div}\,[\text{tr}(\mathbf{A}_1^{(1)2})\mathbf{A}_1^{(1)}] \tag{9.58}$$

As soon as the creeping flow of a Newtonian fluid has been found for an actual case by solving equation (9.57) the right-hand side of (9.58) is known. The secondary flow to be superposed in the case of a viscoelastic fluid results from the solution of this bipotential equation. It would be necessary to specify as boundary conditions for $\Psi^{(1)}$ and $\Psi^{(3)}$ at the fixed walls that their first derivatives should vanish. We explain this in more detail by means of the following example.

9.3.1 Convergent channel flow

As an example we consider the flow in an ideal convergent channel with plane walls which slope towards each other at an angle of $2\varphi_0$ between them. In the intersection line of the two walls there is a slit through which the liquid leaks outwards. The amount of liquid escaping per unit time and unit length (a quantity which has the dimensions [length]^2/time) is denoted by $\dot{F}$. We shall use plane polar coordinates r and φ adapted to the problem. Thus r gives the distance of any arbitrary material point from the intersection line of the two walls, and the angle φ is measured from the line of geometrical symmetry so that the two walls are denoted by $\varphi = \pm\varphi_0$.

From the adhesion condition at the walls it follows that $\partial\Psi/\partial r = 0$, therefore Ψ = constant, and

$$\frac{\partial\Psi}{\partial\varphi} = 0 \text{ for } \varphi = \pm\varphi_0 \tag{9.59}$$

The values for Ψ at the two walls are not equal, but differ by the quantity $\dot{F}$, as one can clearly demonstrate as follows:

$$\begin{aligned}\Psi(r,\varphi_0) - \Psi(r,-\varphi_0) &= \int_{\varphi=-\varphi_0}^{\varphi=\varphi_0} d\Psi = \int_{-\varphi_0}^{\varphi_0} \left(\frac{\partial\Psi}{\partial\varphi}\right)_{r=\text{const}} d\varphi = \\ &= \int_{-\varphi_0}^{\varphi_0} (rv_r)_{r=\text{const}}\, d\varphi = -\dot{F}\end{aligned} \tag{9.60}$$

We can therefore establish without limiting the generality:

$$\Psi = \frac{\dot{F}}{2} \text{ for } \varphi = -\varphi_0, \qquad \Psi = -\frac{\dot{F}}{2} \text{ for } \varphi = \varphi_0 \tag{9.61}$$

We first consider the creeping flow of a Newtonian fluid as a primary flow. Because a characteristic length is lacking, the associated stream function $\Psi^{(1)}$ can, apart from the coordinates r, φ only depend on the constants η_0, $\dot{F}$, φ_0. From the three quantities r, η_0, $\dot{F}$ which have dimensions, no dimensionless one can be formed, i.e. on the basis of dimensional analysis the function $\Psi^{(1)}$ can only have the following form:

$$\frac{\Psi^{(1)}}{\dot{F}} = g(\varphi;\varphi_0) \tag{9.62}$$

The stream function of the primary flow therefore does not at all depend on r. From this it follows that the Newtonian fluid in the convergent channel flows purely radially ($v_\varphi^{(1)} = 0$, cf. (9.54)). This moreover would also apply if inertia were to be considered. Because there is no dependence on r equation (9.57) reduces to an ordinary differential equation, whose general solution can be easily written down. If one determines the free constants so that the boundary conditions (9.59) and (9.61) are met, then one obtains for the stream function of the primary flow

$$\Psi^{(1)} = -\frac{\dot{F}}{2} \cdot \frac{\sin 2\varphi - 2\varphi \cos 2\varphi_0}{\sin 2\varphi_0 - 2\varphi_0 \cos 2\varphi_0} \tag{9.63}$$

Fig. 9.9 shows the shape of this function for various values of the angle of intersection φ_0 between the plane walls. The velocity components $v_\varphi^{(1)} = 0$ and

$$v_r^{(1)} = -\frac{\dot{F}c}{r}(\cos 2\varphi - \cos 2\varphi_0) \tag{9.64}$$

belong to this stream function, whereby the abbreviation

$$c := \frac{1}{\sin 2\varphi_0 - 2\varphi_0 \cos 2\varphi_0} \tag{9.65}$$

has been made.

When the primary flow is known the right-hand side of the differential equation (9.58) which is critical for the secondary flow can be evaluated. It thus turns out that both terms depend in the same way on the position coordinates r and φ, that is they are proportional to $r^{-8} \cdot \sin 2\varphi$, and hence can be summed together. One obtains after a few calculations

$$\Delta\Delta\Psi^{(3)} = \frac{32\beta c^3 \dot{F}^3}{\eta_0} \cdot \frac{\sin 2\varphi}{r^8} \tag{9.66}$$

in which the expression

$$\beta := 3\beta_1 \sin^2 2\varphi_0 + 4(\beta_2 + \beta_3)(1 - 4\cos^2 2\varphi_0) \tag{9.67}$$

is formed from the third order coefficients and the angle of the V-shaped opening. Because this is a third order secondary flow effect, the third power of the flow rate also appears on the right-hand side and c is the constant defined by equation (9.65). For small opening

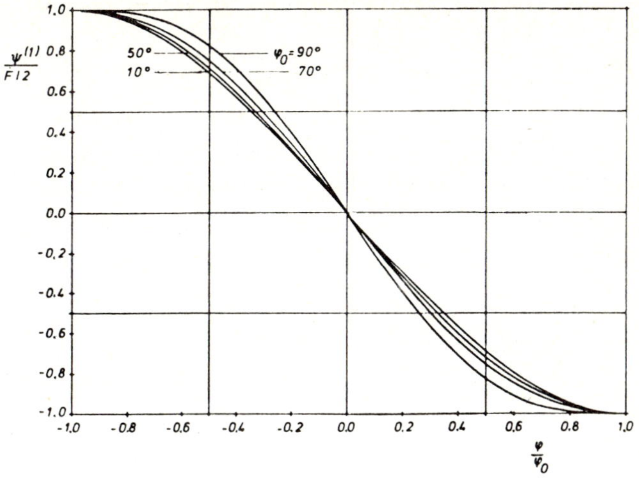

Fig. 9.9 Stream function of the primary flow as a function of the coordinate φ

angles ($2\varphi_0 \ll 1$) β can be approximated by $-12(\beta_2 + \beta_3)$, hence by that coefficient which describes the deviation of the shear viscosity from the Newtonian value. That in this limiting case only the shear viscosity, i.e. the flow property for shear flow enters is understandable when one considers that the space filled by the liquid for small aperture angles is equal to a channel of slowly varying depth. Moreover one can easily prove by use of the expression (9.67) that, independent of the aperture angle of the wedge, $\beta > 0$ holds for fluids which, like most liquid polymers, exhibit shear thinning ($\beta_2 + \beta_3 < 0$) and elongation thickening ($\beta_1 + 2\beta_2 + 2\beta_3 > 0$). For dilatant (shear thickening) liquids β always works out negative, at least for small and large aperture angles ($2\varphi_0 \approx 180^\circ$). For pseudoplastic (shear thinning) liquids $\beta<0$ for moderate aperture angles ($2\varphi_0 \approx 90^\circ$) provided that sufficiently marked elongation thinning is present.

Because with β/η_0 a further constant of dimension $[\text{time}]^2$ is added to the quantities r, φ, η_0, $\dot{F}$, φ_0, a new dimensionless term, $\beta\dot{F}^2/\eta_0 r^4$, enters, i.e. the stream function $\Psi^{(3)}$ made dimensionless with $\dot{F}$ depends on $\beta\dot{F}^2/\eta_0 r^4$ in addition to φ and φ_0. The viscoelastic fluid therefore in contrast to the Newtonian fluid does not flow purely radially towards the sink. Because the theory is linear as regards β the function

$$\Psi^{(3)} = \frac{\beta \dot{F}^3}{\eta_0 r^4} H(\varphi; \varphi_0) \tag{9.68}$$

which gives the dependence on the position coordinate r explicitly, must provide $\Psi^{(3)}$. In fact it reduces equation (9.66) to an ordinary differential equation. The solution, which satisfies the boundary conditions $\Psi^{(3)} = \partial\Psi^{(3)}/\partial\varphi = 0$ for $\varphi = \pm\varphi_0$, is as follows:

$$\frac{12}{c^3} H = \sin 2\varphi + \frac{(-4 \sin 4\varphi_0 + 2 \sin 8\varphi_0) \sin 4\varphi + (3 \sin 2\varphi_0 - \sin 6\varphi_0) \sin 6\varphi}{5 \sin 2\varphi_0 - \sin 10\varphi_0} \tag{9.69}$$

Fig. 9.10 shows the shapes of this function for various values of aperture angle. The superposition of primary and secondary flows leads to the stream line patterns shown in Fig. 9.11 with two oppositely rotating vortices. If the coefficient β defined in equation (9.67) is positive the vortices are located in the middle and the stream lines are deflected outwards to the walls. If however β<0, then the vortices occur at the walls, and the liquid flowing in from outside flows between them to the sink.

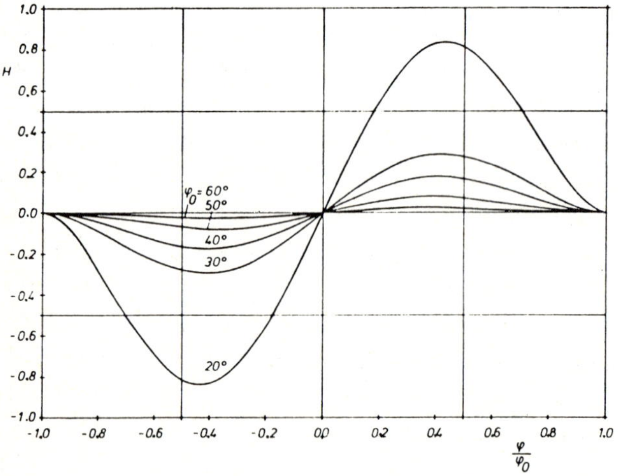

Fig. 9.10 Graphs of the function $H(\varphi,\varphi_0)$ in accordance with equation (9.69)

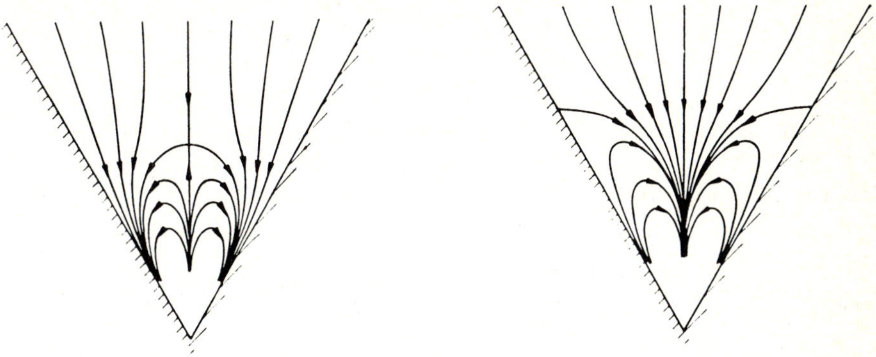

Fig. 9.11 Formation of stream lines in convergent channel flow; result of the third order theory: (a) $\beta > 0$, $\varphi_0 = 30^\circ$, (b) $\beta < 0$, $\varphi_0 = 30^\circ$

A similar appraisal as used for conical nozzle flow yields for the radial extent of the vortex

$$r_* \sim \sqrt[4]{\frac{|\beta|}{\eta_0}\dot{F}^2} \tag{9.70}$$

Fig. 9.10 shows that the amplitude of the secondary flow for small angle of the wedge depends strongly on φ_0, so that the radial extent of the vortex is also considerably affected by the magnitude of φ_0. With a more accurate analysis it turns out that the evaluation (9.70) for small aperture angles has to be corrected by the factor $1/\varphi_0$ on the right-hand side.

It should be pointed out once more that the inertia of the liquid has been omitted. This is permissible so long as the Reynolds number $\rho\dot{F}/\eta_0$ is small compared with unity. The application of the results of the theory therefore assumes that the flow rate is limited in accordance with $\dot{F} \ll \eta_0/\rho$. Moreover the error in the perturbation calculation ought to increase more and more with decreasing distance from the apex of the wedge, because the assumption of a slowly varying previous history of the liquid particles becomes less and less valid the more the particles approach the outlet.

9.4 Steady flow through cylindrical pipes

If an incompressible Newtonian liquid flows in response to an axial pressure gradient in a laminar and steady manner through a straight cylindrical pipe of arbitrary cross-section, then the pressure within a cross-section plane is constant and each material point moves parallel to the pipe axis. For a non-Newtonian liquid this behaviour is true only in the case of a circular section pipe. For other shapes of cross-section the flow has in general more complex kinematic properties because a secondary flow in the plane of the cross-section superposes itself on the main flow in the axial direction, which has already been noted in Section 4.4. This is a normal stress effect in which the second normal stress function is involved.

In an analogous generalisation of the general concept described in Section 9.1 we shall start by stating that the motion of a purely viscous fluid with the real flow function represents a first approximation, and the secondary flow represents a perturbation of this basic state. Therefore it is assumed that the normal stress difference which causes the secondary flow is small compared with the shear stress. The basic state is the axial shear flow discussed in Section 4.4 which has the velocity field $u(y,z)$. In the equation (4.33) k was the constant axial pressure gradient, $\eta(\dot{\gamma},\Theta)$ the viscosity function. Because the shear rate $\dot{\gamma} = |\mathrm{grad}\ u|$ and possibly also the temperature Θ of the liquid varies across the plane of the cross-section, η (and accordingly also the critical normal stress coefficient ν_2) depend ultimately on y,z.

For later use we record the following relationship which satisfies the axial velocity field:

$$D[\eta, u] = \frac{1}{2}\,\frac{\partial\left(\eta, \left(\frac{\partial u}{\partial y}\right)^2 + \left(\frac{\partial u}{\partial z}\right)^2\right)}{\partial(y, z)} \tag{9.71}$$

On the right-hand side there is the Jacobian determinant formed from the partial derivatives of η and $\dot{\gamma}^2$. The expression on the left-hand side is an abbreviated form of

$$D[\varphi, u] := \left(\frac{\partial^2}{\partial y^2} - \frac{\partial^2}{\partial z^2}\right)\left[\varphi\,\frac{\partial u}{\partial y}\frac{\partial u}{\partial z}\right] - \frac{\partial}{\partial y \partial z}\left[\varphi\left(\frac{\partial u}{\partial y}\right)^2 - \varphi\left(\frac{\partial u}{\partial z}\right)^2\right] \tag{9.72}$$

Note that the operator D is linear relative to the first argument. Equation (9.71) is verified when one multiplies equation (4.33) by grad u, adds $0.5\eta\,\mathrm{grad}\,\dot{\gamma}^2$ to both sides, and then performs the operation curl. In the case of an isothermal flow the viscosity function η would only be dependent on $\dot{\gamma}^2$, and in consequence the right-hand side of equation (9.71) would vanish. In the expression thus abbreviated one can recognise equation (4.37).

The extra-stresses occurring in the axial shear flow have been collected together in the matrix (4.31). The elements multiplied by ν_2 are responsible for the secondary flow. The divergence of that part of the extra-stress tensor which consists of the normal stress terms corresponds to a volume force field **f** acting on the liquid, which lies in the plane of the cross-section (with $f_x = 0$). The components f_y and f_z are the expressions on the left-hand side of equations (4.34) and (4.35). The evaluation of the secondary flow therefore proceeds with the aim of determining that motion of a purely viscous fluid with the viscosity function η, which sets in under the influence of this volume force field. As has already been mentioned, this is a steady plane flow within the cross-section of the pipe. The velocity components in the y- and z-directions are denoted by v(y,z) and w(y,z) respectively. With the additional assumption that the Reynolds number for the secondary flow is small compared with unity, the inertia influence can be neglected. The proper Navier-Stokes equations for the plane creeping flow of an incompressible fluid of variable viscosity read

$$\frac{\partial}{\partial y}\left(2\eta\,\frac{\partial v}{\partial y}\right)+\frac{\partial}{\partial z}\left(\eta\,\frac{\partial v}{\partial z}+\eta\,\frac{\partial w}{\partial y}\right)+f_y=\frac{\partial p'}{\partial y} \tag{9.73}$$

$$\frac{\partial}{\partial y}\left(\eta\,\frac{\partial v}{\partial z}+\eta\,\frac{\partial w}{\partial y}\right)+\frac{\partial}{\partial z}\left(2\eta\,\frac{\partial w}{\partial z}\right)+f_z=\frac{\partial p'}{\partial z} \tag{9.74}$$

Because of the incompressibility of the liquid a stream function $\Psi(y,z)$ can be introduced, so that $v = \partial\Psi/\partial z$ and $w = -\partial\Psi/\partial y$ (cf. the comments made in connection with equation (1.103)). In addition if one inserts the above described explicit expressions for f_y and f_z, and eliminates the pressure p' from equations (9.73) and (9.74), then one obtains the following equation for the stream function of the secondary flow:

$$\left(\frac{\partial^2}{\partial y^2}-\frac{\partial^2}{\partial z^2}\right)\left[\eta\left(\frac{\partial^2\Psi}{\partial y^2}-\frac{\partial^2\Psi}{\partial z^2}\right)\right]+4\,\frac{\partial^2}{\partial y\partial z}\left[\eta\,\frac{\partial^2\Psi}{\partial y\partial z}\right]=D[\nu_2,u] \qquad (9.75)$$

One might try to replace the two coefficients η and ν_2 for the sake of simplicity by constant values. But according to the previous treatment in Section 4.4 the secondary flow vanishes in this approximation. Therefore what matters is that it depends just on the variability of the quantities η and ν_2 in the cross-section flowed through, and therefore on the deviation of characteristic constant reference quantities. Because the flow is slow these reference quantities can be identified with the zero values of the coefficients (for $\dot{\gamma} \to 0$) for a temperature $\bar{\Theta}$ characteristic for the actual situation and correspondingly can be denoted by $\bar{\eta}_0$ and $\bar{\nu}_{20}$ respectively. Nevertheless it is possible to simplify equation (9.75) considerably. Of course the right-hand side can also be written in the form $D[\nu_2 - \bar{\nu}_{20}\eta/\bar{\eta}_0, u] + (\bar{\nu}_{20}/\bar{\eta}_0)D[\eta, u]$, in which the second term can be exchanged by use of equation (9.71). Thus the inhomogeneous term of the differential equation (9.75) is expressed so that one can insert for the primary field u the lowest order approximation, namely the isothermal Newtonian flow field, which is denoted by $u^{(1)}$. It satisfies the linear relationship (cf. equation (4.33))

$$\Delta u^{(1)} = -\frac{k}{\bar{\eta}_0} \qquad (9.76)$$

in which Δ represents the Laplace differential operator in the y-z-plane. Similarly one can on the left-hand side of equation (9.75) put $\bar{\eta}_0$ for η in the lowest approximation and obtain therefore for the stream function of the secondary flow the inhomogeneous bipotential equation

$$\bar{\eta}_0\Delta\Delta\Psi = D[\nu_2^{(1)} - \bar{\nu}_{20}\eta^{(1)}/\bar{\eta}_0, u^{(1)}] + \frac{\bar{\nu}_{20}}{2\bar{\eta}_0}\,\frac{\partial(\eta^{(1)}, |\mathrm{grad}\, u^{(1)}|^2)}{\partial(y,z)} \qquad (9.77)$$

Although Ψ is now an approximation of the original stream function, we suppress an additional index as a distinctive criterion. It will turn out that the secondary flow in the actual case can be a phenomenon of the second or fourth order with respect to the pressure gradient.

9.4.1 Isothermal conditions

For the first special case we consider an isothermal pipe flow. It has already been pointed out that in this the second term on the right-hand side of equation (9.77) vanishes, because η depends only on $\dot{\gamma} = |\text{grad } u|$. In the lowest approximation, in which a secondary flow occurs, the two critical constitutive functions $\eta(\dot{\gamma})$ and $\nu_2(\dot{\gamma})$ can be approximated by quadratic expressions:

$$\eta(\dot{\gamma}) = \eta_0 + \eta_2\dot{\gamma}^2, \qquad \nu_2(\dot{\gamma}) = \nu_{20} + \nu_{22}\dot{\gamma}^2 \tag{9.78}$$

The previously used bar can be dispensed with here because temperature differences are not involved. Therefore one obtains as an equation for the stream function Ψ_i under isothermal conditions

$$\Delta\Delta\Psi_i = \frac{\nu_e}{\eta_0} D[\,|\text{grad } u^{(1)}|^2, u^{(1)}] \tag{9.79}$$

ν_e represents a certain combination of the four constants introduced in equation (9.78)

$$\nu_e := \nu_{22} - \frac{\eta_2}{\eta_0}\nu_{20} \equiv \left[\eta \frac{d}{d\dot{\gamma}^2}\left(\frac{\nu_2}{\eta}\right)\right]_{\dot{\gamma}=0} \tag{9.80}$$

It was established in Section 4.4 that a secondary flow is absent if the viscometric functions $\nu_2(\dot{\gamma})$ and $\eta(\dot{\gamma})$ are proportional to each other, which would mean $\nu_e = 0$ here. It is therefore not surprising to meet the expression (9.80) as a factor in the differential equation for Ψ_i. Because the primary field $u^{(1)}$ according to equation (9.76) is proportional to $1/\eta_0$ (so long as the pipe which conducts the flow is at rest), and the right-hand side of equation (9.79) represents a homogeneous fourth order expression in $u^{(1)}$, it follows that $\Psi_i \sim \nu_e/\eta_0^5$. In the context of the *fourth order theory* demonstrated here the isothermal secondary flow is therefore affected by a single combination of coefficients which arise from the viscometric functions $\eta(\dot{\gamma})$ and $\nu_2(\dot{\gamma})$. Note that for the calculation of Ψ_i, a fourth order phenomenon, only the primary flow of the first order $u^{(1)}$ is needed.

If we consider as an example the flow through an elliptical pipe of semi-axes b and c, then the solution to equation (9.76) which satisfies the condition $u^{(1)} = 0$ on the boundary of the ellipse, is:

$$u^{(1)} = \frac{kb^2c^2}{2\eta_0(b^2 + c^2)}\left[1 - \frac{y^2}{b^2} - \frac{z^2}{c^2}\right] \tag{9.81}$$

If one inserts this Newtonian flow field into equation (9.79) and applies the operator (9.72), then the right-hand side reduces to an expression proportional to $y \cdot z$. The solution of the bipotential equation which satisfies the adhesion conditions $\partial\Psi_i/\partial y = \partial\Psi_i/\partial z = 0$ at the wall of the elliptical pipe runs:

$$\Psi_i = -\frac{\nu_e k^4}{4\eta_0^5}\,\frac{(b^2 - c^2)\,b^6c^6}{(5b^4 + 6b^2c^2 + 5c^4)\,(b^2 + c^2)^3}\,yz\left[1 - \frac{y^2}{b^2} - \frac{z^2}{c^2}\right]^2 \tag{9.82}$$

An insignificant additive constant has been made equal to zero.

Fig. 9.12 shows the graphs of Ψ_i = constant, therefore the stream lines of the secondary flow. One recognises, and this is obvious for reasons of symmetry, that the semi-axes of the ellipse are special stream lines ($\Psi_i = 0$), and therefore that there is a vortex in each quadrant. The directions of rotation of the vortices depend on the sign of the constant ν_e. The superposition of the secondary flow on the axial primary flow leads to helically curved path lines of the material points.

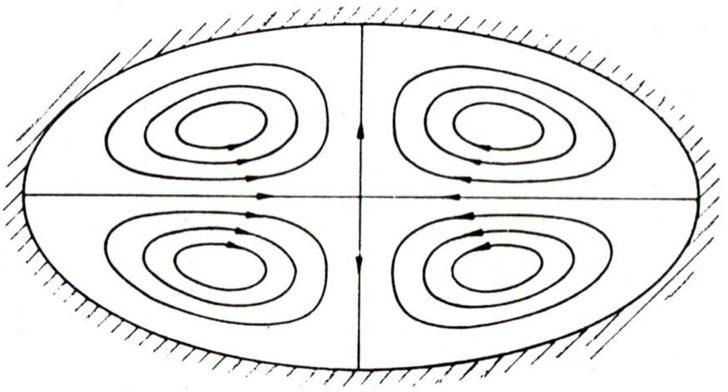

Fig. 9.12 Isothermal secondary flow in an elliptical pipe; $b/c = 2$, $\nu_e > 0$

9.4.2 Effect of dissipation

The flow through a pipe is not in fact isothermal because heat is generated by internal friction, which then flows outwards. The occurrence of a heat flux is however necessarily connected with a temperature gradient. The temperature of the liquid cannot therefore be constant throughout the space. In the case of a circular section pipe the dissipation leads to an axisymmetric temperature distribution, which has been described in Section 3.4. Now the mechanical properties, in particular the normal stress coefficients, are strongly dependent on temperature. Therefore they likewise vary over the cross-section of the flow. For a non-circular section pipe the temperature field, and hence the field of the second normal stress difference too, will not be axisymmetric. One ought therefore to expect a secondary flow to be caused by the dissipation, which is superposed on that considered in Section 9.4.1.

This *thermorheological effect* can be explained in the context of a *second order theory*, hence under the sole consideration of the lower limiting values $\eta_0(\Theta)$ and $\nu_{20}(\Theta)$. For the sake of simplicity we shall assume that the local temperature Θ differs from the reference temperature $\bar{\Theta}$ (say at the wall) by so little that the two constitutive functions can be approximated by the linear expressions

$$\eta_0(\Theta) = \bar{\eta}_0[1 - \kappa(\Theta - \bar{\Theta})] \tag{9.83}$$

$$\nu_{20}(\Theta) = \bar{\nu}_{20}[1 - \mu(\Theta - \bar{\Theta})] \tag{9.84}$$

The coefficients κ and μ, which describe the variability of the viscosity and the normal stress coefficients respectively in terms of the temperature, are in general positive, because η_0 and $|\nu_{20}|$ decrease with increasing temperature (cf. Fig. 2.3). In accordance with equation (2.33) $\mu = 2\kappa$ should apply.

If one now returns to the general relationship (9.77), then one finds that the secondary flow brought about by the dissipation is described by the differential equation

$$\bar{\eta}_0 \Delta\Delta\Psi_d = \bar{\nu}_{20}(\kappa - \mu)\, D[\Theta^{(1)} - \bar{\Theta}, u^{(1)}] - \bar{\nu}_{20}\, \frac{\kappa}{2}\, \frac{\partial(\Theta^{(1)}, |\text{grad}\, u^{(1)}|^2)}{\partial(y, z)} \tag{9.85}$$

Thus $\Theta^{(1)}$ is the temperature field created by the primary flow $u^{(1)}$. We introduce here the Péclet number which is formed like the Reynolds number, in which the kinematic viscosity is replaced by the thermal diffusivity, or what amounts to the same thing, the shear viscosity η_0, by the ratio of the thermal conductivity λ to the specific heat capacity c of the liquid. The Péclet number can therefore also be expressed as the product of the Reynolds number and the Prandtl number $c\eta_0/\lambda$.

If the Péclet number formed with a characteristic secondary flow velocity is small compared with respect to unity, the convective energy transport can be neglected compared with the energy transported by thermal conduction, and the conditional equation for $\Theta^{(1)}$ reads

$$\lambda\Delta\Theta^{(1)} = -\bar{\eta}_0 |\mathrm{grad}\, u^{(1)}|^2 \tag{9.86}$$

in which λ is the thermal conductivity of the liquid. In the case of an elliptical pipe the solution can be given in closed form, which satisfies the condition of constant temperature at the pipe wall ($\Theta^{(1)} = \bar{\Theta}$ at the elliptical boundary). Hence equation (9.85) can also be integrated analytically. The result is

$$\Psi_d = \frac{\bar{\nu}_{20} k^4 b^4 (b^2 - c^2)}{120\,\lambda\,\bar{\eta}_0^4}\left[\kappa M\left(1 + q\frac{y^2}{b^2} + r\frac{z^2}{c^2}\right) + \mu N\left(1 + s\frac{y^2}{b^2} + t\frac{z^2}{c^2}\right)\right] \cdot yz\left(1 - \frac{y^2}{b^2} - \frac{z^2}{c^2}\right)^2 \tag{9.87}$$

The coefficients M, N, q, r, s and t are numerical parameters whose values depend on the semi-axis ratio c/b (Fig. 9.13). The secondary flow consists once more of four vortices which are separated by the semi-axes of the ellipse. The stream line pattern including the flow direction resembles that in Fig. 9.12, although some quantitative differences exist.

In order to examine the relative effect of both phenomena the following evaluations suffice: $u^{(1)} \sim kb^2/\bar{\eta}_0$ according to equation (9.81), $\Theta^{(1)} - \bar{\Theta} \sim k^2 b^4/(\lambda\bar{\eta}_0)$ according to equation (9.86) and finally $\Psi_d \sim \bar{\nu}_{20}\kappa k^4 b^8/(\lambda\bar{\eta}_0^4)$ according to equation (9.85), taking into account the relationship $\mu = 2\kappa$. In order to make the comparison we remember that the strength of the isothermal secondary flow is $\Psi_i \sim \nu_e k^4 b^6/\eta_0^5$ according to equation (9.82). The ratio of the two expressions

$$\frac{\Psi_d}{\Psi_i} \sim \frac{\nu_{20}\eta_0\kappa}{\nu_e\lambda} \cdot b^2 \tag{9.88}$$

is remarkably independent of the pressure gradient and contains apart from material coefficients only the characteristic cross-section dimension b of the pipe which is carrying the flow. The condition that the amplitude ratio of the two secondary flows has the value unity defines a critical pipe radius $b_* \sim \sqrt{\nu_e\lambda/(\nu_{20}\eta_0\kappa)}$ which is dependent only on the thermal and mechanical properties of the fluid. For polymer melts this lies in the centimetre range. For relatively narrow pipes ($b<b_*$) the isothermal secondary flow is dominant, for $b>b_*$ however the effect of dissipation is dominant. If therefore in an elliptical pipe which is not very narrow a secondary flow shown in Fig. 9.12 is observed, then the flow is not so much an isothermal phenomenon, but much more is a thermorheological effect caused by dissipation.

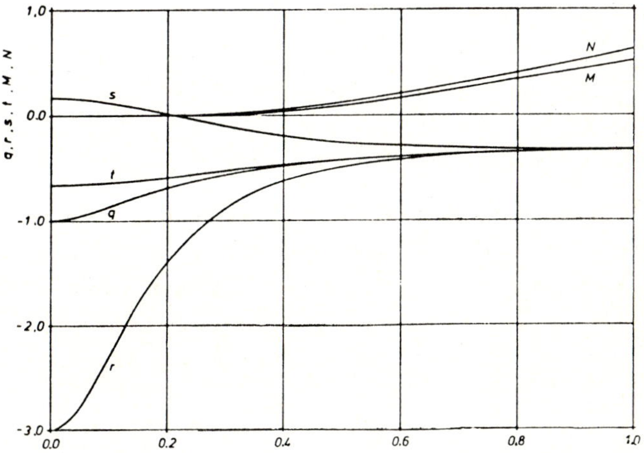

Fig. 9.13 Dependence of the numerical parameters occurring in equation (9.87) on the semi-axis ratio

9.4.3 Effect of a transverse temperature gradient

One should mention here that under the effect of an axial pressure drop even in a circular pipe a thermorheological secondary flow of the second order can occur, if there is a significant temperature gradient perpendicular to the pipe axis. Consider say a flow in a horizontal pipe which is heated above and cooled below. The following considerations may suffice as an explanation. With a non-axisymmetric temperature distribution impressed from the outside on the fluid corresponding viscosity differences exist over the cross-section. Because the shear stress distribution is axisymmetric (equation (3.113) applies as before), the shear rate and the primary velocity field associated with it are necessarily asymmetric. Hence there is also an asymmetric field of normal stress differences, which leads to a secondary flow.

In the quantitative description within a second order theory we can refer to equations (9.76) and (9.85). In the case of a circular pipe of radius r_0 equation (9.76) is solved by the expression $u^{(1)} = (k/4\bar{\eta}_0)(r_0^2 - y^2 - z^2)$. For the sake of simplicity we assume in addition that in the undisturbed fluid there is a linear temperature stratification which is compatible with the laws of heat conduction, but we shall not discuss the possibility of the experimental realisation of this. With suitable orientation of the y-axis, i.e. parallel to the temperature gradient, the following holds for the temperature field

$$\Theta^{(1)} - \bar{\Theta} = \Theta' y \tag{9.89}$$

The constant Θ', which without limitation of the generality can be assumed to be positive, represents the size of the temperature gradient and is determined by the external conditions. Therefore the right-hand side of equation (9.85) reduces to a simple expression proportional to z. The integral, which satisfies the adhesion conditions $\partial\Psi_t/\partial y = \partial\Psi_t/\partial z = 0$ for $y^2 + z^2 = r_0^2$, can be represented in closed form as a polynomial in y,z. To distinguish it from the previously discussed secondary flows we denote the stream function by a new index, t

$$\Psi_t = \frac{\bar{\nu}_{20}(3\kappa - 4\mu)k^2\Theta'}{768\,\bar{\eta}_0^3} z(r_0^2 - y^2 - z^2)^2 \tag{9.90}$$

This expression describes two vortices which are separated by the y-axis, therefore by the pipe diameter parallel to the temperature gradient. Fig. 9.14 shows the stream line pattern.

The directions of the rotation of the vortices depend on the sign of the coefficient $\bar{\nu}_{20}(3\kappa - 4\mu)$. This ought to be generally positive, because the second normal stress coefficient works out mostly negative,

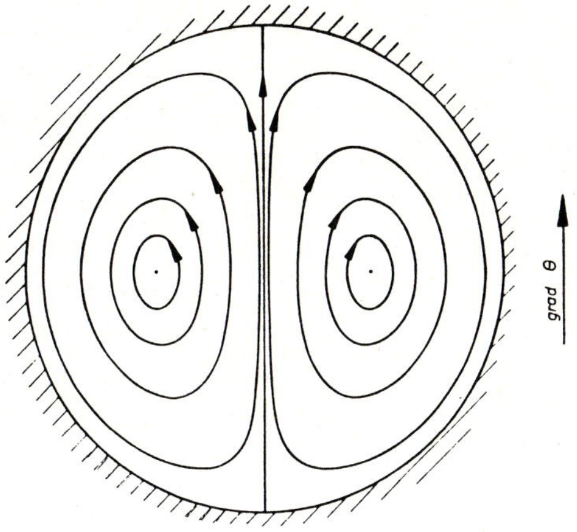

Fig. 9.14 Secondary flow in a circular section pipe under the effect of an axial pressure and a transverse temperature gradient; $\bar{\nu}_{20}(3\kappa - 4\mu)>0$

and according to equation (2.33) $\mu = 2\kappa > 0$, hence $3\kappa - 4\mu < 0$ is to be valid for the constants κ and μ which describe the temperature dependence of the viscosity and normal stress coefficients. Under these conditions the material points move on the symmetry stream line from the cold to the warm side.

Now the secondary flow changes the temperature stratification under some circumstances, and it is therefore significant to make clear once more under what conditions the previous assumptions apply. The theory is based on three independent hypotheses. Because the energy flux by convection was neglected compared with the heat flux, the Péclet number of the secondary flow must be small compared with unity. When furthermore the inertia of the liquid was entirely disregarded, then it was tacitly assumed that also the Reynolds number for the secondary flow is small. Finally it should be remembered that the concept of the calculation of secondary flows is based on the assumption of a sufficiently slow flow, so that the Weissenberg number, which is associated with the primary flow, must be significantly smaller than unity.

9.5 Periodic pipe flow

In addition we now turn to an unsteady flow which can be briefly described as follows. The viscoelastic liquid is in a stationary non-circular cylindrical pipe under the effect of an axial pressure gradient which oscillates sinusoidally with time about a mean value zero, $-\partial p/\partial x = k \sin\omega t$. Thus there naturally arises in the axial direction a periodic flow in which the rate of flow vanishes when averaged over time. More worthy of note however is the fact that there forms in the cross-section plane a time independent (steady) secondary flow. We assume that the amplitude k and the angular velocity ω of the pressure gradient are so small that the flow proceeds slowly as a whole, and the state of the fluid particles within the memory time of the liquid changes only slowly. In obvious generalisation of the earlier stated concept we can regard the actual liquid under such conditions as Newtonian to a first approximation, in which the zero-shear viscosity η_0 is critical, and we shall again designate the associated pure unsteady axial flow as a primary flow. The equation of motion is

$$\rho \frac{\partial u^{(1)}}{\partial t} = \eta_0 \Delta u^{(1)} + k \sin \omega t \tag{9.91}$$

Δ again denotes the two-dimensional Laplace operator. Because of the linearity of the problem the flow velocity $u^{(1)}$ oscillates in the fully developed oscillatory state to which we confine ourselves, likewise sinusoidally with angular velocity ω. This special dependence on time is referred to by the function

$$u^{(1)}(y, z, t) = U_+(y, z) \sin \omega t + U_-(y, z) \cos \omega t \tag{9.92}$$

The primary flow appears here split into two components in which the first proceeds in phase with the pressure gradient, and the second leads the first by a phase angle $\pi/2$. The two amplitude functions U_+ and U_- depend on the position coordinates in the plane of the cross-section. To determine this one has to use the field equations

$$\Delta U_+ = -\frac{\rho\omega}{\eta_0} U_- - \frac{k}{\eta_0} \tag{9.93}$$

$$\Delta U_- = \frac{\rho\omega}{\eta_0} U_+ \tag{9.94}$$

which result from the equation of motion (9.91) if the expression (9.92) is brought in and a comparison of coefficients is made. For a stationary pipe the boundary conditions $U_+ = U_- = 0$ apply at the wall.

To calculate the secondary flow one now determines, as described earlier, the second order extra-stresses caused by the primary flow, $\alpha_1 \mathbf{A}_2^{(1)} + \alpha_2 \mathbf{A}_1^{(1)2}$ and interprets their divergence as a volume force field which drives the secondary flow (cf. equation (9.4)). Because the primary flow which has the axial velocity field $\mathbf{v}^{(1)} = u^{(1)}(y,z,t)\mathbf{e}_x$ is kinematically very simple, this step can be performed without difficulty. The result is

$$\mathbf{f}^{(2)} = \begin{bmatrix} \alpha_1 \dfrac{\partial}{\partial t} \Delta u^{(1)} \\ (2\alpha_1 + \alpha_2)\left[\dfrac{\partial}{\partial y} \left(\dfrac{\partial u^{(1)}}{\partial y} \right)^2 + \dfrac{\partial}{\partial z} \left(\dfrac{\partial u^{(1)}}{\partial y} \dfrac{\partial u^{(1)}}{\partial z} \right) \right] \\ (2\alpha_1 + \alpha_2)\left[\dfrac{\partial}{\partial y} \left(\dfrac{\partial u^{(1)}}{\partial y} \dfrac{\partial u^{(1)}}{\partial z} \right) + \dfrac{\partial}{\partial z} \left(\dfrac{\partial u^{(1)}}{\partial z} \right)^2 \right] \end{bmatrix} \tag{9.95}$$

Because $\mathbf{f}^{(2)}$ has an x-component (the upper element of the column), there arises in the second approximation an axial component of the secondary flow, which however we shall not consider more closely here. The two other components of the volume force are critical for the transverse part of the flow. If one disregards the inertia forces connected with the secondary flow one can use equation (9.1). More appropriately one introduces a stream function $\Psi^{(2)}$ of a type such that $v^{(2)} = \partial\Psi^{(2)}/\partial z$ and $w^{(2)} = -\partial\Psi^{(2)}/\partial y$. Thus one immediately obtains for the equation for the transverse secondary flow

$$\eta_0 \Delta\Delta\Psi^{(2)} = \mathrm{curl}_x\, \mathbf{f}^{(2)} \tag{9.96}$$

The expression on the right-hand side can be represented by considering the special form (9.95) of the volume force field and by use of the operator D defined in equation (9.72) by the primary flow field as follows:

$$\mathrm{curl}_x\, \mathbf{f}^{(2)} = (2\alpha_1 + \alpha_2)\, D[1, u^{(1)}] \tag{9.97}$$

The factor $2\alpha_1 + \alpha_2$ agrees with the lower limit of the second normal stress coefficient and can therefore be replaced by ν_{20} (cf. equation (8.52)). For the expression $D[1, u^{(1)}]$

$$D[1, u^{(1)}] = \frac{\rho}{\eta_0} \frac{\partial(\partial u^{(1)}/\partial t, u^{(1)})}{\partial(y, z)} = \frac{\rho\omega}{\eta_0} \frac{\partial(U_+, U_-)}{\partial(y, z)} \tag{9.98}$$

applies.

The first equality arises from the sole use of the equation of motion (9.91); the second follows because of the special form (9.92). Thus the field equation for finding the secondary flow in the cross-section is

$$\Delta\Delta\Psi^{(2)} = \frac{\omega\rho\nu_{20}}{\eta_0^2} \left[\frac{\partial U_+}{\partial y} \frac{\partial U_-}{\partial z} - \frac{\partial U_+}{\partial z} \frac{\partial U_-}{\partial y} \right] \tag{9.99}$$

Homogeneous boundary conditions prevail at the wall. Note that the right-hand side of equation (9.99) does not in any way depend on time. In consequence the solution $\Psi^{(2)}$ is also independent of time. A sinusoidal oscillatory pressure gradient therefore gives rise to a steady secondary flow in the plane of the cross-section. The factor ν_{20} indicates that this is a normal stress effect.

We do not consider here any special solutions of the differential equations (9.93), (9.94) and (9.99), but restrict ourselves to pointing out that in the special case of an elliptical pipe (with semi-axes b and c) they can be integrated analytically. Fig. 9.15 shows the result. The fourfold symmetry corresponds to the geometrical properties of the ellipse. The direction of rotation of the vortices is controlled by the sign of the coefficient ν_{20}. As regards the strength of the secondary flow one finds $\Psi^{(2)} \sim (\rho^2\nu_{20}/\eta_0^5)\omega^2k^2b^6(b^2 - c^2)$, assuming a small Stokes number, $\rho\omega b^2/\eta_0 \ll 1$. The amplitude of the secondary flow as a second order rheological effect is proportional to the second power of the pressure gradient. Thus enters a single combination of the three constants ρ, η_0 and ν_{20}. It is obvious that the effect vanishes when the semi-axes are of equal length (b = c), i.e. when the ellipse degenerates into a circle.

Finally one should mention that the transverse secondary flow described here also arises in the same form when the pressure in the fluid is constant, but the wall in the axial direction oscillates sinusoidally

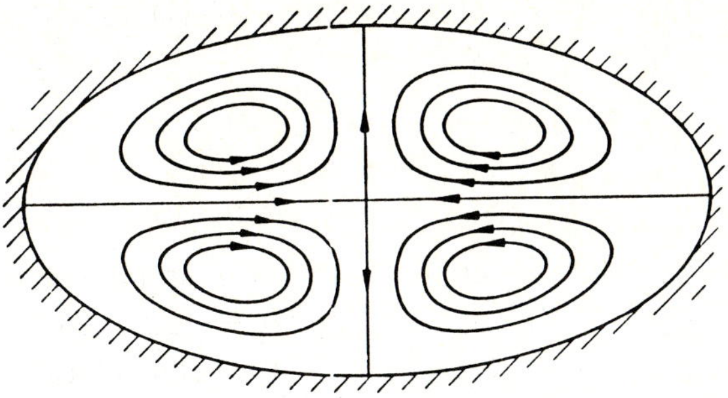

Fig. 9.15 Secondary flow in an elliptical pipe under the effect of an oscillatory axial pressure gradient; $b/c=2$, $\nu_{20}<0$, $\rho\omega b^2/\eta_0<<1$

in time as $x_0\sin\omega t$. An observer moving with the flow, for whom the wall is stationary, finds himself in an accelerated reference system where the acceleration obviously has the value $-\omega^2x_0\sin\omega t$. He therefore records an axial inertia force per unit volume of magnitude $\rho\omega^2x_0\sin\omega t$ which is completely equivalent to a pressure gradient of the same magnitude. Therefore the new situation has been traced back to that previously discussed. The expression $\rho\omega^2x_0$ occurs instead of k. The moving observer sees the same motion as the stationary observer in the case of pressure variation. In the transition to the inertial system the axial velocity field (the primary flow) changes by the value $\omega x_0\cos\omega t$, but the transverse secondary flow remains unchanged. Therefore it is completely the same in both cases.

Problems

9.1 In the calculation of secondary flows one generally starts from the creeping flow of a Newtonian fluid, determines the second (or higher) order forces thereby arising, and then the creeping flow of a Newtonian fluid which is acted on by these forces. According to Section 8.1.1 one can formulate a minimum principle not only for the primary but also for the secondary flow. How do these two minimum principles run in the case of the flow round a rotating body (cf. Section 9.2.2)? *Hint*: One needs the two invariants of the rate of strain tensors associated with the primary flow field $\omega^{(1)}(r,z)$ and the secondary flow field $\Psi^{(2)}(r,z)$, as well as the expression for the volume force field which propels the secondary flow.

9.2 Determine in the context of a second order theory:
(a) The pressure distribution over a rotating sphere in a liquid of infinite extent, hence $p - \tau_{RR}$ (cf. Section 9.2.2),
(b) the pressure distribution over the wall, $p - \tau_{\varphi\varphi}$, for a plane sink flow (cf. Section 9.3.1).

9.3 In a cone-and-plate arrangement like that shown in Fig. 1.12, which has a large angle β between plate and cone, a secondary flow in the meridian plane is created by the rotation of the plate (with angular velocity $\Delta\omega$), in addition to the primary flow in the tangential direction. If one neglects end effects the three flow fields $v_\varphi^{(1)}$, $\Psi^{(2)}$ and $\Psi_I^{(2)}$ can be expressed in the product form $f(R)\cdot g(\vartheta)$; R and ϑ are spherical coordinates. One can determine by dimensional analysis the factors that are dependent on R. In this one should note that besides R and ϑ the constants $\Delta\omega$, η_0, $\alpha_1 + \alpha_2$, ρ, β affect the flow. Hence the equations (9.39), (9.41) and (9.45) reduce to ordinary differential equations for the terms dependent on ϑ. What are these differential equations, and what boundary conditions are involved?

9.4 Is an unsteady uniaxial flow, which has a velocity field given by (9.92) a shear flow; i.e. does the flow field resolve into a set of surfaces which remain undeformed with the passage of time as in the steady case?

Appendix: Formulas for special curvilinear coordinates

The equations of motion $\rho\mathbf{a} = -\text{grad}\ p + \text{div}\ \mathbf{T} + \mathbf{f}$ the velocity gradient tensor $\mathbf{L}$, the first Rivlin-Ericksen tensor $\mathbf{A}_1 = 2\mathbf{D}$, the expressions for div $\mathbf{v}$, curl $\mathbf{v}$, De/Dt and $\Phi := \text{tr}\ (\mathbf{T}\,\mathbf{D})$

and the material time derivative DA/Dt of a symmetrical tensor read with the use of

(a) cylindrical coordinates r, φ, z (indices r, φ, z indicate physical components)

$$\rho\left(\frac{\partial v_r}{\partial t} + v_r\frac{\partial v_r}{\partial r} + \frac{1}{r}v_\varphi\frac{\partial v_r}{\partial \varphi} + v_z\frac{\partial v_r}{\partial z} - \frac{1}{r}v_\varphi^2\right) = -\frac{\partial p}{\partial r} + \frac{\partial \tau_{rr}}{\partial r} + \frac{1}{r}\frac{\partial \tau_{r\varphi}}{\partial \varphi} + \frac{\partial \tau_{zr}}{\partial z} + \frac{1}{r}(\tau_{rr} - \tau_{\varphi\varphi}) + f_r$$

$$\rho\left(\frac{\partial v_\varphi}{\partial t} + v_r\frac{\partial v_\varphi}{\partial r} + \frac{1}{r}v_\varphi\frac{\partial v_\varphi}{\partial \varphi} + v_z\frac{\partial v_\varphi}{\partial z} + \frac{1}{r}v_r v_\varphi\right) = -\frac{1}{r}\frac{\partial p}{\partial \varphi} + \frac{\partial \tau_{r\varphi}}{\partial r} + \frac{1}{r}\frac{\partial \tau_{\varphi\varphi}}{\partial \varphi} + \frac{\partial \tau_{\varphi z}}{\partial z} + \frac{2}{r}\tau_{r\varphi} + f_\varphi$$

$$\rho\left(\frac{\partial v_z}{\partial t} + v_r\frac{\partial v_z}{\partial r} + \frac{1}{r}v_\varphi\frac{\partial v_z}{\partial \varphi} + v_z\frac{\partial v_z}{\partial z}\right) = -\frac{\partial p}{\partial z} + \frac{\partial \tau_{zr}}{\partial r} + \frac{1}{r}\frac{\partial \tau_{\varphi z}}{\partial \varphi} + \frac{\partial \tau_{zz}}{\partial z} + \frac{1}{r}\tau_{zr} + f_z$$

$$\mathbf{L} = \begin{bmatrix} \dfrac{\partial v_r}{\partial r} & \dfrac{1}{r}\dfrac{\partial v_r}{\partial \varphi} - \dfrac{1}{r}v_\varphi & \dfrac{\partial v_r}{\partial z} \\ \dfrac{\partial v_\varphi}{\partial r} & \dfrac{1}{r}\dfrac{\partial v_\varphi}{\partial \varphi} + \dfrac{1}{r}v_r & \dfrac{\partial v_\varphi}{\partial z} \\ \dfrac{\partial v_z}{\partial r} & \dfrac{1}{r}\dfrac{\partial v_z}{\partial \varphi} & \dfrac{\partial v_z}{\partial z} \end{bmatrix} \qquad 2\mathbf{D} = \mathbf{A}_1 = \begin{bmatrix} 2\dfrac{\partial v_r}{\partial r} & \dfrac{\partial v_\varphi}{\partial r} + \dfrac{1}{r}\dfrac{\partial v_r}{\partial \varphi} - \dfrac{1}{r}v_\varphi & \dfrac{\partial v_z}{\partial r} + \dfrac{\partial v_r}{\partial z} \\ \dfrac{\partial v_\varphi}{\partial r} + \dfrac{1}{r}\dfrac{\partial v_r}{\partial \varphi} - \dfrac{1}{r}v_\varphi & 2\left(\dfrac{1}{r}\dfrac{\partial v_\varphi}{\partial \varphi} + \dfrac{1}{r}v_r\right) & \dfrac{1}{r}\dfrac{\partial v_z}{\partial \varphi} + \dfrac{\partial v_\varphi}{\partial z} \\ \dfrac{\partial v_z}{\partial r} + \dfrac{\partial v_r}{\partial z} & \dfrac{1}{r}\dfrac{\partial v_z}{\partial \varphi} + \dfrac{\partial v_\varphi}{\partial z} & 2\dfrac{\partial v_z}{\partial z} \end{bmatrix}$$

$$\operatorname{div} \mathbf{v} = \frac{\partial v_r}{\partial r} + \frac{1}{r} v_r + \frac{1}{r}\frac{\partial v_\varphi}{\partial \varphi} + \frac{\partial v_z}{\partial z}$$

$$\operatorname{curl} \mathbf{v} = \begin{bmatrix} \frac{1}{r}\frac{\partial v_z}{\partial \varphi} - \frac{\partial v_\varphi}{\partial z} \\ \frac{\partial v_r}{\partial z} - \frac{\partial v_z}{\partial r} \\ \frac{\partial v_\varphi}{\partial r} - \frac{1}{r}\frac{\partial v_r}{\partial \varphi} + \frac{1}{r} v_\varphi \end{bmatrix}$$

$$\frac{De}{Dt} = \frac{\partial e}{\partial t} + v_r \frac{\partial e}{\partial r} + \frac{1}{r} v_\varphi \frac{\partial e}{\partial \varphi} + v_z \frac{\partial e}{\partial z}$$

$$\Phi = \tau_{rr}\frac{\partial v_r}{\partial r} + \tau_{\varphi\varphi}\left(\frac{1}{r}\frac{\partial v_\varphi}{\partial \varphi} + \frac{1}{r} v_r\right) + \tau_{zz}\frac{\partial v_z}{\partial z} + \tau_{r\varphi}\left(\frac{\partial v_\varphi}{\partial r} + \frac{1}{r}\frac{\partial v_r}{\partial \varphi} - \frac{1}{r} v_\varphi\right) + \tau_{\varphi z}\left(\frac{1}{r}\frac{\partial v_z}{\partial \varphi} + \frac{\partial v_\varphi}{\partial z}\right) + \tau_{zr}\left(\frac{\partial v_z}{\partial r} + \frac{\partial v_r}{\partial z}\right)$$

$$\left(\frac{DA}{Dt}\right)_{rr} = \frac{\partial A_{rr}}{\partial t} + v_r \frac{\partial A_{rr}}{\partial r} + v_\varphi\left(\frac{1}{r}\frac{\partial A_{rr}}{\partial \varphi} - \frac{2}{r} A_{r\varphi}\right) + v_z \frac{\partial A_{rr}}{\partial z}$$

$$\left(\frac{DA}{Dt}\right)_{r\varphi} = \frac{\partial A_{r\varphi}}{\partial t} + v_r \frac{\partial A_{r\varphi}}{\partial r} + v_\varphi\left(\frac{1}{r}\frac{\partial A_{r\varphi}}{\partial \varphi} + \frac{1}{r}(A_{rr} - A_{\varphi\varphi})\right) + v_z \frac{\partial A_{r\varphi}}{\partial z}$$

$$\left(\frac{DA}{Dt}\right)_{rz} = \frac{\partial A_{rz}}{\partial t} + v_r \frac{\partial A_{rz}}{\partial r} + v_\varphi\left(\frac{1}{r}\frac{\partial A_{rz}}{\partial \varphi} - \frac{1}{r} A_{\varphi z}\right) + v_z \frac{\partial A_{rz}}{\partial z}$$

$$\left(\frac{DA}{Dt}\right)_{\varphi\varphi} = \frac{\partial A_{\varphi\varphi}}{\partial t} + v_r \frac{\partial A_{\varphi\varphi}}{\partial r} + v_\varphi\left(\frac{1}{r}\frac{\partial A_{\varphi\varphi}}{\partial \varphi} + \frac{2}{r} A_{r\varphi}\right) + v_z \frac{\partial A_{\varphi\varphi}}{\partial z}$$

$$\left(\frac{DA}{Dt}\right)_{\varphi z} = \frac{\partial A_{\varphi z}}{\partial t} + v_r \frac{\partial A_{\varphi z}}{\partial r} + v_\varphi\left(\frac{1}{r}\frac{\partial A_{\varphi z}}{\partial \varphi} + \frac{1}{r} A_{rz}\right) + v_z \frac{\partial A_{\varphi z}}{\partial z}$$

$$\left(\frac{DA}{Dt}\right)_{zz} = \frac{\partial A_{zz}}{\partial t} + v_r \frac{\partial A_{zz}}{\partial r} + v_\varphi \frac{1}{r}\frac{\partial A_{zz}}{\partial \varphi} + v_z \frac{\partial A_{zz}}{\partial z}$$

(b) spherical coordinates R, ϑ, φ (indices R, ϑ, φ indicate physical components)

$$\rho\left(\frac{\partial v_R}{\partial t} + v_R\frac{\partial v_R}{\partial R} + \frac{1}{R}v_\vartheta\frac{\partial v_R}{\partial \vartheta} + \frac{1}{R\sin\vartheta}v_\varphi\frac{\partial v_R}{\partial\varphi} - \frac{1}{R}v_\vartheta^2 - \frac{1}{R}v_\varphi^2\right)$$
$$= -\frac{\partial p}{\partial R} + \frac{\partial \tau_{RR}}{\partial R} + \frac{1}{R}\frac{\partial \tau_{R\vartheta}}{\partial\vartheta} + \frac{1}{R\sin\vartheta}\frac{\partial\tau_{\varphi R}}{\partial\varphi} + \frac{1}{R}(2\tau_{RR} - \tau_{\vartheta\vartheta} - \tau_{\varphi\varphi}) + \frac{\cot\vartheta}{R}\tau_{R\vartheta} + f_R$$

$$\rho\left(\frac{\partial v_\vartheta}{\partial t} + v_R\frac{\partial v_\vartheta}{\partial R} + \frac{1}{R}v_\vartheta\frac{\partial v_\vartheta}{\partial \vartheta} + \frac{1}{R\sin\vartheta}v_\varphi\frac{\partial v_\vartheta}{\partial\varphi} + \frac{1}{R}v_R v_\vartheta - \frac{\cot\vartheta}{R}v_\varphi^2\right)$$
$$= -\frac{1}{R}\frac{\partial p}{\partial\vartheta} + \frac{\partial \tau_{R\vartheta}}{\partial R} + \frac{1}{R}\frac{\partial \tau_{\vartheta\vartheta}}{\partial\vartheta} + \frac{1}{R\sin\vartheta}\frac{\partial\tau_{\vartheta\varphi}}{\partial\varphi} + \frac{3}{R}\tau_{R\vartheta} + \frac{\cot\vartheta}{R}(\tau_{\vartheta\vartheta} - \tau_{\varphi\varphi}) + f_\vartheta$$

$$\rho\left(\frac{\partial v_\varphi}{\partial t} + v_R\frac{\partial v_\varphi}{\partial R} + \frac{1}{R}v_\vartheta\frac{\partial v_\varphi}{\partial \vartheta} + \frac{1}{R\sin\vartheta}v_\varphi\frac{\partial v_\varphi}{\partial\varphi} + \frac{1}{R}v_R v_\varphi + \frac{\cot\vartheta}{R}v_\vartheta v_\varphi\right)$$
$$= -\frac{1}{R\sin\vartheta}\frac{\partial p}{\partial\varphi} + \frac{\partial \tau_{\varphi R}}{\partial R} + \frac{1}{R}\frac{\partial \tau_{\vartheta\varphi}}{\partial\vartheta} + \frac{1}{R\sin\vartheta}\frac{\partial\tau_{\varphi\varphi}}{\partial\varphi} + \frac{3}{R}\tau_{\varphi R} + \frac{2\cot\vartheta}{R}\tau_{\vartheta\varphi} + f_\varphi$$

$$\mathbf{L} = \begin{bmatrix} \dfrac{\partial v_R}{\partial R} & \dfrac{1}{R}\dfrac{\partial v_R}{\partial\vartheta} - \dfrac{1}{R}v_\vartheta & \dfrac{1}{R\sin\vartheta}\dfrac{\partial v_R}{\partial\varphi} - \dfrac{1}{R}v_\varphi \\ \dfrac{\partial v_\vartheta}{\partial R} & \dfrac{1}{R}\dfrac{\partial v_\vartheta}{\partial\vartheta} + \dfrac{1}{R}v_R & \dfrac{1}{R\sin\vartheta}\dfrac{\partial v_\vartheta}{\partial\varphi} - \dfrac{\cot\vartheta}{R}v_\varphi \\ \dfrac{\partial v_\varphi}{\partial R} & \dfrac{1}{R}\dfrac{\partial v_\varphi}{\partial\vartheta} & \dfrac{1}{R\sin\vartheta}\dfrac{\partial v_\varphi}{\partial\varphi} + \dfrac{1}{R}v_R + \dfrac{\cot\vartheta}{R}v_\vartheta \end{bmatrix}$$

$$\operatorname{div} \mathbf{v} = \frac{\partial v_R}{\partial R} + \frac{2}{R} v_R + \frac{1}{R}\frac{\partial v_\vartheta}{\partial \vartheta} + \frac{\cot \vartheta}{R} v_\vartheta + \frac{1}{R \sin \vartheta}\frac{\partial v_\varphi}{\partial \varphi}$$

$$\operatorname{curl} \mathbf{v} = \begin{bmatrix} \dfrac{1}{R}\dfrac{\partial v_\varphi}{\partial \vartheta} + \dfrac{\cot \vartheta}{R} v_\varphi - \dfrac{1}{R \sin \vartheta}\dfrac{\partial v_\vartheta}{\partial \varphi} \\ \dfrac{1}{R \sin \vartheta}\dfrac{\partial v_R}{\partial \varphi} - \dfrac{\partial v_\varphi}{\partial R} - \dfrac{1}{R} v_\varphi \\ \dfrac{\partial v_\vartheta}{\partial R} + \dfrac{1}{R} v_\vartheta - \dfrac{1}{R}\dfrac{\partial v_R}{\partial \vartheta} \end{bmatrix}$$

$$\frac{De}{Dt} = \frac{\partial e}{\partial t} + v_R \frac{\partial e}{\partial R} + \frac{1}{R} v_\vartheta \frac{\partial e}{\partial \vartheta} + \frac{1}{R \sin \vartheta} v_\varphi \frac{\partial e}{\partial \varphi}$$

$$\begin{aligned} \Phi = {} & \tau_{RR}\frac{\partial v_R}{\partial R} + \tau_{\vartheta\vartheta}\left(\frac{1}{R}\frac{\partial v_\vartheta}{\partial \vartheta} + \frac{1}{R} v_R\right) + \tau_{\varphi\varphi}\left(\frac{1}{R \sin \vartheta}\frac{\partial v_\varphi}{\partial \varphi} + \frac{1}{R} v_R + \frac{\cot \vartheta}{R} v_\vartheta\right) \\ & + \tau_{R\vartheta}\left(\frac{\partial v_\vartheta}{\partial R} + \frac{1}{R}\frac{\partial v_R}{\partial \vartheta} - \frac{1}{R} v_\vartheta\right) + \tau_{\vartheta\varphi}\left(\frac{1}{R}\frac{\partial v_\varphi}{\partial \vartheta} + \frac{1}{R \sin \vartheta}\frac{\partial v_\vartheta}{\partial \varphi} - \frac{\cot \vartheta}{R} v_\varphi\right) + \tau_{\varphi R}\left(\frac{\partial v_\varphi}{\partial R} + \frac{1}{R \sin \vartheta}\frac{\partial v_R}{\partial \varphi} - \frac{1}{R} v_\varphi\right) \end{aligned}$$

$$2\mathbf{D} = \mathbf{A}_1 = \begin{bmatrix} 2\dfrac{\partial v_R}{\partial R} & \dfrac{\partial v_\vartheta}{\partial R} + \dfrac{1}{R}\dfrac{\partial v_R}{\partial \vartheta} - \dfrac{1}{R} v_\vartheta & \dfrac{\partial v_\varphi}{\partial R} + \dfrac{1}{R \sin \vartheta}\dfrac{\partial v_R}{\partial \varphi} - \dfrac{1}{R} v_\varphi \\ \dfrac{\partial v_\vartheta}{\partial R} + \dfrac{1}{R}\dfrac{\partial v_R}{\partial \vartheta} - \dfrac{1}{R} v_\vartheta & 2\left(\dfrac{1}{R}\dfrac{\partial v_\vartheta}{\partial \vartheta} + \dfrac{1}{R} v_R\right) & \dfrac{1}{R}\dfrac{\partial v_\varphi}{\partial \vartheta} + \dfrac{1}{R \sin \vartheta}\dfrac{\partial v_\vartheta}{\partial \varphi} - \dfrac{\cot \vartheta}{R} v_\varphi \\ \dfrac{\partial v_\varphi}{\partial R} + \dfrac{1}{R \sin \vartheta}\dfrac{\partial v_R}{\partial \varphi} - \dfrac{1}{R} v_\varphi & \dfrac{1}{R}\dfrac{\partial v_\varphi}{\partial \vartheta} + \dfrac{1}{R \sin \vartheta}\dfrac{\partial v_\vartheta}{\partial \varphi} - \dfrac{\cot \vartheta}{R} v_\varphi & 2\left(\dfrac{1}{R \sin \vartheta}\dfrac{\partial v_\varphi}{\partial \varphi} + \dfrac{1}{R} v_R + \dfrac{\cot \vartheta}{R} v_\vartheta\right) \end{bmatrix}$$

$$\left(\frac{DA}{Dt}\right)_{RR} = \frac{\partial A_{RR}}{\partial t} + v_R \frac{\partial A_{RR}}{\partial R} + v_\vartheta\left(\frac{1}{R}\frac{\partial A_{RR}}{\partial \vartheta} - \frac{2}{R}A_{R\vartheta}\right) + v_\varphi\left(\frac{1}{R\sin\vartheta}\frac{\partial A_{RR}}{\partial \varphi} - \frac{2}{R}A_{R\varphi}\right)$$

$$\left(\frac{DA}{Dt}\right)_{R\vartheta} = \frac{\partial A_{R\vartheta}}{\partial t} + v_R \frac{\partial A_{R\vartheta}}{\partial R} + v_\vartheta\left(\frac{1}{R}\frac{\partial A_{R\vartheta}}{\partial \vartheta} + \frac{1}{R}(A_{RR} - A_{\vartheta\vartheta})\right) + v_\varphi\left(\frac{1}{R\sin\vartheta}\frac{\partial A_{R\vartheta}}{\partial \varphi} - \frac{1}{R}A_{\vartheta\varphi} - \frac{\cot\vartheta}{R}A_{R\varphi}\right)$$

$$\left(\frac{DA}{Dt}\right)_{R\varphi} = \frac{\partial A_{R\varphi}}{\partial t} + v_R \frac{\partial A_{R\varphi}}{\partial R} + v_\vartheta\left(\frac{1}{R}\frac{\partial A_{R\varphi}}{\partial \vartheta} - \frac{1}{R}A_{\vartheta\varphi}\right) + v_\varphi\left(\frac{1}{R\sin\vartheta}\frac{\partial A_{R\varphi}}{\partial \varphi} + \frac{1}{R}(A_{RR} - A_{\varphi\varphi}) + \frac{\cot\vartheta}{R}A_{R\vartheta}\right)$$

$$\left(\frac{DA}{Dt}\right)_{\vartheta\vartheta} = \frac{\partial A_{\vartheta\vartheta}}{\partial t} + v_R \frac{\partial A_{\vartheta\vartheta}}{\partial R} + v_\vartheta\left(\frac{1}{R}\frac{\partial A_{\vartheta\vartheta}}{\partial \vartheta} + \frac{2}{R}A_{R\vartheta}\right) + v_\varphi\left(\frac{1}{R\sin\vartheta}\frac{\partial A_{\vartheta\vartheta}}{\partial \varphi} - \frac{2\cot\vartheta}{R}A_{\vartheta\varphi}\right)$$

$$\left(\frac{DA}{Dt}\right)_{\vartheta\varphi} = \frac{\partial A_{\vartheta\varphi}}{\partial t} + v_R \frac{\partial A_{\vartheta\varphi}}{\partial R} + v_\vartheta\left(\frac{1}{R}\frac{\partial A_{\vartheta\varphi}}{\partial \vartheta} + \frac{1}{R}A_{R\varphi}\right) + v_\varphi\left(\frac{1}{R\sin\vartheta}\frac{\partial A_{\vartheta\varphi}}{\partial \varphi} + \frac{\cot\vartheta}{R}(A_{\vartheta\vartheta} - A_{\varphi\varphi}) + \frac{1}{R}A_{R\vartheta}\right)$$

$$\left(\frac{DA}{Dt}\right)_{\varphi\varphi} = \frac{\partial A_{\varphi\varphi}}{\partial t} + v_R \frac{\partial A_{\varphi\varphi}}{\partial R} + v_\vartheta \frac{1}{R}\frac{\partial A_{\varphi\varphi}}{\partial \vartheta} + v_\varphi\left(\frac{1}{R\sin\vartheta}\frac{\partial A_{\varphi\varphi}}{\partial \varphi} + \frac{2}{R}A_{R\varphi} + \frac{2\cot\vartheta}{R}A_{\vartheta\varphi}\right)$$

ACKNOWLEDGEMENTS

Fig. 2.3 Meissner, J.: Deformationsverhalten der Kunststoffe im flüssigen und festen Zustand. Kunststoffe **61**(1971) 576-582

Fig. 2.4,
Fig. 2.9 Laun, H.M.: Das viskoelastische Verhalten von Polyamid-6-Schmelzen. Rheol. Acta **18**(1979) 478-491

Fig. 2.8 Ginn, R.F.; Metzner, A.B.: Measurement of stresses developed in steady laminar shearing flows of viscoelastic media. Trans. Soc. Rheol. **13**(1969) 429-453

Fig. 3.8 Böhme, G.; Nonn, G.: Kennfelder für Schleppströmungspumpen zur Förderung nicht-newtonscher Flüssigkeiten. Rheol. Acta **17**(1978) 115-131

Fig. 5.5,
Fig. 5.6 Carreau, P.J.: Rheological equations from molecular network theories. Trans. Soc. Rheol. **16**(1972) 99-127

Fig. 5.9a Ashare, E.: Rheological properties of narrow distribution polystyrene solutions. Trans. Soc. Rheol. **12**(1968) 535-557

Fig. 5.9b Han, C.D.; Kim, K.U.; Siskoviic, N.; Huang, C.R.: An appraisal of rheological models as applied to polymer melt flow. Rheol. Acta **14**(1975) 533-549

Fig. 6.1 Macdonald, I.F.: Parallel superposition of simple shearing and small amplitude oscillatory motions. Trans. Soc. Rheol. **17**(1973) 537-555

Fig. 7.1 Laun, H.M.; Münstedt, M.: Elongational behaviour of a low density polyethylene melt. Rheol. Acta **17**(1978) 415-425

REFERENCES

1 Books

Astarita, G.; Marucci, G.: Principles of non-Newtonian fluid mechanics. London: McGraw Hill 1974

Becker, E.; Bürger, W.: Kontinuumsmechanik. Stuttgart: Teubner 1975.

Bird, R.B.; Armstrong, R.C.; Hassager, O.: Dynamics of polymeric liquids. Vol. I, Fluid mechanics. New York: Wiley 1977

Bird, R.B.; Hassager, O.; Armstrong, R.C.; Curtiss, Ch. F.: Dynamics of polymeric liquids. Vol. II, Kinetic theory. New York: Wiley 1977

Bird, R.B.; Stewart, W.E.; Lightfoot, E.N.: Transport phenomena. New York: Wiley 1960

Brydson, J.A.: Flow properties of polymer melts. London: Iliffe 1970

Christensen, R.M.: Theory of viscoelasticity. New York: Academic Press 1971

Coleman, B.D.; Markovitz, H.; Noll, W.: Viscometric flows of non-Newtonian fluids. Berlin-Heidelberg-New York: Springer 1966

Crochet, M.J.; Davies, A.R.; Walters, K.: Numerical simulation of non-Newtonian flow. Amsterdam-Oxford-New York-Tokyo: Elsevier 1984

Darby, R.: Viscoelastic fluids. New York: Marcel Dekker 1976

Ebert, F.: Strömung nicht-newtonscher Medien. Brunswick: Vieweg 1980

Eirich, F.R. (Ed.): Rheology. Theory and applications. Vols. 1-5. New York: Academic Press 1956-1969

Ferry, J.D.: Viscoelastic properties of polymers. 2nd. Ed. New York: Wiley 1970

Fredrickson, A.G.: Principles and applications of rheology. Englewood Cliffs: Prentice Hall 1964

Han, C.D.: Rheology in polymer processing. New York: Academic Press 1976

Happel, J.; Brenner, H.: Low Reynolds number hydrodynamics. Leyden: Noordhoff 1973

Harris, J.: Rheology and non-Newtonian flow. New York: Longman 1977

Hughes, W.F.: An introduction to viscous flow. Washington: McGraw Hill 1979

Huilgol, R.R.: Continuum mechanics of viscoelastic liquids. New York: Wiley 1975

Hutton, J.F.; Pearson, J.R.A.; Walters, K.(Ed.): Theoretical rheology. New York: Wiley 1974

Lenk, R.S.: Rheologie der Kunststoffe. Munich: Hanser 1971

Lodge, A.S.: Elastic liquids. London: Academic Press 1964

McKelvey, J.M.: Polymer processing. New York: Wiley 1962

Middleman, S.: The flow of high polymers. New York: Interscience 1968

Nielsen, L.E.: Polymer rheology. New York: Marcel Dekker 1977

Pearson, J.R.A.: Mechanical principles of polymer melt processing. 2nd. Ed. M.I.T. Press 1975

Petrie, C.J.S.: Elongational flows. London: Pitman 1979

Pipkin, A.C.: Lectures on viscoelasticity theory. New York: Springer 1972

Reiner, M.: Rheologie in elementarer Darstellung. 2nd. Ed. Munich: Hanser 1969

Reiner, M.: Advanced rheology. London: Lewis 1971

Rivlin, R.S. (Ed.): The mechanics of viscoelastic fluids. New York: The American Society of Mechanical Engineering 1977

Schowalter,W.R.: Mechanics of non-Newtonian fluids. Oxford: Pergamon Press 1978

Skelland, A.H.P.: Non-Newtonian flow and heat transfer. New York: Wiley 1967

Tadmor, Z.; Gogos, C.G.: Principles in polymer processing.New York: Wiley 1979

Tanner, R.I.: Engineering rheology. Oxford: Clarendon Press 1985

Torner, R.V.: Grundprozesse der Verarbeitung von Polymeren. Leipzig: VEB Deutscher Verlag für Grundstoffindustrie 1974

Truesdell, C.; Noll, W.: Die nichtlinearen Feldtheorien der Mechanik. In: Handbuch der Physik, Vol. III/3. Berlin-Heidelberg-New York: Springer 1965

VDI-Gesellschaft Kunststofftechnik (Ed.): Praktische Rheologie der Kunststoffe. Düsseldorf: VDI-Verlag 1978

Walters, K.: Rheometry. London: Chapman and Hall 1975

Walters, K. (Ed.): Rheometry - Industrial applications. Chichester-New York-Brisbane-Toronto: Wiley 1980

Wazer, van, J.R.; Lyons, J.W.; Kim, K.Y.; Colwell, R.E.: Viscosity and flow measurement. A laboratory handbook of rheology. New York: Interscience 1963

Wilkinson, W.L.: Non-Newtonian fluids. New York: Pergamon Press 1960

Williams, D.J.: Polymer science and engineering. Englewood Cliffs: Prentice Hall 1971

2 Summary articles

Becker, E.: Simple non-Newtonian fluid flows. Advances in Applied Mechanics **20**(1980) 177-226

Bird, R.B.: Useful non-Newtonian models. Annual Review of Fluid Mechanics **8**(1976) 13-34

Goddard, J.D.: Polymer fluid mechanics. Advances in Applied Mechanics **19** (1979) 143-219

Pipkin, A.C.; Tanner, R.I.: A survey of theory and experiment in viscometric flows of viscoelastic liquids. Mechanics Today **1**(1973) 262-321

Pipkin, A.C.; Tanner, R.I.: Steady non-viscometric flows of viscoelastic liquids. Annual Review of Fluid Mechanics **9**(1977) 13-32

Truesdell, C.: The meaning of viscometry in fluid dynamics. Annual Review of Fluid Mechanics **6**(1974) 111-146

INDEX

Absolute temperature 40
Acceleration gradient tensor 16
Acceleration vector 6
Axial shear flow 156, 264

Bingham fluid 59, 67
Bipotential equation 314
Boundary layer 219, 225, 273
Boundary layer equation 225
Boundary layer simplification 222
Boundary layer thickness 184
Brinkman number 74

Capillary viscometer 130
Cauchy strain tensor 20
Cauchy stress formulas 36
Channel flow 81
Coextrusion 138
Complex shear modulus 173
Complex shear viscosity 174, 214
Cone-and-plate flow 30, 70, 141
Constant density fluid 32
Constitutive equation, finite linear viscoelasticity 267
Constitutive equation for slow and slowly varying processes 204, 281
Constitutive equation for steady extensional flows 249
Constitutive equation for viscometric flows 66
Continuity equation 31
Control volume 31
Convectional flow, free 284
Convective derivative 6
Couette flow 27, 69, 146
Couette flow, unsteady 192
Creep recovery 172
Creep test 171
Creeping flow 142, 233

Dead time 48, 208
Deformation gradient, relative 19
Deformation history 19, 47
Density field 31
Derivative, convective 6
Derivative, Jaumann's 17
Derivative, local 5
Derivative, material 5
Derivative, Oldroyd's 17
Derivative, substantial 5
Determinism 47
Die-swell 150
Differential models 276
Differential viscosity 54
Dilatant fluid 51
Dimensional analysis 126, 275, 296
Dispersion 183

Dispersion law 41
Displacement thickness 191, 227, 275
Dissipation 40, 72, 132, 325
Dissipation function 40, 263
Dynamic shear modulus 174
Dynamic viscosity 174
Dynamic vorticity number 14

Efficiency of a friction pump 89
Efficiency of a screw extruder 101
Eigenvalue of a symmetrical tensor 13
Eigenvector 13
Ellis model 59
Elongation rate 10, 245
Elongational flow, plane 4
Elongational viscosity 248
Energy, dissipated 40
Energy equation 38
Energy, internal 38
Energy, kinetic 38
Entropy production rate 261
Equation of continuity 31
Equation of motion 37
Expanding sphere 254
Extensional flow 245
Extensional flow, biaxial, plane 246
Extensional flow, simple 42, 245
Extensional flow, uniaxial 245
Extruder 94

Fading memory 163
Fibre extrusion 250
Field 1
Film blowing 254
Finite linear viscoelasticity 267
Flow characteristic 86, 100
Flow, creeping 142
Flow function 50
Flow function, second 218
Flow function, universal representation 51
Flow function, usual models 59
Flow, isochoric 5, 32
Flow law 50
Flow, nearly viscometric 213
Flow, partly controllable 141
Flow, plane 33
Flow, rotational symmetric 292
Flow, steady 3
Flow, unsteady 3
Flow, viscometric 24
Flow with constant stretch history 26

Fluid 35
Fluid, constant density 32
Fluid, dilatant 51
Fluid, generalised Newtonian 262
Fluid, incompressible 32
Fluid, Newtonian 45
Fluid, non-Newtonian 46
Fluid particles 1
Fluid, second order 281
Fluid, shear thickening 51
Fluid, shear thinning 51
Fluid, simple 47
Fluid, third order 281
Fluid, viscoelastic 163
Fluids without memory 260
Form parameter 228
Free jet 150
Friction coefficient 117
Friction pump 81
Friction pump characteristic 86
Frozen wave (propagation) speed 210

Generalised Newtonian fluid 262
Gravity wave 41
Green's strain tensor 21
Gümbel curve 119

Heat capacity, specific 41
Heat flux density 38
Helical flow 28, 135
Homogeneous deformation 42, 245
Hydrodynamic lubrication 81, 119, 124
Hydrostatic pressure 35

Incompressible fluid 32
Indifference curve 241
Instantaneous reaction 46
Integral model 265
Internal energy 38
Irreversible thermodynamics 261
Isochoric flow 5, 32

Jaumann's derivative 17
Journal bearing 111, 230
Jump relationship 189

K-BKZ model 268
Kinematic vorticity number 14
Kinetic energy 38

Law of conservation of mass 31
Lethersich model 208
Limiting viscosity 50, 53

Linear viscoelasticity 162, 267
Local action 46
Local derivative 5
Lubricating film 219, 230
Lubricating gap 111, 231

Mass flux 31
Material coefficients of the second order 149, 282
Material coefficients of the third order 283
Material coordinate 47
Material derivative 5
Material functional 47, 161, 247
Material objectivity 49, 266, 277
Material point 1
Materially steady 25, 247
Maxwell body 164, 275
Maxwell-Chartoff rheometer 42, 272
Maxwell-Oldroyd model 279
Memory, fading 47, 163
Memory range 205, 216
Memory, short 216
Minimum principle for generalised Newtonian fluids 262
Motion with constant stretch history 48, 247

Natural basis 61, 156, 223
Navier-Stokes equations 287, 321
Nearly viscometric flow 213
Newtonian flow 45, 281
Non-Newtonian flow 46
Normal stress 34
Normal stress coefficients 62
Normal stress coefficients, lower limiting values 62
Normal stress functions 61
Nozzle flow 294

Oldroyd's 8-constant model 279
Orthogonality principle 262
Oscillatory stress and deformation 173
Ostwald-de Waele model 59

Partly controllable flow 141
Path line 2
Péclet number 325
Pipe flow 126, 320
Pipe flow, periodic 330
Pipe flow, unsteady 196
Plane elongational flow 4
Plane flow 33
Plane shear flow 11
Plane stagnation point flow 3
Poiseuille flow 26
Poisson's equation 158
Position vector 2

Potential 262
Potential flow 14, 245
Potential vortex 148
Power 89, 94
Prandtl number 284
Prandtl-Eyring model 59
Pressure 35, 36
Pressure cushion 119
Pressure distribution 104
Pressure-drag flow 81
Pressure drop 127, 156
Pressure gradient 38, 82
Pressure parameter 87, 128
Pressure, piezometric 289
Primary flow 287
Principal axis system 12
Principal invariants 13
Propulsion power 104
Pseudoplastic fluid 51

Rabinowitsch model 59
Rate of shear 12
Rate of strain tensor 9
Rayleigh problem 184
Reiner model 59
Reiner-Philippoff model 59
Reiner-Rivlin fluid 250
Relative deformation gradient 19
Relaxation function 164
Relaxation spectrum 166
Relaxation test 164
Relaxation time 165
Reversible shear 172
Reynolds number 118, 131, 239
Rheological laws (see also constitutive equation) 259
Rheometer 247, 272
Right-Cauchy-Green tensor 20
Rivlin-Ericksen tensors 17
Rolling 104
Rotational symmetric flow 292
Rotational viscometer 69

Screw extruder 94
Second flow function 218
Secondary flow 149, 159, 287
Secondary flow, in cylindrical pipes 320
Secondary flow, plane 312
Secondary flow, rotational symmetric 292
Shear flow 11
Shear flow, axial 156, 264
Shear flow, plane 11, 46
Shear flow, steady 24, 45
Shear flow, steady, stability 237

Shear flow, unsteady 45, 161
Shear modulus, complex 173
Shear modulus, dynamic 174
Shear stress 34
Shear thickening fluid 51
Shear thinning fluid 51
Shear viscosity 45, 50
Shock absorber 177, 209
Shock front 185
Similarity assumption 192
Similarity solution 227
Simple extensional flow 245
Simple fluid 47, 266
Simple shear flow 11
Slenderness condition 222
Slit nozzle 160
Slow and slowly varying processes 204, 281
Sommerfeld number 114
Specific heat capacity 41
Specific internal energy 38
Spherical symmetric state of stress 36
Squeezing flow 124
Stability limit 240
Stagnation point flow 3, 225
Steady flow 3
Steady shear flow 24, 49
Steady shear flow, natural basis 61, 156
Stokes number 199, 332
Strain tensor 17
Strain tensor, Cauchy 20
Strain tensor, Green 21
Stream function 33, 292, 314
Stream line 2
Stream surface 2
Stress 34
Stress growth experiment 167, 269
Stress tensor 36
Stress vector 34
Substantial derivative 5
Suction 273
Surface force 34
Symmetry of the stress tensor 36

Temperature 40
Temperature gradient 40
Temperature profile 79, 133
Tensor of the extra-stresses 36
Theorem of momentum 126
Theory of the fourth order 323
Theory of the second order 289
Theory of the third order 313
Thermal conductivity 41
Thermodynamic pressure 36

Thermorheological effect 325
Torsional flow 30, 42
Trace of a tensor 13
Transverse wave 183
Tuning of a shock absorber 177
Turbulence, viscoelastic 237

Unsteady flow 3
Useful power of a friction pump 89

Velocity field 31
Velocity gradient tensor 6
Velocity profile 83, 102, 127
Velocity vector 2
Viscoelasticity 162, 267
Viscometer 69, 130
Viscometric flow 24
Viscosity 45
Viscosity, complex 174, 214
Viscosity, dependence on concentration 52
Viscosity, dependence on molecular weight 52
Viscosity, dependence on temperature 50
Viscosity, differential 54
Viscosity, dynamic 174
Viscosity in the stress growth experiment 169
Volume flux 83, 128
Volume force 34
Volume force, conservative 38
Vorticity number 14
Vorticity tensor 9
Vorticity vector 8

Wall shear stress 82, 127, 275
Wave number 239
Wave speed 185, 239
Wave speed, equilibrium value 210
Wave speed, frozen 210
Wavefront 185
Weissenberg effect 146
Weissenberg number 74, 227, 240, 282
Wire coating tool 136

Yield point 59
Yield stress 59

Zero-shear viscosity 50